S0-AEB-738

Solutions Manual

SAXON Math™ HOMESCHOOL 6/5

Stephen Hake
John Saxon

SAXON™
PUBLISHERS

Saxon Publishers gratefully acknowledges the contributions of the following individuals in the completion of this project:

Authors: Stephen Hake, John Saxon

Editorial: Chris Braun, Bo Björn Johnson, Brooke Butner, Brian E. Rice

Editorial Support Services: Christopher Davey, Jay Allman, Shelley Turner, Jean Van Vleck, Darlene Terry

Production: Alicia Britt, Karen Hammond, Donna Jarrel, Brenda Lopez, Adriana Maxwell, Cristi D. Whiddon

Project Management: Angela Johnson, Becky Cavnar

© 2005 Saxon Publishers, Inc., and Stephen Hake

All rights reserved. No part of ***Saxon Math 6/5—Homeschool,*** **Third Edition, Solutions Manual** may be reproduced, stored in a retrieval system, or transmitted in any form or by any means, electronic, mechanical, photocopying, recording, or otherwise, without the prior written permission of the publisher.

Printed in the United States of America

ISBN: 978-1-591-41326-4

32 0928 21

4500826117

Solutions for

Lessons and Investigations

Lesson 1, Warm-Up

a. **6**

b. **60**

c. **600**

d. **90**

e. **800**

f. **100**

g. **60**

h. **1500**

Problem Solving
9

Lesson 1, Lesson Practice

a. The rule is **"Count up by twos."**
12, 14, 16

b. The rule is **"Count up by sevens."**
28, 35, 42

c. The rule is **"Count up by fours."**
16, 20, 24

d. The rule is **"Count down by threes."**
12, 9, 6

e. The rule is **"Count down by fives."**
30, 25, 20

f. The rule is **"Count up by sixes."**
30, 36, 42

g. **5 digits**

h. **4 digits**

i. **7 digits**

j. **7**

k. **6**

l. **1**

Lesson 1, Mixed Practice

1. The rule is "Count up by fives."
25

2. The rule is "Count down by sevens."
35

3. The rule is "Count up by eights."
32

4. The rule is "Count up by nines."
54

5. The rule is "Count down by threes."
15

6. The rule is "Count down by fours."
16

7. The rule is "Count up by sevens."
21

8. The rule is "Count down by fives."
35

9. The rule is "Count up by fours."
24

10. The rule is "Count up by eights."
40

11. The rule is "Count down by sixes."
42

12. The rule is "Count up by sevens."
35

13. The rule is **"Count up by threes."**
15, 18, 21

14. The rule is **"Count up by eights."**
32, 40, 48

15. The rule is **"Count up by sixes."**
24, 30, 36

16. The rule is **"Count down by fives."**
25, 20, 15

17. The rule is **"Count up by threes."**
27, 30, 33

18. The rule is **"Count up by nines."**
36, 45, 54

19. **Sequence**

20. **6 digits**

21. **5 digits**

22. **8 digits**

23. **8**

24. **7**

25. **0**

Lesson 2, Warm-Up

a. **12**

b. **120**

c. **1200**

d. **130**

e. **150**

f. **900**

g. **180**

h. **200**

Problem Solving
90

Lesson 2, Activity

Counting number	11	12	13	14	15	16	17	18	19	20
Half of number	$5\frac{1}{2}$	6	$6\frac{1}{2}$	7	$7\frac{1}{2}$	8	$8\frac{1}{2}$	9	$9\frac{1}{2}$	10

Lesson 2, Lesson Practice

a. **Even**

b. **Even**

c. **Odd**

d. **Odd**

e. **Even**

f. **Even**

g. **Even number**

h. **$3\frac{1}{2}$ cookies**

Lesson 2, Mixed Practice

1. **Odd**

2. **0**

3. **7**

4. **Odd**

5. **Even**

6. **Odd**

7. **3716**

8. **56,789**

9. **333,456**

10. **323**

11. The rule is "Count up by threes."
18, 21, 24

12. The rule is "Count up by eights."
40, 48, 56

13. The rule is "Count down by tens."
90, 80, 70

14. The rule is "Count down by fours."
16, 12, 8

15. The rule is "Count down by fives."
40, 35, 30

16. The rule is "Count up by nines."
45, 54, 63

17. The rule is "Count down by threes."
27, 24, 21

18. The rule is "Count up by sixes."
36, 42, 48

19. The rule is "Count up by sevens."
35, 42, 49

20. The rule is "Count down by sixes."
48, 42, 36

21. The rule is "Count down by fours."
36, 32, 28

22. The rule is "Count down by nines."
72, 63, 54

23. The rule is "Count down by eights."
64, 56, 48

24. The rule is "Count down by sevens."
63, 56, 49

25. **B. 31**

26. $2\frac{1}{2}$

27. **B. Half of 12**

Lesson 3, Warm-Up

a. **100**

b. **320**

c. **340**

d. **540**

e. **290**

f. **650**

g. **180**

h. **720**

Problem Solving
900

Lesson 3, Activity

1. **1 hundred + 2 tens + 0 ones**

2. **1 hundred + 3 tens + 2 ones**

Lesson 3, Lesson Practice

a. **6**

b. **350**

c. **$215**

Lesson 3, Mixed Practice

1. **578**

2. **250**

3. **6**

4. **5**

5. **$100 bill**

6. **C. 536**

7. **B. 2345**

8. **98**

9. **B. 25**

10. **C. 5's**

11. The rule is "Count up by threes."
18, 21, 24

12. The rule is "Count down by sixes."
36, 30, 24

13. The rule is "Count up by eights."
32, 40, 48

14. The rule is "Count down by eights."
56, 48, 40

15. The rule is "Count up by fours."
28, 32, 36

16. The rule is "Count down by fours."
28, 24, 20

17. The rule is **"Count up by nines."**
54, 63, 72

18. The rule is **"Count down by nines."**
54, 45, 36

19. The rule is **"Count up by tens."**
40, 50, 60

20. The rule is **"Count down by threes."**
24, 21, 18

21. **40**

22. **500**

23. **$334**

24. **Hundreds**

25. **36**

26. **$5**

27. **$2.50 or $2** $\frac{1}{2}$

28. **Odd number**

Lesson 4, Warm-Up

a. **640**

b. **300**

c. **650**

d. **640**

e. **700**

f. **310**

g. **940**

h. **560**

Problem Solving
123, 132, 231, 312, 321

Lesson 4, Lesson Practice

a. **243, 324, 423**

b. 36 ⓐ $<$ 632

c. 110 $>$ 101

d. 90 $=$ 90

e. 112 $<$ 121

f. **20 $<$ 30**

g. **12 $>$ 8**

Lesson 4, Mixed Practice

1. **4 < 10**

2. **15 > 12**

3. 97 (<) 101

4. 34 (<) 43

5. **365**

6. **6**

7. **3**

8. **$10 bill**

9. **Odd**

10. **Even**

11. **Even**

12. **999**

13. **B. 4's**

14. **354, 435, 543**

15. 6, 12, 18, 24, 30, 36, 42, 48, 54
54

16. **$640**

17. The rule is "Count up by fours."
32, 36, 40

18. The rule is "Count down by twos."
100, 98, 96

19. The rule is "Count up by sixes."
18, 24, 30

20. The rule is "Count up by sevens."
21, 28, 35

21. The rule is "Count down by eights."
16, 8, 0

22. The rule is "Count down by nines."
18, 9, 0

23. **90**

24. **110**

25. 8, 16, 24, 32, 40, 48, 56
56

26. 2, 4, 6, 8, 10, 12, 14, 16, 18, 20, 22
Even

27. $4\frac{1}{2}$

28. **Odd**

Lesson 5, Warm-Up

a. **560**

b. **650**

c. **770**

d. **500**

e. **760**

f. **460**

g. **890**

h. **520**

Problem Solving
234, 243, 324, 342, 423, 432

Lesson 5, Lesson Practice

a. **Five hundred sixty-three dollars and forty-five cents**

b. **One hundred one**

c. **One hundred eleven**

d. **245**

e. **420**

f. **$503.50**

Lesson 5, Mixed Practice

1. **$374.20**

2. **Six hundred twenty-three dollars and fifteen cents**

3. **205**

4. **One hundred nine**

5. **150 > 115**

6. 346 ⓒ 436

7. **579**

8. **246, 426, 462, 624**

9. **6**

10. **100**

11. **Even**

12. **Odd**

13. **Even**

14. **998**

15. **B. 3's**

16. **B. 29**

17. **$702**

18. The rule is "Count up by nines."
27, 36, 45, 54

19. The rule is "Count up by fives."
40, 45, 50, 55

20. The rule is "Count up by sixes."
24, 30, 36, 42

21. The rule is **"Count down by tens."**
70, 60, 50, 40

22. The rule is **"Count down by nines."**
63, 54, 45, 36

23. The rule is **"Count down by eights."**
64, 56, 48, 40

24. The rule is **"Count up by sevens."**
28, 35, 42, 49

25. 3, 6, 9, 12, 15, 18, 21, 24, 27
27

26. 1, 3, 5, 7, 9, 11, 13, 15, 17, 19
Odd

27. **Odd**

28. **$3.50** $\left(\text{or } \$3\frac{1}{2}\right)$

Lesson 6, Warm-Up

a. **790**

b. **620**

c. **350**

d. **650**

e. **680**

f. **940**

g. **290**

h. **900**

Problem Solving
345, 354, 435, 453, 534, 543

Lesson 6, Lesson Practice

a.

```
  8 \
  6  > 10
+ 2 /
 16
```

b.

```
        4 \
10 <    7  \
        3  / > 10
      + 6 /
       20
```

c.

```
  9
  6 \
     > 10
+ 4 /
 19
```

d.

```
        4 \
        5 / > 10
12 <    6
      + 7
       22
```

e.

```
  7 \
  3 / > 10
+ 4
 14
```

f.

```
  2 \
  6 / > 8
  3 \
+ 5 / > 8
 16
```

g.

```
  6 \
  7  > 11
+ 5 /
 18
```

h.

```
        8
13 <    7 \
        5 / > 10
      + 3
       23
```

i. **B. Less than 50**

j.

```
   1 1
  $463
+ $158
  $621
```

k.

```
  1
  674
+ 555
 1229
```

l.

```
   1 1
  $323
  $142
+ $365
  $830
```

m.

```
  1 1
  543
+  98
  641
```

n.

```
     1
   $ 47
     1
+ $485
  $532
```

Lesson 6, Mixed Practice

1.

```
    1
  $520
+  $86
  $606
```

2. **Two hundred twelve dollars and fifty cents**

3. **2**

4. **Even**

5. **Odd**

6. **Odd**

7. **$508**

8. **Five hundred eighty**

9.

```
  1 \
  6  > 10
+ 9 /
 16
```

10.

```
  7
  6 \
     > 10
+ 4 /
 17
```

11.

$$9 \begin{cases} 8 \\ 3 \\ 1 \end{cases} 10$$
$$+\ 7$$
$$\overline{19}$$

12.

$$12 \begin{cases} 4 \\ 5 \\ 6 \end{cases} 10$$
$$+\ 7$$
$$\overline{22}$$

13.
$$\overset{1}{\$436}$$
$$+\ \$527$$
$$\overline{\$963}$$

14.
$$\overset{1\ 1}{592}$$
$$+\ 408$$
$$\overline{1000}$$

15.
$$\overset{1\ 1}{963}$$
$$+\ 79$$
$$\overline{1042}$$

16.
$$\overset{1}{\$180}$$
$$+\ \$747$$
$$\overline{\$927}$$

17. A. 28

18. 10, 20, 30, 40, 50, 60, 70, 80
80

19. 6, 12, 18, 24, 30, 36, 42, 48
48

20. 7, 14, 21, 28, 35, 42, 49, 56
56

21. 8, 16, 24, 32, 40, 48, 56, 64
64

22. 916 $<$ 960

23. 690 $>$ 609

24.

$$\begin{array}{c} 5 \\ 5 \end{array} 10 \quad\quad \begin{array}{c} 4 \\ 5 \end{array} 10$$
$$+\ 5 \quad\quad\quad +\ 6$$
$$\overline{15} \quad\quad\quad \overline{15}$$

$$15 = 15$$

25. 11

26. Even

27. Even

28. $4\frac{1}{2}$ pages

LESSON 7, WARM-UP

a. $50

b. $750

c. $500

d. 480

e. $100

f. $1000

g. $400

h. 720

Problem Solving

$$\begin{array}{r} 123 \\ 132 \\ 213 \\ 231 \\ 312 \\ +\ 321 \\ \hline 1332 \end{array}$$

12; The sum of each column is the same because the same digits appear in each column (two 1's, two 2's, and two 3's).

Lesson 7, Lesson Practice

a. 36,420
Thirty-six thousand, four hundred twenty

b. $12,300
Twelve thousand, three hundred dollars

c. 4,567
Four thousand, five hundred sixty-seven

d. 63,117

e. 256,700

f. 50,924

g. $750,000

h.

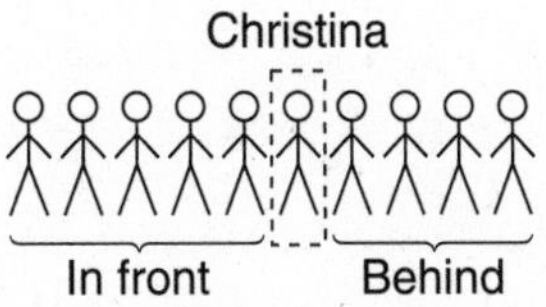

Five people were in front of Christina and four people were behind her.

Lesson 7, Mixed Practice

1.
$$\begin{array}{r} {}^{1\,1} \\ \$462 \\ +\ \ \$88 \\ \hline \mathbf{\$550} \end{array}$$

2. 6

3. 707

4. Twenty-nine thousand, thirty-five feet

5.
$$\begin{array}{r} {}^{1} \\ {}_{1}54 \\ +\ 246 \\ \hline \mathbf{300} \end{array}$$

6.
$$\begin{array}{r} {}^{1} \\ \$463 \\ +\ \$364 \\ \hline \mathbf{\$827} \end{array}$$

7.
$$\begin{array}{r} {}^{1\,1} \\ \$286 \\ +\ \$414 \\ \hline \mathbf{\$700} \end{array}$$

8.
$$\begin{array}{r} {}^{1} \\ 709 \\ +\ 314 \\ \hline \mathbf{1023} \end{array}$$

9. 10, 20, 30, 40, 50, 60, 70
70

10. 5, 10, 15, 20, 25, 30, 35
35

11. 6, 12, 18, 24, 30, 36, 42
42

12. 7, 14, 21, 28, 35, 42, 49
49

13. 8, 16, 24, 32, 40, 48, 56
56

14. 9, 18, 27, 36, 45, 54, 63
63

15. 250 ⓹ 215

16. 365 + 366 ⓹ 365 + 365

17.
$$\begin{array}{r} {}^{1\,1} \\ \$436 \\ \$72 \\ +\ \ \$54 \\ \hline \mathbf{\$562} \end{array}$$

18.
$$\begin{array}{r} {}^{2\,1} \\ 361 \\ 493 \\ +\ 147 \\ \hline \mathbf{1001} \end{array}$$

19.
$$\begin{array}{r} {}^{1\,1} \\ 506 \\ 79 \\ +\ 434 \\ \hline \mathbf{1019} \end{array}$$

20. 408 < 480

21. B. 5's

22. **Odd**

23. **Even**

24. **Even**

25. **101**

26. **4 people in front; 7 people behind**

27. **Odd**

28. **No. Half of 5 is $2\frac{1}{2}$, and $2\frac{1}{2}$ birds cannot fly away.**

Lesson 8, Warm-Up

a. **6000**

b. **10,000**

c. **800**

d. **950**

e. **$300**

f. **$850**

g. **1500**

h. **1000**

Problem Solving

$$\begin{array}{r} 246 \\ 264 \\ 426 \\ 462 \\ 624 \\ +\ 642 \\ \hline \mathbf{2664} \end{array}$$

The greater sum (2664) is double the smaller sum (1332).

Lesson 8, Lesson Practice

a. $$\begin{array}{r} 17 \\ -\ 9 \\ \hline \mathbf{8} \end{array}$$

b. $$\begin{array}{r} 12 \\ -\ 8 \\ \hline \mathbf{4} \end{array}$$

c. $$\begin{array}{r} 15 \\ -\ 9 \\ \hline \mathbf{6} \end{array}$$

d. $$\begin{array}{r} 11 \\ -\ 5 \\ \hline \mathbf{6} \end{array}$$

e. $$\begin{array}{r} 17 \\ -\ 8 \\ \hline \mathbf{9} \end{array}$$

f. $$\begin{array}{r} 16 \\ -\ 8 \\ \hline \mathbf{8} \end{array}$$

g. **7 + 8 = 15**
8 + 7 = 15
15 − 7 = 8
15 − 8 = 7

h. **5 + 7 = 12**
7 + 5 = 12
12 − 5 = 7
12 − 7 = 5

Lesson 8, Mixed Practice

1. **3**

2. **1, 3, 5, 7, 9**

3. $$\begin{array}{r} 15 \\ -\ 7 \\ \hline \mathbf{8} \end{array}$$

4. $$\begin{array}{r} {}^{1}\ \ \\ 560 \\ +\ 56 \\ \hline \mathbf{616} \end{array}$$

5.
$$\begin{array}{r} 7 \\ -\ 4 \\ \hline \mathbf{3} \end{array}$$

6.
$$\begin{array}{r} {}^{1} \\ 64 \\ +\ 206 \\ \hline \mathbf{270} \end{array}$$

7. **Eight hundred twelve thousand dollars**

8. **802**

9. **65**

10. **Four hundred forty-four**

11. 6, 12, 18, 24, 30, 36, 42, 48, 54
The rule is **"Count up by sixes."; 54**

12. 3, 6, 9, 12, 15, 18, 21, 24, 27
The rule is **"Count up by threes."; 27**

13.
$$\begin{array}{r} \mathbf{4 + 8 = 12} \\ \mathbf{8 + 4 = 12} \\ \mathbf{12 - 4 = 8} \\ \mathbf{12 - 8 = 4} \end{array}$$

14. **Even**
Example: $3 + 3 = 6$

15.
$$\begin{array}{r} 18 \\ -\ 9 \\ \hline \mathbf{9} \end{array}$$

16.
$$\begin{array}{r} 15 \\ -\ 7 \\ \hline \mathbf{8} \end{array}$$

17.
$$\begin{array}{r} 12 \\ -\ 5 \\ \hline \mathbf{7} \end{array}$$

18.
$$\begin{array}{r} 11 \\ -\ 8 \\ \hline \mathbf{3} \end{array}$$

19.
$$\begin{array}{r} 14 \\ -\ 6 \\ \hline \mathbf{8} \end{array}$$

20.
$$\begin{array}{r} 13 \\ -\ 9 \\ \hline \mathbf{4} \end{array}$$

21.
$$\begin{array}{r} {}^{1} \\ \$_{1}36 \\ \$403 \\ +\ \ \$97 \\ \hline \mathbf{\$536} \end{array}$$

22.
$$\begin{array}{r} {}^{1\,1} \\ 572 \\ 386 \\ +\ \ 38 \\ \hline \mathbf{996} \end{array}$$

23.
$$\begin{array}{r} {}^{1} \\ {}_{1}47 \\ 135 \\ +\ \ 70 \\ \hline \mathbf{252} \end{array}$$

24.
$$\begin{array}{r} {}^{1\,1} \\ \$590 \\ \$306 \\ +\ \ \$75 \\ \hline \mathbf{\$971} \end{array}$$

25.
$$\begin{array}{r} {}^{1\,1} \\ 287 \\ +\ 364 \\ \hline \mathbf{651} \end{array}$$

26. **1000; Even**

27. **9 girls**

28. **$2\frac{1}{2}$ miles**

LESSON 9, WARM-UP

a. **$500**

b. **12,000**

c. **$200**

d. **1500**

e. **40**

f. **400**

g. **4000**

h. **930**

Problem Solving

PTA TPA ATP

(PAT) (TAP) (APT)

Lesson 9, Lesson Practice

a.
$$\begin{array}{r} {\scriptstyle 8} \quad \\ \$4\,\cancel{9}{}^{1}6 \\ -\ \$1\,5\,7 \\ \hline \mathbf{\$3\,3\,9} \end{array}$$

b.
$$\begin{array}{r} {\scriptstyle 39} \quad \\ \cancel{40}{}^{1}0 \\ -\ 13\,6 \\ \hline \mathbf{26\,4} \end{array}$$

c.
$$\begin{array}{r} {\scriptstyle 2} \qquad \\ \$\cancel{3}{}^{1}1\,5 \\ -\ \$2\,6\,4 \\ \hline \mathbf{\$5\,1} \end{array}$$

d.
$$\begin{array}{r} {\scriptstyle 4\ 9} \quad \\ \$\cancel{5}\,\cancel{0}{}^{1}0 \\ -\ \$6\,3 \\ \hline \mathbf{\$4\,3\,7} \end{array}$$

e.
$$\begin{array}{r} {\scriptstyle 3\ {}^{1}2} \quad \\ \cancel{4}\,\cancel{3}{}^{1}5 \\ -\ 7\,6 \\ \hline \mathbf{3\,5\,9} \end{array}$$

f.
$$\begin{array}{r} {\scriptstyle 79} \quad \\ \cancel{80}{}^{1}0 \\ -\ 40\,6 \\ \hline \mathbf{39\,4} \end{array}$$

g.
$$\begin{array}{r} {\scriptstyle 7} \quad \\ \cancel{8}{}^{1}6 \\ -\ 4\,8 \\ \hline \mathbf{3\,8} \end{array}$$

h.
$$\begin{array}{r} {\scriptstyle 0} \qquad \\ \$\cancel{1}{}^{1}3\,2 \\ -\ \$4\,0 \\ \hline \mathbf{\$9\,2} \end{array}$$

i.
$$\begin{array}{r} {\scriptstyle 19} \quad \\ \cancel{20}{}^{1}3 \\ -\ 4\,7 \\ \hline \mathbf{15\,6} \end{array}$$

Lesson 9, Mixed Practice

1.
$$\begin{array}{r} {\scriptstyle 4\ {}^{1}4} \quad \\ \$\cancel{5}\,\cancel{5}{}^{1}0 \\ -\ \$7\,5 \\ \hline \mathbf{\$4\,7\,5} \end{array}$$

2. **0, 2, 4, 6, 8**

3. **9**

4. **10**

5.
$$\begin{array}{r} 15 \\ -\ 7 \\ \hline \mathbf{8} \end{array}$$

6.
7 + 8 = 15
8 + 7 = 15
15 − 7 = 8
15 − 8 = 7

7.
$$\begin{array}{r} {\scriptstyle 1} \quad \\ 190 \\ +\ 119 \\ \hline \mathbf{309} \end{array}$$

8. **540 > 514**

9. **Seven hundred sixty-one thousand, two hundred sixty-six acres**

10. **132**

11.
$$\begin{array}{r} {\scriptstyle 2\ {}^{1}3} \quad \\ \$\cancel{3}\,\cancel{4}{}^{1}6 \\ -\ \$1\,7\,8 \\ \hline \mathbf{\$1\,6\,8} \end{array}$$

12.
$$\begin{array}{r} {\scriptstyle 4} \quad \\ \cancel{5}{}^{1}6 \\ -\ 3\,8 \\ \hline \mathbf{1\,8} \end{array}$$

13.
$$\begin{array}{r} {\scriptstyle 1} \qquad \\ \$\cancel{2}{}^{1}1\,9 \\ -\ \$7\,3 \\ \hline \mathbf{\$1\,4\,6} \end{array}$$

14.
$$\begin{array}{r} {\scriptstyle 59} \quad \\ \cancel{60}{}^{1}0 \\ -\ 32\,1 \\ \hline \mathbf{27\,9} \end{array}$$

15.
$$\begin{array}{r} \overset{29}{\cancel{30}}{}^{1}0 \\ -\ 124 \\ \hline \mathbf{176} \end{array}$$

16.
$$\begin{array}{r} \$\overset{49}{\cancel{50}}{}^{1}0 \\ -\ \$246 \\ \hline \mathbf{\$254} \end{array}$$

17.
$$\begin{array}{r} \overset{5}{\cancel{6}}{}^{1}08 \\ -\ 314 \\ \hline \mathbf{294} \end{array}$$

18.
$$\begin{array}{r} \overset{3}{\cancel{4}}\,\overset{10}{\cancel{1}}{}^{1}5 \\ -\ 378 \\ \hline \mathbf{37} \end{array}$$

19.
$$\begin{array}{r} {}^{1\,1} \\ \$787 \\ \$156 \\ +\ \$324 \\ \hline \mathbf{\$1267} \end{array}$$

20.
$$\begin{array}{r} {}^{2\,1} \\ 573 \\ 90 \\ +\ 438 \\ \hline \mathbf{1101} \end{array}$$

21.
$$\begin{array}{r} {}^{1\,1} \\ \$645 \\ \$489 \\ +\ \$65 \\ \hline \mathbf{\$1199} \end{array}$$

22.
$$\begin{array}{r} {}^{1\,1} \\ 429 \\ 85 \\ +\ 671 \\ \hline \mathbf{1185} \end{array}$$

23. 7, 14, 21, 28, 35, 42, 49, 56, 63
63

24. 9, 18, 27, 36, 45, 54, 63, 72, 81
81

25. 8, 16, 24, 32, 40, 48, 56, 64, 72
72

26. 370
Even number

27. 31 − 12 $\bigcirc\!\!\!\!>$ 31 − 15

$$\begin{array}{r} \overset{2}{\cancel{3}}{}^{1}1 \\ -\ 12 \\ \hline 19 \end{array} \qquad \begin{array}{r} \overset{2}{\cancel{3}}{}^{1}1 \\ -\ 15 \\ \hline 16 \end{array}$$

28. $\mathbf{10\frac{1}{2}}$

Lesson 10, Warm-Up

a. **9500**

b. **2000**

c. **$1000**

d. **470**

e. **100**

f. **250**

g. **900**

h. **15**

Problem Solving

RTA	**(TAR)**	**(ART)**
(RAT)	**TRA**	**ATR**

Lesson 10, Lesson Practice

a.
$$\begin{array}{r} 67 \\ -\ 35 \\ \hline 32 \end{array} \nearrow \begin{array}{r} 35 \\ +\ 32 \\ \hline 67 \end{array}$$
m = **32**

b.
$$\begin{array}{r} \overset{3}{\cancel{4}}{}^{1}0 \\ -\ 27 \\ \hline 13 \end{array} \nearrow \begin{array}{r} {}^{1} \\ 13 \\ +\ 27 \\ \hline 40 \end{array}$$
n = **13**

c.

$$
\begin{array}{r} 5 \\ 7\!\!\rightarrow\!21 \\ 9 \\ +\;f \\ \hline 30 \end{array}
\qquad
\begin{array}{r} \overset{2}{3}{}^{1}0 \\ -\;2\,1 \\ \hline 9 \end{array}
\qquad
\begin{array}{r} 5 \\ 7 \\ 9 \\ +\;9 \\ \hline 30 \end{array}
$$

$f = \mathbf{9}$

d.

$$
\begin{array}{r} 15 \\ k \\ 10 \\ +\;25 \\ \hline 70 \end{array}\!\!\!\searrow 50
\qquad
\begin{array}{r} 70 \\ -\;50 \\ \hline 20 \end{array}
\qquad
\begin{array}{r} \overset{1}{1}5 \\ 20 \\ 10 \\ +\;25 \\ \hline 70 \end{array}
$$

$k = \mathbf{20}$

Lesson 10, Mixed Practice

1.
$$\begin{array}{r} \$200 \\ +\;\$467 \\ \hline \mathbf{\$667} \end{array}$$

2.
$4 + 5 = 9$
$5 + 4 = 9$
$9 - 4 = 5$
$9 - 5 = 4$

3. **645**

4. $613 < 630$

5.
$$\begin{array}{r} \overset{5}{6}{}^{1}1 \\ -\;3\,4 \\ \hline 2\,7 \end{array}
\qquad
\begin{array}{r} \overset{1}{3}4 \\ +\;27 \\ \hline 61 \end{array}$$
$m = \mathbf{27}$

6.
$$\begin{array}{r} \overset{4}{5}\,\overset{1}{1}{}^{0}0 \\ -\;\;\;5\,1 \\ \hline \mathbf{4\,5\,9} \end{array}$$

7. **9**

8. **B. 4's**

9. **Odd**

Example: $3 + 4 = 7$

10. $100 - 10 \;\bigcirc\!\!> \; 100 - 20$
$$\begin{array}{r} \overset{0}{\cancel{1}}\,{}^{1}0\,0 \\ -\;\;1\,0 \\ \hline 9\,0 \end{array}
\qquad
\begin{array}{r} \overset{0}{\cancel{1}}\,{}^{1}0\,0 \\ -\;\;2\,0 \\ \hline 8\,0 \end{array}$$

11.
$$\begin{array}{r} \$\overset{2}{3}\,\overset{15}{6}\,{}^{1}3 \\ -\;\$1\,7\,9 \\ \hline \mathbf{\$1\,8\,4} \end{array}$$

12.
$$\begin{array}{r} \overset{39}{4}\overset{}{0}{}^{1}0 \\ -\;1\,7\,6 \\ \hline \mathbf{2\,2\,4} \end{array}$$

13.
$$\begin{array}{r} \$\overset{4}{5}\,\overset{16}{7}{}^{1}0 \\ -\;\;\$9\,1 \\ \hline \mathbf{\$4\,7\,9} \end{array}$$

14.
$$\begin{array}{r} \overset{49}{5}\overset{}{0}{}^{1}4 \\ -\;1\,7\,5 \\ \hline \mathbf{3\,2\,9} \end{array}$$

15.
$$\begin{array}{r} \overset{1\,2}{\$3}67 \\ \$48 \\ +\;\$135 \\ \hline \mathbf{\$550} \end{array}$$

16.
$$\begin{array}{r} \overset{1\,1}{1}79 \\ 484 \\ +\;201 \\ \hline \mathbf{864} \end{array}$$

17.
$$\begin{array}{r} \overset{1\,1}{\$305} \\ \$897 \\ +\;\$725 \\ \hline \mathbf{\$1927} \end{array}$$

18.
$$\begin{array}{r} \overset{1}{}32 \\ _{1} \\ 248 \\ +\;165 \\ \hline \mathbf{445} \end{array}$$

19.
$$\begin{array}{r} \$\overset{3}{4}\,\overset{15}{6}{}^{1}3 \\ -\;\;\$8\,5 \\ \hline \mathbf{\$3\,7\,8} \end{array}$$

20.
$$\begin{array}{r} ^{1\,1}\\ 432 \\ 84 \\ +\ 578 \\ \hline \mathbf{1094} \end{array}$$

21.
$$\begin{array}{r} \overset{3}{\cancel{4}}{}^{1}2 \\ -\ 1\,8 \\ \hline 2\,4 \end{array} \nearrow \begin{array}{r} \overset{1}{1}8 \\ +\ 24 \\ \hline 42 \end{array}$$
w = **24**

22.
$$\begin{array}{r} \overset{7}{\cancel{8}}{}^{1}0 \\ -\ 1\,2 \\ \hline 6\,8 \end{array} \nearrow \begin{array}{r} \overset{1}{1}2 \\ +\ 68 \\ \hline 80 \end{array}$$
r = **68**

23. The rule is "Count up by threes."
15, 18, 21, 24

24. The rule is "Count up by fours."
20, 24, 28, 32

25. The rule is "Count up by sixes."
30, 36, 42, 48

26. **Ten $100 bills**

27. **$5\frac{1}{2}$ inches**

28. **Half of 37,295 is not a whole number, because 37,295 is an odd number.**

Investigation 1

(a) **How many miles did the troop hike in the morning?**

(b) **How many miles did the troop hike in the afternoon?**

(c) **Altogether, how many miles did the troop hike?**

(d) **How much money did Jack have when he went to the store?**

(e) **How much money did Jack spend at the store?**

(f) **How much money did Jack have when he left the store?**

(g) **How many pens are there at the Lazy W Ranch?**

(h) **How many cows are in each pen?**

(i) **Altogether, how many cows are at the Lazy W Ranch?**

(j) **How old is Abe?**

(k) **How old is Gabe?**

(l) **Abe is how much younger than Gabe?**
Gabe is how much older than Abe?

Lesson 11, Warm-Up

a. **9200**

b. **2000**

c. **500**

d. **820**

e. **50**

f. **400**

g. **14**

h. **24**

Problem Solving

Sample answers:
two, tow;
ten, net;
eat, ate, tea;
tar, rat, art;
apt, pat, tap

Lesson 11, Lesson Practice

a. **\$24 + m = \$41**

$$\begin{array}{r} \$\overset{3}{\not 4}{}^{1}1 \\ -\ \$24 \\ \hline \$17 \end{array} \quad \rightarrow \quad \begin{array}{r} \overset{1}{\$24} \\ +\ \$17 \\ \hline \$41 \end{array}$$

\$17

b. **S + 16 = 30**

$$\begin{array}{r} \overset{2}{\not 3}{}^{1}0 \\ -\ 16 \\ \hline 14 \end{array} \quad \rightarrow \quad \begin{array}{r} \overset{1}{14} \\ +\ 16 \\ \hline 30 \end{array}$$

14 laps

c. **See student work. Answer to the question is \$36.**

Lesson 11, Mixed Practice

1. **13 + p = 21**

$$\begin{array}{r} \overset{1}{\not 2}{}^{1}1 \text{ points} \\ -\ 13 \text{ points} \\ \hline \mathbf{8 \text{ points}} \end{array}$$

2. **S + 29 = 62**

$$\begin{array}{r} \overset{5}{\not 6}{}^{1}2 \text{ points} \\ -\ 29 \text{ points} \\ \hline \mathbf{33 \text{ points}} \end{array}$$

3. **397 + 406 = t**

$$\begin{array}{r} \overset{1\,1}{397} \text{ miles} \\ +\ 406 \text{ miles} \\ \hline \mathbf{803 \text{ miles}} \end{array}$$

4.
8 + 9 = 17
9 + 8 = 17
17 − 8 = 9
17 − 9 = 8

5. **312**

6. $\underbrace{8 + 7 + 6}_{21} \ \circledequal\ \underbrace{6 + 7 + 8}_{21}$

7. **80,000 > 18,000**

8. **40 − 14 = 26**

9. **Even**

Example:

$$\begin{array}{r} 3 \\ 5 \\ +\ 2 \\ \hline 10 \end{array}$$

10. **\$408.70**

11.

$$\begin{array}{r} \$8\ \overset{6}{\not 7}{}^{1}2 \\ -\ \$5\ 6 \\ \hline \mathbf{\$8\ 1\ 6} \end{array}$$

12.

$$\begin{array}{r} \overset{6}{\not 7}{}^{1}0\ 6 \\ -\ 1\ 3\ 4 \\ \hline \mathbf{5\ 7\ 2} \end{array}$$

13.

$$\begin{array}{r} \$\overset{79}{\not 8\not 0}{}^{1}0 \\ -\ \$13\ 9 \\ \hline \mathbf{\$66\ 1} \end{array}$$

14.

$$\begin{array}{r} \overset{2}{\not 3}{}^{1}6\ 5 \\ -\ 2\ 8\ 5 \\ \hline \mathbf{8\ 0} \end{array}$$

15.

$$\begin{array}{r} \overset{59}{\not 6\not 0}{}^{1}0 \\ -\ 57\ 8 \\ \hline 2\ 2 \end{array}$$

A = **22**

16.

$$\begin{array}{r} \overset{1\,1}{\$640} \\ \$152 \\ +\ \$749 \\ \hline \mathbf{\$1541} \end{array}$$

17.

$$\begin{array}{r} \overset{1\,1}{365} \\ 294 \\ +\ 716 \\ \hline \mathbf{1375} \end{array}$$

18.

$$\begin{array}{r} \overset{1\,1}{\$475} \\ \$233 \\ +\ \$76 \\ \hline \mathbf{\$784} \end{array}$$

19.
```
   2 ¹0
  $3̸ 1̸ ¹7
 −  $5 8
  $2 5 9
```
$2 5 9

20.
```
 433
  56 >489        8
+  Q           4 9̸ ¹7
 497         − 4 8 9
                   8
```
Q = **8**

21.
```
  15
−  7
   8
```
w = **8**

22.
```
  6
  7̸ ¹0
− 1 5
  5 5
```
y = **55**

23. The rule is "Count up by nines."
45, 54, 63, 72

24. The rule is "Count up by eights."
40, 48, 56, 64

25. 7, 14, 21, 28, 35, 42, 49, 56, 63, 70
70

26. **A.** 

27. **$1\frac{1}{2}$ miles**

28. **5; $2\frac{1}{2}$**

29. **The total amount in the story should be $33. See student work.**

Lesson 12, Warm-Up

a. **7000**

b. **500**

c. **150**

d. **1500**

e. **60**

f. **200**

g. **45**

h. **11**

Problem Solving

Lance, Molly, José;
Lance, José, Molly;
Molly, Lance, José;
Molly, José, Lance;
José, Lance, Molly;
José, Molly, Lance

Lesson 12, Lesson Practice

a. **A.** ____________

b. 

c. ____________

d. **Ray should be neither horizontal nor vertical.**
One possibility: 

e. 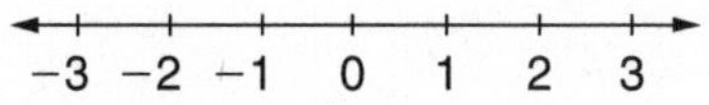

f. **5 unit segments**

g. The rule is "Count down by twos."
4, 2, 0, −2, −4

h. **−3 < 3** or **3 > −3**

i. **2**

j. **14**

Lesson 12, Mixed Practice

1. 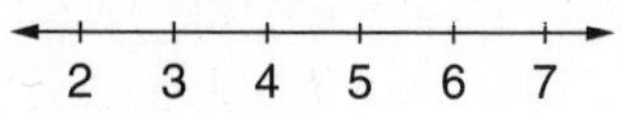

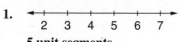

5 unit segments

2. 

3. **94 + 86 = t**

$$\begin{array}{r} \scriptstyle 1 \\ 94 \text{ pounds} \\ +\ 86 \text{ pounds} \\ \hline \textbf{180 pounds} \end{array}$$

4. **86 + p = 110**

$$\begin{array}{r} \scriptstyle 0\ {}^{1}0 \\ \cancel{1}\,\cancel{1}{}^{1}0 \text{ pounds} \\ -\ \ 8\ 6 \text{ pounds} \\ \hline \textbf{2 4 pounds} \end{array}$$

5.
$$\begin{array}{r} \scriptstyle 7\ {}^{1}5 \\ \cancel{8}\,\cancel{6}{}^{1}2 \\ -\ \ 7\ 9 \\ \hline \textbf{7 8 3} \end{array}$$

6.
$$\begin{array}{r} \scriptstyle 3\ {}^{1}1 \\ \$\cancel{4}\,\cancel{2}{}^{1}0 \\ -\ \$1\ 3\ 7 \\ \hline \textbf{\$2 8 3} \end{array}$$

7.
$$\begin{array}{r} \scriptstyle 4 \\ \cancel{5}{}^{1}0\ 8 \\ -\ \ 9\ 6 \\ \hline \textbf{4 1 2} \end{array}$$

8.
$$\begin{array}{r} \scriptstyle 49 \\ \$\cancel{50}{}^{1}0 \\ -\ \$13\ 6 \\ \hline \textbf{\$36 4} \end{array}$$

9.
$$\begin{array}{r} \scriptstyle 2 \\ \$248 \\ \$514 \\ +\ \ \$18 \\ \hline \textbf{\$780} \end{array}$$

10.
$$\begin{array}{r} \scriptstyle 1\,1 \\ 907 \\ 45 \\ +\ 653 \\ \hline \textbf{1605} \end{array}$$

11.
$$\begin{array}{r} \scriptstyle 1\,1 \\ \$367 \\ \$425 \\ +\ \$740 \\ \hline \textbf{\$1532} \end{array}$$

12.
$$\begin{array}{r} 568 \\ -\ 427 \\ \hline 141 \end{array}$$
w = **141**

13.
$$\begin{array}{r} 38 \\ 427 \\ +\quad P \\ \hline 475 \end{array} \quad (38 + 427 = 465) \qquad \begin{array}{r} 475 \\ -\ 465 \\ \hline 10 \end{array}$$
P = **10**

14.
$$\begin{array}{r} \scriptstyle 4\ {}^{1}7 \\ \$\cancel{5}\,\cancel{8}{}^{1}0 \\ -\ \ \$9\ 4 \\ \hline \textbf{\$4 8 6} \end{array}$$

15. **B. 50 and 60**

16. **18,000 < 80,000**

17. **4 + 6 = 10**
6 + 4 = 10
10 − 4 = 6
10 − 6 = 4

18. **Odd**
Example: 7 − 4 = 3

19.
$$\begin{array}{r} \scriptstyle 4 \\ 1\ \cancel{5}{}^{1}0 \\ -\ \ 1\ 8 \\ \hline 1\ 3\ 2 \end{array}$$
m = **132**

20.
$$\begin{array}{r} \scriptstyle 4 \\ \cancel{5}{}^{1}1 \\ -\ 1\ 2 \\ \hline 3\ 9 \end{array}$$
y = **39**

21. $x + y$ ⓧ 19, with the circled symbol being $<$

22. The rule is "Count up by twos."
8, 10, 12, 14, 16, 18

23. The rule is "Count up by threes."
12, 15, 18, 21, 24, 27

24. The rule is "Count up by fours."
16, 20, 24, 28, 32, 36

25. The rule is "Count down by fives."
15, 10, 5, 0, −5, −10

26. Five thousand, two hundred eighty

27. Even

28. The total amount in the story should be **$11.37.**
See student work.

29.
$$\begin{array}{r} \overset{2\ 1}{\$1.50} \\ \$0.65 \\ \$0.65 \\ +\ \$1.79 \\ \hline \$4.59 \end{array}$$

LESSON 13, WARM-UP

a. **6000**

b. **1900**

c. **450**

d. **1500**

e. **160**

f. **300**

g. **75**

h. **18**

Problem Solving

We write a 9 in the ones place of the bottom addend, because $4 + 9 = 13$. The 1 carries to the tens column. We write a 6 in the tens place of the top addend because $1 + 3 + 6 = 10$. The one carries to the hundreds column. We write a 6 in the hundreds place of the sum because $1 + 3 + 2 = 6$.

$$\begin{array}{r} 3\underline{6}4 \\ +\ 239 \\ \hline \underline{6}03 \end{array}$$

LESSON 13, LESSON PRACTICE

a. 4×8

b. 3×25

c. 3×6 or 6×3

d.
$$\begin{array}{r} \overset{1}{\$5.26} \\ +\ \$8.92 \\ \hline \$14.18 \end{array}$$

e.
$$\begin{array}{r} \$\overset{2}{3}.\overset{1}{2}7 \\ -\ \$2.65 \\ \hline \$0.62 \end{array}$$

f.
$$\begin{array}{r} \$10.00 \\ \$3.75 \\ +\ \$2.00 \\ \hline \$15.75 \end{array}$$

g.
$$\begin{array}{r} \$\overset{4}{5}.\overset{9}{\cancel{0}}{}^{1}0 \\ -\ \$1.87 \\ \hline \$3.13 \end{array}$$

LESSON 13, MIXED PRACTICE

1.
$$\xleftarrow{\quad}\underset{\substack{0\ \ 1\ \ 2\ \ 3\ \ 4\ \ 5\ \ 6\ \ 7\ \ 8}}{\rule{0pt}{0pt}}\xrightarrow{\quad}\ ;$$
4 unit segments

2. 卌 ||||

3. $M + 14 = 33$

$$\begin{array}{r} \overset{2}{\cancel{3}}{}^{1}3 \text{ miles} \\ -\ 14 \text{ miles} \\ \hline 19 \text{ miles} \end{array}$$

4. $11 + g = 23$

$$\begin{array}{r} 23 \text{ children} \\ -\ 11 \text{ boys} \\ \hline 12 \text{ girls} \end{array}$$

5. $3 + 7 = 10$
$7 + 3 = 10$
$10 - 7 = 3$
$10 - 3 = 7$

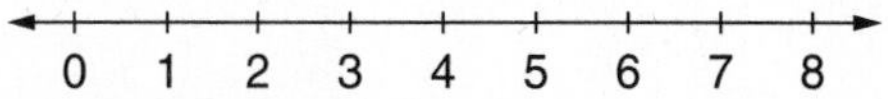

6.
$$\begin{array}{r} \overset{2}{2}5 \text{ miles} \\ 25 \text{ miles} \\ 25 \text{ miles} \\ 25 \text{ miles} \\ +\ 25 \text{ miles} \\ \hline \textbf{125 miles} \end{array}$$

7.
$$\begin{array}{r} \overset{29}{\not{3}\not{0}}{}^{1}0 \\ -\ 11\,4 \\ \hline \textbf{18 6} \end{array}$$

8.
$$\begin{array}{r} \$\overset{4}{\not{5}}.\overset{^{1}5}{\not{6}}{}^{1}0 \\ -\ \$2.\,8\,4 \\ \hline \textbf{\$2. 7 6} \end{array}$$

9.
$$\begin{array}{r} \overset{19}{\not{2}\not{0}}{}^{1}3 \\ -\ 8\,7 \\ \hline \textbf{11 6} \end{array}$$

10.
$$\begin{array}{r} \$\overset{4}{\not{5}}\,\overset{^{1}0}{\not{1}}{}^{1}2 \\ -\ \$1\,2\,3 \\ \hline \textbf{\$3 8 9} \end{array}$$

11.
$$\begin{array}{r} 683 \\ -\ 432 \\ \hline 251 \end{array}$$

B = **251**

12.
$$\begin{array}{r} \$\overset{1}{2}.\overset{1}{5}4 \\ \$5.36 \\ +\ \$0.75 \\ \hline \textbf{\$8.65} \end{array}$$

13.
$$\begin{array}{r} \overset{2}{3}\overset{1}{8}7 \\ 496 \\ +\ 874 \\ \hline \textbf{1757} \end{array}$$

14.
$$\begin{array}{r} \$_{2}\overset{2}{9}7 \\ \$436 \\ +\ \$468 \\ \hline \textbf{\$1001} \end{array}$$

15. 15 − 5 ⓶ 15 − 6

10   9

(15 − 5 **>** 15 − 6, circled)

16. Odd

Example:
$$\begin{array}{r} 2 \\ 4 \\ +\ 3 \\ \hline 9 \end{array}$$

17.
$$\begin{array}{r} \$\overset{1}{4}.\overset{1}{5}6 \\ +\ \$13.76 \\ \hline \textbf{\$18.32} \end{array}$$

18.
$$\begin{array}{r} \$5\,\overset{0}{\not{1}}\,\overset{^{1}1}{\not{2}}{}^{1}7 \\ -\ \$4\,9 \\ \hline \textbf{\$5 0 7 8} \end{array}$$

19.
$$\begin{array}{r} N \\ 27 \\ +\ 123 \\ \hline 153 \end{array} > 150 \qquad \begin{array}{r} 153 \\ -\ 150 \\ \hline 3 \end{array}$$

N = **3**

20.
$$\begin{array}{r} 2\,\overset{4}{\not{5}}\,\overset{^{1}0}{\not{1}}{}^{1}0 \\ -\ 4\,3\,2 \\ \hline \textbf{2 0 7 8} \end{array}$$

21.
$$\begin{array}{r} \$\overset{4}{\not{5}}.\,\overset{9}{\not{0}}{}^{1}0 \\ -\ \$3.\,3\,6 \\ \hline \textbf{\$1. 6 4} \end{array}$$

22.
$$\begin{array}{r} \$5.\overset{1}{0}0 \\ \$3.36 \\ +\ \$0.54 \\ \hline \textbf{\$8.90} \end{array}$$

23. The rule is "Count up by sixes."
24, 30, 36, 42, 48, 54

24. The rule is "Count up by sevens."
28, 35, 42, 49, 56, 63

25. The rule is "Count up by eights."
32, 40, 48, 56, 64, 72

26. The rule is "Count up by nines."
36, 45, 54, 63, 72, 81

27. 3 + 3 + 3 + 3 ⊜ 4 + 4 + 4

4 × 3   3 × 4

28. **6 × 7**

29. **4 × 5 or 5 × 4**

Lesson 14, Warm-Up

a. **3000**

b. **2000**

c. **840**

d. **$10.00**

e. **20**

f. **50¢; 75¢; $1.00**

Problem Solving

We write an 8 in the ones place of the top addend because 8 + 4 = 12. The 1 carries to the tens column. We write a 2 in the tens place of the sum because 2 + 9 + 1 = 12. The 1 carries to the hundreds column. We write a 4 in the hundreds place of the bottom addend because 1 + 5 + 4 = 10. The 1 carries to the thousands column, where we place it in the sum.

$$\begin{array}{r} \mathbf{52\underline{8}} \\ \mathbf{+\ \underline{4}94} \\ \hline \mathbf{\underline{1}0\underline{2}2} \end{array}$$

Lesson 14, Lesson Practice

a. **22 + 12 = 34**

b.
$$\begin{array}{r} \mathbf{27} \\ \mathbf{+\ 29} \\ \hline \mathbf{56} \end{array}$$

c. $6 + 8 = 14$

$w =$ **14**

d. $17 + y = 23$

$$\begin{array}{r} \overset{1}{\cancel{2}}{}^{1}3 \\ -\ 1\ 7 \\ \hline 6 \end{array}$$

$y =$ **6**

e.
$$\begin{array}{r} \overset{1}{4}8 \\ +\ 24 \\ \hline 72 \end{array}$$

$N =$ **72**

f.
$$\begin{array}{r} 63 \\ -\ 20 \\ \hline 43 \end{array}$$

$P =$ **43**

g.
$$\begin{array}{r} \overset{1}{1}4 \\ +\ 36 \\ \hline 50 \end{array}$$

$Q =$ **50**

h.
$$\begin{array}{r} \overset{3}{\cancel{4}}{}^{1}2 \\ -\ 2\ 4 \\ \hline 1\ 8 \end{array}$$

$R =$ **18**

Lesson 14, Mixed Practice

1. 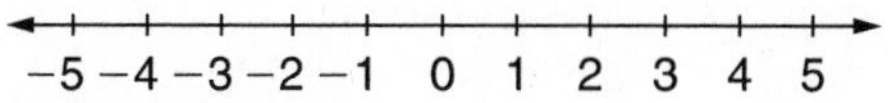

;

4 unit segments

2. **Four dollars and forty-eight cents**

3. **818,080**

4. 

5. **85 + *p* = 260**

$$\begin{array}{r} \overset{1}{\cancel{2}}\ \overset{15}{\cancel{6}}{}^{1}0 \text{ pages} \\ -\quad 8\ 5 \text{ pages} \\ \hline \mathbf{1\ 7\ 5\ pages} \end{array}$$

6. **32 + 24 = *t***

$$\begin{array}{r} 32 \text{ ounces} \\ +\ 24 \text{ ounces} \\ \hline \mathbf{56\ ounces} \end{array}$$

7. **56 < 65**

8. **312**

9.
$$\begin{array}{r} \$4\,\overset{2}{\cancel{3}}.\overset{\overset{10}{\cancel{1}}}{1}0 \\ -\ \ \$1.\,5\,4 \\ \hline \$4\,1.\,5\,6 \end{array}$$

10.
$$\begin{array}{r} \$\overset{2}{\cancel{3}}.\,\overset{9}{\cancel{0}}\,\overset{11}{1} \\ -\ \ \$1.\,0\,3 \\ \hline \$1.\,9\,8 \end{array}$$

11.
$$\begin{array}{r} \overset{59}{\cancel{6}}\cancel{0}\,\overset{10}{0} \\ -\ 3\,6\,4 \\ \hline 2\,3\,6 \end{array}$$

$M = \mathbf{236}$

12.
$$\begin{array}{r} 4\,\overset{5}{\cancel{6}}\,\overset{11}{\cancel{2}}\,5 \\ -\ 1\,3\,8\,7 \\ \hline 3\,2\,3\,8 \end{array}$$

13.
$$\begin{array}{r} \overset{1}{}\$3.67 \\ \$4.12 \\ +\ \$5.01 \\ \hline \$12.80 \end{array}$$

14.
$$\begin{array}{r} \overset{1\,1}{}\$573 \\ \$96 \\ +\ \$427 \\ \hline \$1096 \end{array}$$

15.
$$\begin{array}{r} \overset{1}{}\overset{1}{}68 \\ 532 \\ +\ 176 \\ \hline 776 \end{array}$$

16.
$$\begin{array}{r} \overset{5}{\cancel{6}}\,\overset{12}{\cancel{3}}\,4 \\ -\ 4\,3\,6 \\ \hline 1\,9\,8 \end{array}$$

$Y = \mathbf{198}$

17.
$$\begin{array}{r} \overset{9}{\cancel{10}}\,\overset{10}{0} \\ -\ \ 4\,8 \\ \hline 5\,2 \end{array}$$

$N = \mathbf{52}$

18.
$$\begin{array}{r} \$\overset{2}{\cancel{3}}\overset{1}{1}.4\,0 \\ -\ \$1\,3.4\,0 \\ \hline \$1\,8.0\,0 \end{array}$$

19.
$$\begin{array}{r} 6 \\ 48 \diagdown 63 \\ 9 \\ +\ \ W \\ \hline 100 \end{array} \qquad \begin{array}{r} \overset{9}{\cancel{10}}\,\overset{10}{0} \\ -\ \ 6\,3 \\ \hline 3\,7 \end{array}$$

$W = \mathbf{37}$

20.
$$\begin{array}{r} \overset{1\,1\,1}{}3714 \\ 56 \\ +\ \ 459 \\ \hline 4229 \end{array}$$

21. $32 + 18 = \mathbf{50}$

22. $3, 0, -3, -6, -9, -12$

23.
$2 + 8 = 10$
$8 + 2 = 10$
$10 - 2 = 8$
$10 - 8 = 2$

24.
$$\begin{array}{r} 12 \\ +\ 17 \\ \hline 29 \end{array}$$

$N = \mathbf{29}$

25.
$$\begin{array}{r} \overset{1\,1}{}125 \\ +\ 175 \\ \hline 300 \end{array}$$

$P = \mathbf{300}$

26. 4×10

27. **More than half**

28.

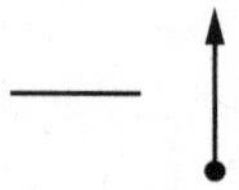

29.
$$\begin{array}{r} {}_2 6 \text{ points} \\ 13 \text{ points} \\ 7 \text{ points} \\ +\ 6 \text{ points} \\ \hline 32 \text{ points} \end{array}$$

LESSON 15, WARM-UP

a. **150**

b. **1500**

c. **50**

d. **500**

e. **280**

f. **45**

g. **550**

h. **$3.25**

i. **75¢; $1.25; $1.50**

j. **30**

Problem Solving

Billy: B
Ricardo: R
Sherry: S

1st	2nd	3rd
B	**R**	**S**
B	**S**	**R**
R	**B**	**S**
R	**S**	**B**
S	**B**	**R**
S	**R**	**B**

Lesson 15, Activity

	0	1	2	3	4	5	6	7	8	9	10
0	0	0	0	0	0	0	0	0	0	0	0
1	0	1	2	3	4	5	6	7	8	9	10
2	0	2	4	6	8	10	12	14	16	18	20
3	0	3	6	9	12	15	18	21	24	27	30
4	0	4	8	12	16	20	24	28	32	36	40
5	0	5	10	15	20	25	30	35	40	45	50
6	0	6	12	18	24	30	36	42	48	54	60
7	0	7	14	21	28	35	42	49	56	63	70
8	0	8	16	24	32	40	48	56	64	72	80
9	0	9	18	27	36	45	54	63	72	81	90
10	0	10	20	30	40	50	60	70	80	90	100

Lesson 15, Lesson Practice

a. **20**

b. **12**

c. **18**

d. **80**

e. $6 \times 7 =$ **42**

f. $8 \times 9 =$ **72**

g. $8 \times 4 =$ **32**

h. $3 \times 10 =$ **30**

i. $50 \times 0 =$ **0**

j. $25 \times 1 =$ **25**

k. **Factors**

Lesson 15, Mixed Practice

1. 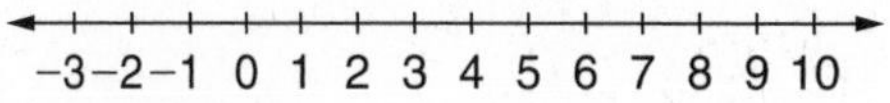

5 unit segments

2. **8 people**

3. 

4.
1 + 9 = 10
9 + 1 = 10
10 − 1 = 9
10 − 9 = 1

5. **194 + *p* = 429**

$$\begin{array}{r} \overset{3}{\not{4}}\,{}^{1}2\,9 \text{ pages} \\ -\ 1\,9\,4 \text{ pages} \\ \hline \mathbf{2\,3\,5 \text{ pages}} \end{array}$$

6. **172 + 168 + 189 = *t***

$$\begin{array}{r} {}^{2\,1} \\ 172 \text{ pages} \\ 168 \text{ pages} \\ +\ 189 \text{ pages} \\ \hline \mathbf{529 \text{ pages}} \end{array}$$

7. $3 \times 6 =$ **18**

8. $4 \times 8 =$ **32**

9. $7 \times 9 = \mathbf{63}$

10. $9 \times 10 = \mathbf{90}$

11.
$$\begin{array}{r} 100 \\ +\ 819 \\ \hline 919 \end{array}$$
$A = \mathbf{919}$

12.
$$\begin{array}{r} {\scriptstyle 5\ \ 9\ \ } \\ \$6.0^{1}0 \\ -\ \$5.\,4\,3 \\ \hline \mathbf{\$0.\,5\,7} \end{array}$$

13.
$$\begin{array}{r} {\scriptstyle 4\ \ 9\ \ } \\ \$5\,0^{1}1 \\ -\ \$2\,5\,6 \\ \hline \mathbf{\$2\,4\,5} \end{array}$$

14.
$$\begin{array}{r} {\scriptstyle 4\ {}^{1}0\ \ } \\ 5\,1^{1}0 \\ -\ 2\,5\,6 \\ \hline 2\,5\,4 \end{array}$$
$Q = \mathbf{254}$

15.
$$\begin{array}{r} {\scriptstyle 2\,1\ } \\ \$564 \\ \$796 \\ +\ \$287 \\ \hline \mathbf{\$1647} \end{array}$$

16.
$$\begin{array}{r} {\scriptstyle 3\ {}^{1}2\ \ } \\ 4\,3^{1}2 \\ -\ \ \ 9\,6 \\ \hline 3\,3\,6 \end{array}$$
$N = \mathbf{336}$

17.
$$\begin{array}{r} {\scriptstyle 1\,1\ } \\ 608 \\ 930 \\ +\ 762 \\ \hline \mathbf{2300} \end{array}$$

18.
$$\begin{array}{r} {\scriptstyle 1\ 1\ } \\ \$4.36 \\ \$2.18 \\ +\ \$3.94 \\ \hline \mathbf{\$10.48} \end{array}$$

19.
$$\begin{array}{r} 360 \\ 47 \\ +\ \ B \\ \hline 518 \end{array} \quad (360 + 47 = 407) \qquad \begin{array}{r} 518 \\ -\ 407 \\ \hline 111 \end{array}$$
$B = \mathbf{111}$

20.
$$\begin{array}{r} {\scriptstyle 9\ \ \ 9\ \ } \\ \$10.0^{1}0 \\ -\ \ \$9.\,1\,8 \\ \hline \mathbf{\$0.\,8\,2} \end{array}$$

21. **132**

22. $5 + 5 + 5$ ⊜ 3×5

23. $\mathbf{12 = 10 + 2}$

24. The rule is "Count up by eighths."
56

25. **$880.08**

26. 346,129 ⧀ 346,132

27. Half of 12 is 6. Half of 6 is **3.**

28. $\mathbf{4 \times 6}$ **or** $\mathbf{6 \times 4}$

29. $\mathbf{-1 < 4}$ **or** $\mathbf{4 > -1}$

Lesson 16, Warm-Up

a. **15**

b. **150**

c. **1500**

d. **30**

e. **300**

f. **3000**

g. **$1.25**

h. **156**

i. **42**

j. **44**

Problem Solving

The digits in the tens column indicate that the sum of the ones column must be 18. We write a 9 in the ones place of the top and bottom addends, because 9 + 9 = 18. The one carries to the tens column. We write an 8 in the hundreds place of the top addend because 1 + 8 + 3 = 12. The one carries to the thousands column, where we place it in the sum.

$$\begin{array}{r} \underline{8}6\underline{9} \\ +\ 37\underline{9} \\ \hline \mathbf{\underline{1}248} \end{array}$$

Lesson 16, Lesson Practice

a. **500 − *A* = 293**

$$\begin{array}{r} \overset{49}{\cancel{50}}{}^{1}0 \text{ runners} \\ -\ 29\,3 \text{ runners} \\ \hline \mathbf{20\,7 \text{ runners}} \end{array}$$

b. **S − $85 = $326**

$$\begin{array}{r} \overset{1\,1}{\$326} \\ +\ \ \$85 \\ \hline \mathbf{\$411} \end{array}$$

c. **26 − 14 = *L***

$$\begin{array}{r} 26 \text{ members} \\ -\ 14 \text{ members} \\ \hline \mathbf{12 \text{ members}} \end{array}$$

d. **See student work; answer to the question is $8.**

Lesson 16, Mixed Practice

1. **$26 + *D* = $32**

$$\begin{array}{r} \$\overset{2}{\cancel{3}}{}^{1}2 \\ -\ \$2\,6 \\ \hline \mathbf{\$6} \end{array}$$

2. 𝍸 𝍸 𝍸

3. **Two hundred five dollars and fifty cents**

4. **6 + 8 = 14**
   8 + 6 = 14
   14 − 6 = 8
   14 − 8 = 6

5. **400 + 600 = *t***

$$\begin{array}{r} 400 \text{ ink spots} \\ +\ 600 \text{ ink spots} \\ \hline \mathbf{1000 \text{ ink spots}} \end{array}$$

6. **C − 24 = 52**

$$\begin{array}{r} 52 \text{ chairs} \\ +\ 24 \text{ chairs} \\ \hline \mathbf{76 \text{ chairs}} \end{array}$$

7. **$24 − $8 = *L***

$$\begin{array}{r} \$\overset{1}{\cancel{2}}{}^{1}4 \\ -\ \ \$8 \\ \hline \mathbf{\$1\,6} \end{array}$$

8. 3 × 7 = **21**

9. 6 × 7 = **42**

10. 3 × 8 = **24**

11. 7 × 10 = **70**

12.

$$\begin{array}{r} {}_{1}\overset{1}{5}6 \\ +\ 256 \\ \hline 312 \end{array}$$

B = **312**

13.

$$\begin{array}{r} \overset{8}{\cancel{9}}{}^{1}0\,0 \\ -\ \ 9\,0 \\ \hline 8\,1\,0 \end{array}$$

C = **810**

14.

$$\begin{array}{r} \$\overset{3}{\cancel{4}}.{}^{1}1\,8 \\ -\ \$2.\,8\,8 \\ \hline \mathbf{\$1.\,3\,0} \end{array}$$

15.

$$\begin{array}{r} \$\overset{39}{\cancel{40}}{}^{1}6 \\ -\ \ \$27\,8 \\ \hline \mathbf{\$12\,8} \end{array}$$

16.

$$\begin{array}{r} \overset{1\,1}{\$357} \\ \$946 \\ +\ \ \$130 \\ \hline \mathbf{\$1433} \end{array}$$

17. $\begin{array}{r} \overset{9}{\cancel{10}}\,\overset{9}{\cancel{0}}{}^{1}0 \\ -\quad 8\,4\,3 \\ \hline 1\,5\,7 \end{array}$

$G = \mathbf{157}$

18. $\begin{array}{r} \overset{1\,1}{365} \\ 52 \\ +\ 548 \\ \hline \mathbf{965} \end{array}$

19. $\begin{array}{r} \overset{2\ 2}{\$3.15} \\ \$2.87 \\ +\ \$1.98 \\ \hline \mathbf{\$8.00} \end{array}$

20. **Odd**
Example:
$3 \times 5 = 15$

21. **C.** ⟷

22. **840 > 814**

23. The rule is "Count by sixes."
42

24. $\begin{array}{ccc} 4 \times 3 & \textcircled{=} & 2 \times 6 \\ 12 & & 12 \end{array}$

25. $\begin{array}{r} \overset{5}{\cancel{6}}{}^{1}3 \\ -\ 3\,6 \\ \hline 2\,7 \end{array}$

$y = \mathbf{27}$

26. **Vertical**

27. **50¢; 25¢**

28. **6 × 8 or 8 × 6**

29. **The amount that remains should be $17. See student work.**

Lesson 17, Warm-Up

a. **24**

b. **240**

c. **2400**

d. **40**

e. **400**

f. **4000**

g. **92**

h. **184**

i. **520**

j. **49**

Problem Solving

$40 \times \$0.25 = \$10.00$
$50 \times \$0.10 = \$5.00$
One roll of quarters has the same value as 2 rolls of dimes.

Lesson 17, Lesson Practice

a. $\begin{array}{r} \overset{3}{\$36} \\ \times\quad 5 \\ \hline \mathbf{\$180} \end{array}$

b. $\begin{array}{r} 50 \\ \times\ 8 \\ \hline \mathbf{400} \end{array}$

c. $\begin{array}{r} \overset{3\ 2}{\$0.43} \\ \times\quad 7 \\ \hline \mathbf{\$3.01} \end{array}$

d. $\begin{array}{r} \overset{3}{340} \\ \times\quad 8 \\ \hline \mathbf{2720} \end{array}$

e. $\begin{array}{r} \overset{2\ 3}{\$7.68} \\ \times\quad 4 \\ \hline \mathbf{\$30.72} \end{array}$

f. $\begin{array}{r} \overset{3}{506} \\ \times\quad 6 \\ \hline \mathbf{3036} \end{array}$

g.
$$\begin{array}{r} {\scriptstyle 6\,2} \\ \$394 \\ \times \quad\; 7 \\ \hline \mathbf{\$2758} \end{array}$$

h.
$$\begin{array}{r} {\scriptstyle 6\;\;} \\ 607 \\ \times \quad 9 \\ \hline \mathbf{5463} \end{array}$$

i.
$$\begin{array}{r} {\scriptstyle 2\;2\;} \\ \$9.68 \\ \times \qquad 3 \\ \hline \mathbf{\$29.04} \end{array}$$

j.
$$\begin{array}{r} \mathbf{\$42} \\ \mathbf{\$42} \\ \mathbf{\$42} \\ + \;\mathbf{\$42} \\ \hline \mathbf{\$168} \end{array} \qquad \begin{array}{r} \mathbf{\$42} \\ \times \quad \mathbf{4} \\ \hline \mathbf{\$168} \end{array}$$

k. $\mathbf{5 \times \$0.25 = \$1.25}$

$$\begin{array}{r} {\scriptstyle 1\;2\;} \\ \$0.25 \\ \times \qquad 5 \\ \hline \mathbf{\$1.25} \end{array}$$

l.
$$\begin{array}{r} {\scriptstyle 1\;\;} \\ \$2.14 \\ \times \qquad 3 \\ \hline \mathbf{\$6.42} \end{array}$$

Lesson 17, Mixed Practice

1. |

2.
$$\begin{array}{r} \text{120 pages} \\ \text{120 pages} \\ + \;\text{120 pages} \\ \hline \textbf{360 pages} \end{array} \qquad \begin{array}{r} \text{120 pages} \\ \times \quad 3 \qquad\quad \\ \hline \textbf{360 pages} \end{array}$$

3. $\mathbf{6 + \mathit{h} = 14}$

$$\begin{array}{r} \text{14 hours} \\ - \;\; \text{6 hours} \\ \hline \textbf{8 hours} \end{array}$$

4. $\mathbf{\mathit{D} - \$1.45 = \$2.65}$

$$\begin{array}{r} {\scriptstyle 1\;1\;} \\ \$2.65 \\ + \;\$1.45 \\ \hline \mathbf{\$4.10} \end{array}$$

5.
$$\begin{array}{r} {\scriptstyle 1\;} \\ 24 \\ \times \;\; 3 \\ \hline \mathbf{72} \end{array}$$

6.
$$\begin{array}{r} {\scriptstyle 2\;} \\ \$36 \\ \times \quad 4 \\ \hline \mathbf{\$144} \end{array}$$

7.
$$\begin{array}{r} {\scriptstyle 2\;} \\ 45 \\ \times \;\; 5 \\ \hline \mathbf{225} \end{array}$$

8.
$$\begin{array}{r} {\scriptstyle 3\;} \\ \$56 \\ \times \quad 6 \\ \hline \mathbf{\$336} \end{array}$$

9.
$$\begin{array}{r} {\scriptstyle 1\;3\;} \\ \$3.25 \\ \times \qquad 6 \\ \hline \mathbf{\$19.50} \end{array}$$

10.
$$\begin{array}{r} {\scriptstyle 2\,1\;} \\ 432 \\ \times \quad 9 \\ \hline \mathbf{3888} \end{array}$$

11.
$$\begin{array}{r} {\scriptstyle 3\;4\;} \\ \$2.46 \\ \times \qquad 7 \\ \hline \mathbf{\$17.22} \end{array}$$

12.
$$\begin{array}{r} {\scriptstyle 5\,3\;} \\ 364 \\ \times \quad 8 \\ \hline \mathbf{2912} \end{array}$$

13.
$$\begin{array}{r} {\scriptstyle 2\;\;{}^{1}0\;\;} \\ \cancel{3}\,\cancel{1}\,{}^{1}6 \\ -\;1\;4\;7 \\ \hline 1\;6\;9 \end{array}$$

$C = \mathbf{169}$

14.
$$\begin{array}{r} {\scriptstyle 3\;{}^{1}1\;\;} \\ \$\cancel{4}.\cancel{2}\,{}^{1}0 \\ -\;\;\$3.7\;5 \\ \hline \mathbf{\$0.4\;5} \end{array}$$

15.
$$\begin{array}{r} {\scriptstyle 59\;\;} \\ \cancel{60}\,{}^{1}4 \\ -\;40\;6 \\ \hline 19\;8 \end{array}$$

$W = \mathbf{198}$

16.

$$\begin{array}{r} 800 \\ +\ \ 73 \\ \hline 873 \end{array}$$

M = **873**

17.

$$\begin{array}{r} 3 \\ N \\ 15 \\ +\ \ 9 \\ \hline 60 \end{array} \quad (N + 15 \to 27) \qquad \begin{array}{r} \overset{5}{\cancel{6}}{}^{1}0 \\ -\ 2\,7 \\ \hline 3\,3 \end{array}$$

N = **33**

18.

$$\begin{array}{r} \overset{1\,1\ \ 1}{\$90.00} \\ \$6.75 \\ \$7.98 \\ +\ \ \$0.02 \\ \hline \mathbf{\$104.75} \end{array}$$

19.

$$\begin{array}{r} \overset{1\ 2}{\$0.24} \\ \times\ \ \ \ 5 \\ \hline \mathbf{\$1.20} \end{array}$$

20.

$$\begin{array}{r} \overset{4}{26} \\ \times\ \ 7 \\ \hline \mathbf{182} \end{array}$$

21. **Even**
Example:

$$\begin{array}{r} 2 \\ \times\ 4 \\ \hline 8 \end{array}$$

22. 12 × 1 (>) 24 × 0
12 0

23. **504,000 < 514,000**

24. The rule is "Count up by sevens."
42

25. **3**

26. **13**

27. **Odd**

28. **3 × 8 or 8 × 3**

29.

$$\begin{array}{r} \overset{2}{\mathbf{24}} \\ \mathbf{\times\ \ 5} \\ \hline \mathbf{120} \end{array}$$

Lesson 18, Warm-Up

a. **96**

b. **92**

c. **170**

d. **4200**

e. **900**

f. **352**

g. **$2.50**

h. **33**

Problem Solving

The only two possibilities for the hundreds digit of the sum are 8 (if 1 is carried from the tens column) and 7. We try 8 as the hundreds digit of the sum, meaning that the sum of the tens column would be greater than 10. Thus, the possible sums of the tens column are 11, 12, 13, and 14. The only unused number less than 4 is 1. So we know that the tens digit of the sum is 1. We write a 6 in the tens place of the top addend because 6 + 5 = 11. We write a 7 in the ones place of the top addend and a 9 in the ones place of the sum (since 7 + 2 = 9).

$$\begin{array}{r} \mathbf{367} \\ \mathbf{+\ 452} \\ \hline \mathbf{819} \end{array}$$

Lesson 18, Lesson Practice

a. 5 × 7 × 6
30 × 7 = **210**

b. 10 × 9 × 8
90 × 8 = **720**

c. 3 × 4 × 25
3 × 100 = **300**

d. $4 \times 3 \times 2 \times 1 \times 0$

$12 \times 2 \times 0$

$24 \times 0 = \mathbf{0}$

e. **16 blocks; Student's multiplication problem should contain the factors 2, 2, and 4 (for example, 2 × 4 × 2 = 16).**

f. $5 \times 6 = 30$
$M = \mathbf{6}$

g. $3 \times 7 = 21$
$B = \mathbf{7}$

h. $3 \times 4 = N \times 2$
$12 = N \times 2$
$12 = 6 \times 2$
$N = \mathbf{6}$

i.
$$\begin{array}{r} 6 \\ \times\ 4 \\ \hline 24 \end{array}$$
$P = \mathbf{6}$

j.
$$\begin{array}{r} 9 \\ \times\ 9 \\ \hline 81 \end{array}$$
$Q = \mathbf{9}$

k.
$$\begin{array}{r} 0 \\ \times\ 9 \\ \hline 0 \end{array}$$
$W = \mathbf{0}$

Lesson 18, Mixed Practice

1. ↔ **horizontal**

↕ **vertical**

2. **223 − *P* = 195**

$$\begin{array}{rl} \overset{1}{\not 2}\,\overset{1}{\not 2}{}^{1}3 & \text{pounds} \\ -\ 1\ 9\ 5 & \text{pounds} \\ \hline \mathbf{2\ 8} & \textbf{pounds} \end{array}$$

3. **14 + *g* = 33**

$$\begin{array}{rl} \overset{2}{\not 3}{}^{1}3 & \text{children} \\ -\ 1\ 4 & \text{boys} \\ \hline \mathbf{1\ 9} & \textbf{girls} \end{array}$$

4. **17 + 14 = *t***

$$\begin{array}{rl} \overset{1}{1}7 & \text{boys} \\ +\ 14 & \text{girls} \\ \hline \mathbf{31} & \textbf{children} \end{array}$$

5. $6 \times 4 \times 5$

$30 \times 4 = \mathbf{120}$

6. $5 \times 6 \times 12$

$60 \times 6 = \mathbf{360}$

7. $5 \times 10 \times 6$

$50 \times 6 = \mathbf{300}$

8. $9 \times 7 \times 10$

$63 \times 10 = \mathbf{630}$

9.
$$\begin{array}{r} \$4\overset{5}{0}7 \\ \times\quad 8 \\ \hline \mathbf{\$3256} \end{array}$$

10.
$$\begin{array}{r} \overset{4}{3}\overset{3}{7}5 \\ \times\quad 6 \\ \hline \mathbf{2250} \end{array}$$

11.
$$\begin{array}{r} \$\overset{7}{4}.\overset{5}{8}6 \\ \times\quad 9 \\ \hline \mathbf{\$43.74} \end{array}$$

12.
$$\begin{array}{r} \overset{5}{3}08 \\ \times\quad 7 \\ \hline \mathbf{2156} \end{array}$$

13. $9 \times 4 = 36$
$G = \mathbf{4}$

14. $\begin{array}{r} {}^{6\,2} \\ \$573 \\ \times \quad 9 \\ \hline \mathbf{\$5157} \end{array}$

15. $8 \times 6 = 48$
$H = \mathbf{6}$

16. $\begin{array}{r} {}^{2\,3} \\ \$7.68 \\ \times \quad 4 \\ \hline \mathbf{\$30.72} \end{array}$

17. $\begin{array}{r} 456 \\ 78 \\ + \quad F \\ \hline 904 \end{array} \; > 534 \qquad \begin{array}{r} {}^{8} \\ \not{9}\,{}^{1}0\,4 \\ -\,5\,3\,4 \\ \hline 3\,7\,0 \end{array}$
$F = \mathbf{370}$

18. $\begin{array}{r} {}^{2} \\ 34 \\ {}_{1}75 \\ 123 \\ + \quad 9 \\ \hline \mathbf{241} \end{array}$

19. $\begin{array}{r} {}^{2\;\;{}^{1}5\;\;{}^{1}6} \\ \$\not{3}\,\not{6}.\,\not{7}\,{}^{1}0 \\ - \quad \$7.\,9\,3 \\ \hline \mathbf{\$2\,8.\,7\,7} \end{array}$

20. $\begin{array}{r} {}^{1} \\ {}_{1}46 \\ +\,354 \\ \hline 400 \end{array}$
$H = \mathbf{400}$

21. $\begin{array}{r} {}^{1\;2} \\ \$0.37 \\ \times \quad 4 \\ \hline \mathbf{\$1.48} \end{array}$

22. **Even**
Example:
$\begin{array}{r} 3 \\ \times\, 4 \\ \hline 12 \end{array}$

23. $6 \times 4 = 8 \times N$
$24 = 8 \times 3$
$N = \mathbf{3}$

24. $\mathbf{8 \times 8 > 9 \times 7}$

25. $\mathbf{7 + 8 = 15}$
$\mathbf{8 + 7 = 15}$
$\mathbf{15 - 7 = 8}$
$\mathbf{15 - 8 = 7}$

26. $\mathbf{3 \times 6 = 18}$ **or** $\mathbf{6 \times 3 = 18}$

27. **One possibility:** $2 \times 4 \times 3$
$2 \times 12 = 24$
$\mathbf{2 \times 4 \times 3 = 24}$

28. The rule is "Count down by twos."
0, −2, −4

29. 3 feet × 2 = **6 feet**

LESSON 19, WARM-UP

a. **60**

b. **90**

c. **192**

d. **260**

e. **200**

f. **664**

g. **$4.50**

h. **100**

Problem Solving

(a) soup: s
eggs: e
ham: h

B	L	D
s	e	h
s	h	e
e	s	h
e	h	s
h	e	s
h	s	e

(b) **4 arrangements**

Lesson 19, Lesson Practice

a. $\begin{array}{r} \mathbf{8} \\ 2\overline{)16} \end{array}$

b. $\begin{array}{r} \mathbf{6} \\ 4\overline{)24} \end{array}$

c. $\begin{array}{r} \mathbf{5} \\ 6\overline{)30} \end{array}$

d. $\begin{array}{r} \mathbf{7} \\ 8\overline{)56} \end{array}$

e. $\begin{array}{r} \mathbf{7} \\ 3\overline{)21} \end{array}$

f. $\begin{array}{r} \mathbf{3} \\ 10\overline{)30} \end{array}$

g. $\begin{array}{r} \mathbf{4} \\ 7\overline{)28} \end{array}$

h. $\begin{array}{r} \mathbf{4} \\ 9\overline{)36} \end{array}$

i. $\mathbf{3 \times 8 = 24}$
$\mathbf{8 \times 3 = 24}$
$\begin{array}{r} \mathbf{8} \\ \mathbf{3\overline{)24}} \end{array} \quad \begin{array}{r} \mathbf{3} \\ \mathbf{8\overline{)24}} \end{array}$

Lesson 19, Mixed Practice

1. $\mathbf{\$45 - D = \$29}$

$$\begin{array}{r} \$\overset{3}{\cancel{4}}{}^{1}5 \\ -\ \$2\,9 \\ \hline \mathbf{\$1\,6} \end{array}$$

2. $\mathbf{243 + 487 + 608 = t}$

$$\begin{array}{r} {}^{1\,1} \\ 243 \text{ cans} \\ 487 \text{ cans} \\ +\ 608 \text{ cans} \\ \hline \mathbf{1338 \text{ cans}} \end{array}$$

3.
$$\begin{array}{r} 6 \text{ desks} \\ 6 \text{ desks} \\ 6 \text{ desks} \\ 6 \text{ desks} \\ +\ 6 \text{ desks} \\ \hline \mathbf{30 \text{ desks}} \end{array} \qquad \begin{array}{r} 6 \text{ desks} \\ \times\ 5 \phantom{\text{ desks}} \\ \hline \mathbf{30 \text{ desks}} \end{array}$$

4. **Four thousand, five hundred eighty-seven dollars and twenty cents**

5. $\mathbf{7 \times 8 = 56}$
$\mathbf{8 \times 7 = 56}$
$\begin{array}{r} \mathbf{8} \\ \mathbf{7\overline{)56}} \end{array} \quad \begin{array}{r} \mathbf{7} \\ \mathbf{8\overline{)56}} \end{array}$

6. $\begin{array}{r} \mathbf{8} \\ 3\overline{)24} \end{array}$

7. $\begin{array}{r} \mathbf{3} \\ 6\overline{)18} \end{array}$

8. $\begin{array}{r} \mathbf{8} \\ 4\overline{)32} \end{array}$

9. $\begin{array}{r} \mathbf{4} \\ 10\overline{)40} \end{array}$

10.
$$\begin{array}{r} {}^{5\,2} \\ \$4.83 \\ \times\ 7 \\ \hline \mathbf{\$33.81} \end{array}$$

11.
$$\begin{array}{r} {}^{4\,7} \\ 659 \\ \times\ 8 \\ \hline \mathbf{5272} \end{array}$$

12.
$$\begin{array}{r} {}^{2} \\ \$706 \\ \times\ 4 \\ \hline \mathbf{\$2824} \end{array}$$

13. $9 \times 6 = 54$
$M = \mathbf{6}$

14. $8 \times 10 \times 7$
$56 \times 10 = \mathbf{560}$

15. $9 \times 8 \times 5$
$9 \times 40 = \mathbf{360}$

16.
$$\begin{array}{r} \$\overset{5}{\cancel{6}}{}^{1}5.\,\overset{3}{\cancel{4}}{}^{1}0 \\ -\ \$1\,9.\,1\,8 \\ \hline \mathbf{\$4\,6.\,2\,2} \end{array}$$

17. $$\begin{array}{r} \overset{39}{\cancel{40}}\ \overset{9}{\cancel{0}}{}^{1}0 \\ -\ 13\ 5\ 7 \\ \hline 26\ 4\ 3 \end{array}$$
T = **2643**

18. $$\begin{array}{r} \overset{1}{{}_{1}269} \\ +\ 1915 \\ \hline 2184 \end{array}$$
R = **2184**

19. $$\begin{array}{r} \overset{1}{907} \\ 415 \\ +\ 653 \\ \hline \mathbf{1975} \end{array}$$

20. $$\begin{array}{r} \overset{1\ 1}{\$3.67} \\ \$4.25 \\ +\ \$7.40 \\ \hline \mathbf{\$15.32} \end{array}$$

21. $$\begin{array}{r} \overset{7}{\cancel{8}}\ \overset{{}^{1}0}{\cancel{1}}{}^{1}3 \\ -\ 4\ 2\ 7 \\ \hline 3\ 8\ 6 \end{array}$$
K = **386**

22. $$\begin{array}{r} 356 \\ L \\ +\ 67 \\ \hline 500 \end{array} \Big\rangle 423 \qquad \begin{array}{r} \overset{49}{\cancel{50}}{}^{1}0 \\ -\ 42\ 3 \\ \hline 7\ 7 \end{array}$$
L = **77**

23. $$\begin{array}{r} \overset{1}{\cancel{2}}\ \overset{{}^{1}4}{\cancel{5}}{}^{1}0 \\ -\ \ 8\ 6 \\ \hline 1\ 6\ 4 \end{array}$$
w = **164**

24. $6 \times 6 = 4N$
$36 = 4 \times 9$
N = **9**

25. **8 × 6 < 7 × 7**

26. **$7\frac{1}{2}$ inches**

27. **3 × 7 = 21 or 7 × 3 = 21**

28. $2 \times 2 \times 2$
$4 \times 2 = 8$
2 × 2 × 2 = 8

29. **Colorado, Missouri, Mississippi**

LESSON 20, WARM-UP

a. **128**

b. **120**

c. **120**

d. **92**

e. **216**

f. **10**

g. **100**

Problem Solving

The hundreds digit of the top addend must be 9 because the sum is greater than 1000. The thousands digit of the sum must be 1, and the hundreds digit of the sum must be 0. The tens digit of the bottom addend must be 9, because the sum of the tens column must be 13. The sum of the ones column must be 10. The ones digit of the sum must be 0, and the ones digit of the top addend must be 9.

$$\begin{array}{r} \mathbf{939} \\ \mathbf{+\ \ 91} \\ \hline \mathbf{1030} \end{array}$$

LESSON 20, LESSON PRACTICE

a. **$2\overline{)10}$, 10 ÷ 2, $\frac{10}{2}$**

b. **$6\overline{)24}$, 24 ÷ 6, $\frac{24}{6}$**

c. **Twenty-one divided by three**

d. **Twelve divided by six**

e. **Thirty divided by five**

f. **$7\overline{)63}$**

g. **$6\overline{)42}$**

h. **$6\overline{)30}$**

i. **Quotient, 7; dividend, 63; divisor, 9**

j. **6**

k. **6**

l. **7**

m. **6**

n. $24 \div 4 \; \textcircled{>} \; 24 \div 6$
   6 　　 4

Lesson 20, Mixed Practice

1. 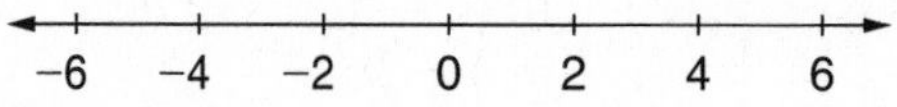

2.
4 × 9 = 36
9 × 4 = 36
36 ÷ 4 = 9
36 ÷ 9 = 4

3. 𝍸 𝍸 𝍸 |

4.
40 pages
40 pages
40 pages
+ 40 pages
160 pages

40 pages
× 4
160 pages

5. **397 + *b* = 806**

79
~~80~~¹6 children
− 39 7 girls
40 9 boys

6.
1 1
526
+ 684
1210

7. **Twenty-four divided by six**

8. **Fifteen divided by three**

9. $\frac{15}{3} \; \textcircled{>} \; \frac{15}{5}$
5 　　 3

10. $8 \times 3 = 24$
M = **3**

11. $10\overline{)90}$ = **9**

12. $\frac{27}{3}$ = **9**

13.
1 1 4
$23.18
× 6
$139.08

14.
5 2 4
4726
× 8
37,808

15.
2 6
$34.09
× 7
$238.63

16. 5 × 6 × 7 ⊜ 7 × 6 × 5
30 × 7 　　 7 × 30
210 　　 210

17.
2 1
$3.52
× 5
$17.60

18.
39 9
$~~40~~. ~~0~~¹0
− $24. 6 8
$15. 3 2

19.
0 ¹1
~~1~~ ~~2~~¹0 7
− 9 4 3
2 6 4
R = **264**

20.
$$\begin{array}{r} \scriptstyle 1\,1 \\ 4444 \\ +\ 1358 \\ \hline 5802 \end{array}$$
Z = **5802**

21.
$$\begin{array}{r} \scriptstyle 1\,1\,1 \\ 3426 \\ 1547 \\ +\ 2684 \\ \hline \mathbf{7657} \end{array}$$

22.
$$\begin{array}{r} 4\,3\,\overset{3}{\not{4}}\,{}^{1}3 \\ -\ 4\,3\,1\,8 \\ \hline 2\,5 \end{array}$$
M = **25**

23.
$$\begin{array}{r} \scriptstyle 1\ 1 \\ \$13.06 \\ \$4.90 \\ +\ \$60.75 \\ \hline \mathbf{\$78.71} \end{array}$$

24. **10 × 2 > 10 + 2**

25. The rule is "Count down by sixes."
0, −6, −12

26. **3**

27. **4 × 5 = 20 or 5 × 4 = 20**

28. **The amount that "went away" should be $26. See student work.**

29. **1492, 1620, 1776, 1789**

INVESTIGATION 2

1. 20 pumpkins ÷ 4 = **5 pumpkins**

2. 20 pumpkins ÷ 10 = **2 pumpkins**

3. 20 pumpkins ÷ 2 = **10 pumpkins**

4. The number of ripe pumpkins is 5 + 2 + 10 = 17. So there are 20 − 17, or **3 pumpkins** that are not yet ripe.

5. 80 pounds ÷ 4 = **20 pounds**

6. 80 pounds ÷ 2 = **40 pounds**

7. 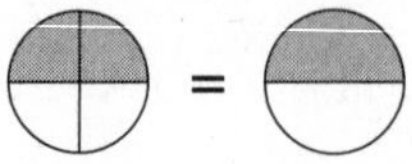

8. 25% + 25% = **50%**

9. **five tenths**

10. **less than one half**

11. 25% + 10% + 10% = **45%**

12. 50% + 10% + 10% + 10% + 10% = **90%**

13. Possibilities include:
$\mathbf{\frac{10}{10} = 1, \frac{1}{2} + \frac{5}{10} = 1,}$ **and** $\mathbf{\frac{2}{4} + \frac{5}{10} = 1.}$

14. 25% + 10% = **35%**

15. 25% + 25% + 10% + 10% = **70%**

16. 50% + 25% + 10% = **85%**

17. $\mathbf{\frac{1}{4}}$

18. 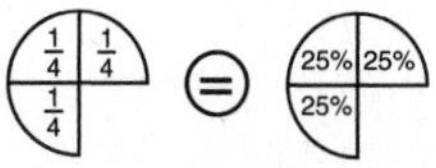

$\frac{3}{4}$ ⊜ 75%

19. 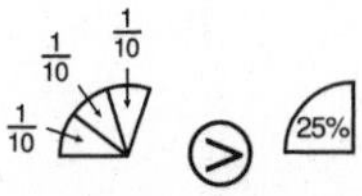

$\frac{3}{10}$ ⊘ 25% (circled >)

20. 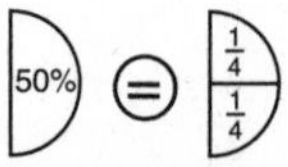

50% ⊜ $\frac{1}{4} + \frac{1}{4}$

21.

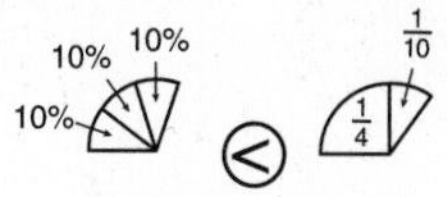

$30\% < \frac{1}{4} + \frac{1}{10}$

22.

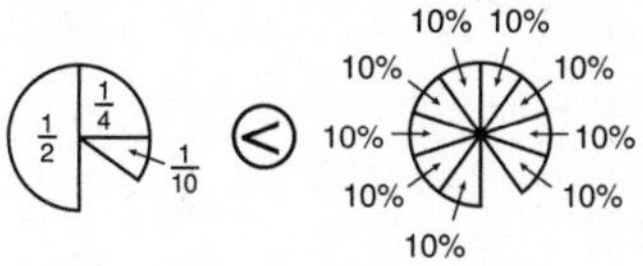

$\frac{1}{2} + \frac{1}{4} + \frac{1}{10} < 90\%$

23.

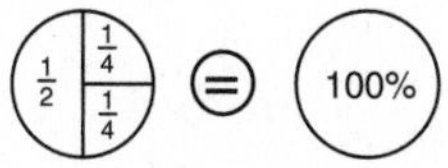

$\frac{1}{2} + \frac{1}{4} + \frac{1}{4} = 100\%$

24.

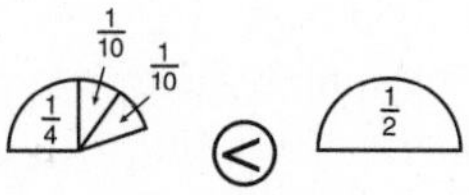

$\frac{1}{4} + \frac{1}{10} + \frac{1}{10} < \frac{1}{2}$

25.

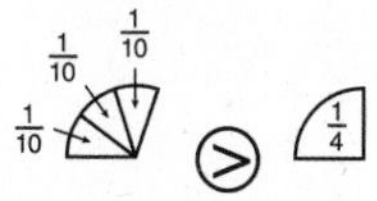

$\frac{1}{10} + \frac{1}{10} + \frac{1}{10} > \frac{1}{4}$

Lesson 21, Warm-Up

a. **135**

b. **216**

c. **180**

d. **216**

e. **190**

f. **360**

g. **245**

h. **3**

Problem Solving

Uppercase: **A, B, D, O, P, Q, R**

Lowercase: **a, b, d, e, g, o, p, q**

Lesson 21, Lesson Practice

a. **4 · 12 = *T***

$$\begin{array}{r} 12 \text{ eggs} \\ \times \quad 4 \text{ cartons} \\ \hline \mathbf{48 \text{ eggs}} \end{array}$$

b. **6*N* = 30**

30 ÷ 6 = **5 desks**

c. **7*G* = 21**

21 ÷ 7 = **3 piles**

d. **7*N* = 56**

56 ÷ 7 = **8 zebus**

e. **See student work.**

$$\begin{array}{r} \overset{4\ 3}{\$0.75} \\ \times \quad 6 \\ \hline \mathbf{\$4.50} \end{array}$$

Lesson 21, Mixed Practice

1. **8*p* = 56**

56 ÷ 8 = **7 scouts**

2. **32 − 8 = *t***

$$\begin{array}{r} \overset{2}{\cancel{3}}{}^{1}2 \text{ ounces} \\ - \quad 8 \text{ ounces} \\ \hline \mathbf{24 \text{ ounces}} \end{array}$$

3. **\$487 + *M* = \$800**

$$\begin{array}{r} \$\overset{7}{\cancel{8}}\overset{9}{\cancel{0}}{}^{1}0 \\ - \ \$487 \\ \hline \mathbf{\$313} \end{array}$$

4.

5. **6 × 7 = 42**
7 × 6 = 42
42 ÷ 6 = 7
42 ÷ 7 = 6

6. $8\overline{)72}$ with quotient **9**

7. $42 \div 6 = 7$
$N = \mathbf{7}$

8. $9\overline{)36}$ = **4**

9. $48 \div 6 = 8$
$N = \mathbf{8}$

10. $56 \div 7 = \mathbf{8}$

11. $\frac{70}{10} = \mathbf{7}$

12. $24 \div 4$ (>) $30 \div 6$
6 5

13.
5 5
367
× 8
2936

14.
2
$5.04
× 7
$35.28

15.
3 6
837
× 9
7533

16. 6 × 8 × 10
48 × 10 = **480**

17. 7 × 20 × 4
7 × 80 = **560**

18.
39 9
~~$40~~. ~~0~~¹0
− $29. 3 4
$10. 6 6

19.
1
4568
+ 6318
10,886
R = **10,886**

20.
49 9
~~50~~ ~~0~~¹3
− 8 7 6
41 2 7
W = **4127**

21.
7
6 ~~8~~¹7
− 2 6 8
4 1 9
M = **419**

22.
1 1
$9.65
$2.43
+ $1.45
$13.53

23.
2 1
382
96
+ 182
660

24. $12 \div 2 =$ **6 items**

25. The rule is "Count up by tens."
100, 110, 120

26. **Ten divided by two**

27. **60**

28.
1
12 books
× 5 boxes
60 books
There were 60 books in all 5 boxes.

29. **50%**

Lesson 22, Warm-Up

a. **50**

b. **250**

c. **285**

d. **224**

e. **168**

f. **175**

g. **210**

h. **700**

i. **4**

Problem Solving

To calculate the ones digit of the subtrahend, 1 must be borrowed from the tens column. We write a 9 in the ones place of the subtrahend because 16 − 7 = 9. We write a 5 in the tens place of the minuend because 1 + 1 + 3 = 5. We write a 2 in the hundreds place of the subtrahend because 4 − 2 = 2.

$$\begin{array}{r} \mathbf{456} \\ -\ \mathbf{219} \\ \hline \mathbf{237} \end{array}$$

h. $\begin{array}{r} \mathbf{5\ R\ 5} \\ 9\overline{)50} \\ -45 \\ \hline 5 \end{array}$

i. $\begin{array}{r} \mathbf{3\ R\ 7} \\ 9\overline{)34} \\ -27 \\ \hline 7 \end{array}$

j. $\mathbf{5\overline{)44}}$

k. **30**

LESSON 22, LESSON PRACTICE

a. $\begin{array}{r} \mathbf{4\ R\ 3} \\ 5\overline{)23} \\ -20 \\ \hline 3 \end{array}$

b. $\begin{array}{r} \mathbf{8\ R\ 2} \\ 6\overline{)50} \\ -48 \\ \hline 2 \end{array}$

c. $\begin{array}{r} \mathbf{4\ R\ 5} \\ 8\overline{)37} \\ -32 \\ \hline 5 \end{array}$

d. $\begin{array}{r} \mathbf{5\ R\ 3} \\ 4\overline{)23} \\ -20 \\ \hline 3 \end{array}$

e. $\begin{array}{r} \mathbf{7\ R\ 1} \\ 7\overline{)50} \\ -49 \\ \hline 1 \end{array}$

f. $\begin{array}{r} \mathbf{6\ R\ 4} \\ 6\overline{)40} \\ -36 \\ \hline 4 \end{array}$

g. $\begin{array}{r} \mathbf{4\ R\ 2} \\ 10\overline{)42} \\ -40 \\ \hline 2 \end{array}$

LESSON 22, MIXED PRACTICE

1. 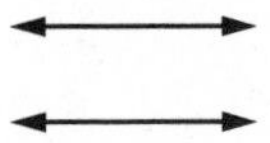

2. $\mathbf{4N = 32}$

$32 \div 4 =$ **8 night crawlers**

3. $\mathbf{26 - 9 = M}$

$$\begin{array}{r} \overset{1}{\cancel{2}}{}^{1}6 \text{ miles} \\ -\quad 9 \text{ miles} \\ \hline 1\ 7 \text{ miles} \end{array}$$

Approximately 17 miles

4. $\mathbf{840 + 418 = t}$

$$\begin{array}{r} 840 \text{ mice} \\ +\ 418 \text{ mice} \\ \hline \mathbf{1258\ mice} \end{array}$$

5. $\begin{array}{r} \mathbf{5\ R\ 6} \\ 10\overline{)56} \\ -50 \\ \hline 6 \end{array}$

6. $\begin{array}{r} \mathbf{6\ R\ 2} \\ 3\overline{)20} \\ -18 \\ \hline 2 \end{array}$

7. $\begin{array}{r} \mathbf{4\ R\ 2} \\ 7\overline{)30} \\ -28 \\ \hline 2 \end{array}$

8. $3 \times 7 \times 10$

$21 \times 10 = \mathbf{210}$

9. $2 \times 3 \times 4 \times 5$

$6 \times 20 = \mathbf{120}$

10.
$$\begin{array}{r} {}^{7\,3} \\ \$394 \\ \times \quad 8 \\ \hline \mathbf{\$3152} \end{array}$$

11.
$$\begin{array}{r} {}^{3\,3} \\ 678 \\ \times \quad 4 \\ \hline \mathbf{2712} \end{array}$$

12.
$$\begin{array}{r} {}^{4\,8} \\ \$6.49 \\ \times \quad 9 \\ \hline \mathbf{\$58.41} \end{array}$$

13. $\frac{63}{7} = \mathbf{9}$

14. $\frac{56}{8} = \mathbf{7}$

15. $\frac{42}{6} = \mathbf{7}$

16.
$$\begin{array}{r} {}^{5} \\ \$4.08 \\ \times \quad 7 \\ \hline \mathbf{\$28.56} \end{array}$$

17.
$$\begin{array}{r} {}^{3\,2\,3} \\ 3645 \\ \times \quad 6 \\ \hline \mathbf{21{,}870} \end{array}$$

18.
$$\begin{array}{r} {}^{3\;\;1} \\ 3904 \\ \times \quad 4 \\ \hline \mathbf{15{,}616} \end{array}$$

19. $8 \times 0 = 4N$

$0 = 4 \times 0$

$N = \mathbf{0}$

20.
$$\begin{array}{r} {}^{1\,1} \\ 548 \\ +\ 462 \\ \hline 1010 \end{array}$$

$C = \mathbf{1010}$

21.
$$\begin{array}{r} {}^{2\ \ {}^{1}5} \\ \$\not{3}\,\not{6}.{}^{1}1\ 5 \\ -\ \$2\,9.\,8\ 1 \\ \hline \mathbf{\$6.\,3\ 4} \end{array}$$

22.
$$\begin{array}{r} {}^{59\ \ 9} \\ \not{6}\not{0}\ \not{0}{}^{1}0 \\ -\quad 9\ 6\ 3 \\ \hline 50\ 3\ 7 \end{array}$$

$a = \mathbf{5037}$

23. **Twelve divided by four**

24. **No**

Example:
$$\begin{array}{r} 3 \\ \times\ 2 \\ \hline 6 \end{array}$$

$6 \div 2 = 3$

25. The rule is "Count down by ten."

0, −10, −20

26.
$$\begin{array}{r} 3 \\ 3\overline{)10} \\ -9 \\ \hline \end{array}$$

1 quarter

27. 46,208 (>) 46,028

28. **Two $\frac{1}{4}$ circles**

29. **25%**

Lesson 23, Warm-Up

a. **70**

b. **750**

c. **245**

d. **175**

e. **210**

f. **294**

g. **252**

h. **820**

i. **3**

Problem Solving

car: c

boat: b

pogo stick: p

A	B	C
c	**b**	**p**
c	**p**	**b**
b	**p**	**c**
b	**c**	**p**
p	**b**	**c**
p	**c**	**b**

Lesson 23, Lesson Practice

a. **Answers may vary.**

b. **B.** $\mathbf{\frac{8}{15}}$

c. $\frac{5}{8} \ (>) \ \frac{5}{12}$

d. $\frac{12}{24} \ (=) \ \frac{6}{12}$

Lesson 23, Mixed Practice

1. **$5.00 − $3.48 = *M***

$$\begin{array}{r} \overset{4}{\cancel{5}}.\overset{9}{\cancel{0}}{}^{1}0 \\ -\ \$3.48 \\ \hline \mathbf{\$1.52} \end{array}$$

2. **$1.45 + $0.95 = *t***

$$\begin{array}{r} \overset{1}{\$}\overset{1}{1}.45 \\ +\ \$0.95 \\ \hline \mathbf{\$2.40} \end{array}$$

3. **7 × 52 = *D***

$$\begin{array}{r} \overset{1}{5}2 \\ \times\ \ 7 \\ \hline \mathbf{364\ days} \end{array}$$

4. $3d = \mathbf{\$24}$

$\$24 \div 3 = \mathbf{\$8}$

5. (a) **10 pigs**

(b) **5 pigs**

6. $\frac{3}{10} \ (<) \ \frac{3}{6}$

7.
$$\begin{array}{r} \mathbf{6\ R\ 4} \\ 6\overline{)40} \\ -36 \\ \hline 4 \end{array}$$

8.
$$\begin{array}{r} \mathbf{6\ R\ 2} \\ 3\overline{)20} \\ -18 \\ \hline 2 \end{array}$$

9. $60 \div 10 = 6$

$N = \mathbf{6}$

10.
$$\begin{array}{r} \$3.\overset{5}{0}8 \\ \times\ \ \ \ 7 \\ \hline \mathbf{\$21.56} \end{array}$$

11.
$$\begin{array}{r} \overset{1}{2}5\overset{1}{1}4 \\ \times\ \ \ 3 \\ \hline \mathbf{7542} \end{array}$$

12.
$$\begin{array}{r} \overset{7}{6}\overset{5}{9}7 \\ \times\ \ 8 \\ \hline \mathbf{5576} \end{array}$$

13. **Thirty-five divided by seven**

14. $4 \times 3 \times 10$

$12 \times 10 = \mathbf{120}$

15. $12 \times 2 \times 10$

$24 \times 10 = \mathbf{240}$

16.
$$\begin{array}{r} \overset{39}{\cancel{40}}\ \overset{{}^{1}2}{\cancel{3}}{}^{1}5 \\ -\ 35\ 8\ 7 \\ \hline 4\ 4\ 8 \end{array}$$

$S = \mathbf{448}$

17.
$$\begin{array}{r} \scriptstyle 1\,1 \\ 5694 \\ +\ 1056 \\ \hline 6750 \end{array}$$
M = **6750**

18.
$$\begin{array}{r} \scriptstyle 69\ \ 9 \\ \$\not{7}\not{0}.\ \not{0}{}^{1}0 \\ -\ \ \$7.\ 5\ 3 \\ \hline \mathbf{\$62.\ 4\ 7} \end{array}$$

19.
$$\begin{array}{r} \scriptstyle 1\ 1 \\ \$5.00 \\ \$_{1}8.75 \\ \$10.00 \\ +\ \ \ \$0.35 \\ \hline \mathbf{\$24.10} \end{array}$$

20.
$$\begin{array}{r} \$6.25 \\ \$0.85 \\ \$4.00 \\ +\ \ \ \ D \\ \hline \$20.00 \end{array} \rightarrow \$11.10 \qquad \begin{array}{r} \scriptstyle 19 \\ \$\not{2}\not{0}.{}^{1}0\ 0 \\ -\ \$11.\ 1\ 0 \\ \hline \$8.\ 9\ 0 \end{array}$$
D = **\$8.90**

21. **7 × 9 = 63**
9 × 7 = 63
63 ÷ 7 = 9
63 ÷ 9 = 7

22. **52, 48, 16**

23.
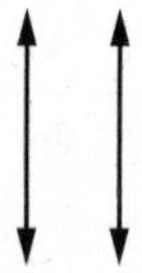

24. **Two hundred twelve thousand, five hundred**

25. **7 + 9 = 16**
9 + 7 = 16
16 − 7 = 9
16 − 9 = 7

26. **D.** $\mathbf{\frac{80}{40}}$

27. **75%**

28.
$$\begin{array}{r} \scriptstyle 2\ 4 \\ \$0.25 \\ \times\ \ \ \ \ 9 \\ \hline \$2.25 \end{array}$$
9 × \$0.25 = \$2.25

29. **Sample answer: There are three cartons with a dozen eggs in each. How many eggs are there total?**
$$\begin{array}{r} \mathbf{12} \\ \times\ \mathbf{3} \\ \hline \mathbf{36} \end{array}$$ **eggs**

Lesson 24, Warm-Up

a. **120**

b. **200**

c. **336**

d. **294**

e. **252**

f. **6**

g. **3**

h. **3**

i. **900**

j. **4**

Problem Solving

The ones column must borrow 1 from the tens column. We write a 0 in the ones place of the minuend because 10 − 2 = 8. The tens column must borrow 1 from the hundreds column. We write an 8 in the tens place of the subtrahend because 13 − 8 = 5. We write a 4 in the hundreds place of the minuend because 3 + 1 = 4.

$$\begin{array}{r} \mathbf{4\underline{4}\underline{0}} \\ -\ \mathbf{3\underline{8}2} \\ \hline \mathbf{58} \end{array}$$

LESSON 24, LESSON PRACTICE

a. $6 - (4 - 2) =$
$\quad 6 - \quad 2 \quad = \textbf{4}$

b. $(6 - 4) - 2 =$
$\quad\quad 2 \quad - 2 = \textbf{0}$

c. $(8 \div 4) \div 2 =$
$\quad\quad 2 \quad \div 2 = \textbf{1}$

d. $8 \div (4 \div 2) =$
$\quad 8 \div \quad 2 \quad = \textbf{4}$

e. $12 \div (4 - 1) =$
$\quad 12 \div \quad 3 \quad = \textbf{4}$

f. $(12 \div 4) - 1 =$
$\quad\quad 3 \quad - 1 = \textbf{2}$

g. Addition, subtraction, multiplication, and division

h. $(8 \div 4) \div 2 \;\textcircled{<}\; 8 \div (4 \div 2)$
$\quad 2 \div 2 \quad\quad\quad\quad 8 \div 2$
$\quad\quad 1 \quad\quad\quad\quad\quad\quad 4$
No, the associative property does not apply.

i. $(8 - 4) - 2 \;\textcircled{<}\; 8 - (4 - 2)$
$\quad 4 - 2 \quad\quad\quad\quad 8 - 2$
$\quad\quad 2 \quad\quad\quad\quad\quad\quad 6$
No, the associative property does not apply.

j. $(8 \times 4) \times 2 \;\textcircled{=}\; 8 \times (4 \times 2)$
$\quad 32 \times 2 \quad\quad\quad\quad 8 \times 8$
$\quad\quad 64 \quad\quad\quad\quad\quad\quad 64$
Yes, the associative property applies.

LESSON 24, MIXED PRACTICE

1.
$$\begin{array}{r} \$0.50 \\ + \ \$0.25 \\ \hline \$0.75 \end{array}$$

2. $25 \times 4 = t$

$$\begin{array}{r} \overset{2}{2}5 \text{ horses} \\ \times \ \ 4 \text{ horseshoes} \\ \hline 100 \text{ horseshoes} \end{array}$$

3. $12 - E = 9$

$$\begin{array}{r} 12 \text{ eggs} \\ - \ \ 9 \text{ eggs} \\ \hline 3 \text{ eggs} \end{array}$$

4. $98 + S = 956$

$$\begin{array}{r} \overset{8}{\cancel{9}}\,\overset{14}{\cancel{5}}6 \text{ seats} \\ - \ \ \ 9\,8 \text{ seats} \\ \hline 8\,5\,8 \text{ seats} \end{array}$$

Addition pattern

5.
$5 \times 10 = 50$
$10 \times 5 = 50$
$50 \div 5 = 10$
$50 \div 10 = 5$

6. $3 \times (4 + 5) \;\textcircled{>}\; (3 \times 4) + 5$
$\quad 3 \times 9 \quad\quad\quad\quad 12 + 5$
$\quad\quad 27 \quad\quad\quad\quad\quad 17$

7. $30 - (20 + 10) =$
$\quad 30 - \quad 30 \quad = \textbf{0}$

8. $(30 - 20) + 10 =$
$\quad\quad 10 \quad + 10 = \textbf{20}$

9. $4 \times (6 \times 5) \;\textcircled{=}\; (4 \times 6) \times 5$
$\quad 4 \times 30 \quad\quad\quad\quad 24 \times 5$
$\quad\quad 120 \quad\quad\quad\quad\quad 120$

10.
$$\begin{array}{r} \textbf{8 R 4} \\ 7\overline{)60} \\ -56 \\ \hline 4 \end{array}$$

11.
$$\begin{array}{r} \textbf{8 R 2} \\ 6\overline{)50} \\ -48 \\ \hline 2 \end{array}$$

12.
$$\begin{array}{r} \textbf{4 R 4} \\ 10\overline{)44} \\ -40 \\ \hline 4 \end{array}$$

13.
$$\begin{array}{r} \overset{1\ 2}{\$50.36} \\ \times \quad\quad 4 \\ \hline \$201.44 \end{array}$$

14.
$$\begin{array}{r} {}^{2\ 4} \\ 7408 \\ \times \quad 6 \\ \hline \mathbf{44{,}448} \end{array}$$

15.
$$\begin{array}{r} {}^{536} \\ 4637 \\ \times \quad 9 \\ \hline \mathbf{41{,}733} \end{array}$$

16.
$$\begin{array}{r} {}^{1\ \ 1} \\ \$14.08 \\ + \quad \$9.62 \\ \hline \$23.70 \end{array}$$
W = **\$23.70**

17.
$$\begin{array}{r} 4\ 7\ \not{3}^{2}\ {}^{1}0 \\ -\ 2\ 7\ 1\ 2 \\ \hline 2\ 0\ 1\ 8 \end{array}$$
J = **2018**

18.
$$\begin{array}{r} \$\not{30}^{29}.\ \not{0}^{9}\ {}^{1}0 \\ -\quad \$0.\ 5\ 6 \\ \hline \mathbf{\$29.\ 4\ 4} \end{array}$$

19.
$$\begin{array}{r} {}^{1\ 1} \\ \$3.54 \\ \$12.00 \\ +\ \ \$1.66 \\ \hline \mathbf{\$17.20} \end{array}$$

20.
$$\begin{array}{r} \$\not{20}^{19}.\ \not{0}^{9}\ {}^{1}0 \\ -\ \$16.\ 4\ 5 \\ \hline \mathbf{\$3.\ 5\ 5} \end{array}$$

21. **5 + 9 = 14**
9 + 5 = 14
14 − 9 = 5
14 − 5 = 9

22. **2**

23.
$$\begin{array}{r} {}^{1\ 2} \\ \mathbf{\$0.35} \\ \times \quad \mathbf{4} \\ \hline \mathbf{\$1.40} \end{array}$$

24. The rule is "Count up by threes."
3, 6, 9, 12, 15, 18, 21, 24, 27, 30
30

25. **B. 25**
25 ÷ 5 = 5

26. 

27. **7 × 8 = 56**
8 × 7 = 56
56 ÷ 8 = 7
56 ÷ 7 = 8

28. (8 + 4) + 2 ⊜ 8 + (4 + 2)
12 + 2   8 + 6
14   14

Yes

29. (a) **7**

(b) $\mathbf{\frac{7}{14}}$

LESSON 25, WARM-UP

a. **100**

b. **1000**

c. **144**

d. **252**

e. **294**

f. **20**

g. **10**

h. **4**

i. **265**

j. **7**

Problem Solving

Counting by 3's: 3, 6, 9, 12, 15, 18, 21, 24, 27, 30, 33, 36, 39, (42), 45, 48, …

Counting by 7's: 7, 14, 21, 28, 35, (42), 49, 56, 63, 70, …

Tom's two-digit even number is **42.**

Lesson 25, Lesson Practice

a. Factors of 4: **1, 2, 4**

b. Factors of 3: **1, 3**

c. Factors of 6: **1, 2, 3, 6**

d. Factors of 5: **1, 5**

e. Factors of 8: **1, 2, 4, 8**

f. Factors of 11: **1, 11**

g. Factors of 9: **1, 3, 9**

h. Factors of 12: **1, 2, 3, 4, 6, 12**

i. Factors of 1: **1**

j. Factors of 14: **1, 2, 7, 14**

k. Factors of 2: **1, 2**

l. Factors of 15: **1, 3, 5, 15**

m. **D. 263**

n. **D. 501**

o. **C. 6**

Lesson 25, Mixed Practice

1. $\mathbf{9 \times 24 = t}$

$$\begin{array}{r} \overset{3}{2}4 \text{ trees} \\ \times \ \ 9 \text{ rows} \\ \hline \mathbf{216 \text{ trees}} \end{array}$$

2. $\mathbf{\$10.00 - \$6.75 = m}$

$$\begin{array}{r} \$\overset{9}{\not{1}\not{0}}.\overset{9}{\not{0}}{}^{1}0 \\ - \ \$6.75 \\ \hline \mathbf{\$3.25} \end{array}$$

3. $\mathbf{4 \times \$1.12 = t}$

$$\begin{array}{r} \$1.12 \\ \times \quad 4 \\ \hline \mathbf{\$4.48} \end{array}$$

4. Factors of 30: **1, 2, 3, 5, 6, 10, 15, 30**

5. Factors of 13: **1, 13**

6. $4 \times (6 \times 10)$ (=) $(4 \times 6) \times 10$

$4 \times 60 \qquad 24 \times 10$

$240 \qquad 240$

7. **Associative property**

8. $6 \times (7 + 8) =$

$6 \times 15 = \mathbf{90}$

9. $(6 \times 7) + 8 =$

$42 + 8 = \mathbf{50}$

10.
$\mathbf{10 \times 12 = 120}$
$\mathbf{12 \times 10 = 120}$
$\mathbf{120 \div 10 = 12}$
$\mathbf{120 \div 12 = 10}$

11. $54 \div 9 = 6$

$n = \mathbf{6}$

12.

$$\begin{array}{r} \mathbf{6\ R\ 7} \\ 8\overline{)55} \\ -48 \\ \hline 7 \end{array}$$

13.

$$\begin{array}{r} {}^{112} \\ 1234 \\ \times \quad 5 \\ \hline \mathbf{6170} \end{array}$$

14.

$$\begin{array}{r} {}^{5\ 5} \\ \$5.67 \\ \times \quad 8 \\ \hline \mathbf{\$45.36} \end{array}$$

15.

$$\begin{array}{r} {}^{54} \\ 987 \\ \times \quad 6 \\ \hline \mathbf{5922} \end{array}$$

16.

$$\begin{array}{r} \$13.55 \\ + \ \$5.00 \\ \hline \$18.55 \end{array}$$

$W = \mathbf{\$18.55}$

17.

$$\begin{array}{r} \overset{19}{\not{2}\not{0}}\ \overset{9}{\not{0}}{}^{1}1 \\ - \ 10\ 0\ 2 \\ \hline 9\ 9\ 9 \end{array}$$

$R = \mathbf{999}$

18.
$$\begin{array}{r} {}^{2\,1} \\ 4387 \\ 124 \\ +\quad 96 \\ \hline \mathbf{4607} \end{array}$$

19.
$$\begin{array}{r} {}^{2\,1\,1} \\ 3715 \\ 987 \\ +\ 850 \\ \hline \mathbf{5552} \end{array}$$

20.
$$\begin{array}{r} \$6.75 \\ \$8.00 \\ \$1.36 \\ +\qquad P \\ \hline \$20.00 \end{array} \quad \$6.75 + \$8.00 + \$1.36 \rightarrow \$16.11 \qquad \begin{array}{r} {}^{19\ \ \ 9} \\ \$\not{2}\not{0}.\not{0}{}^{1}0 \\ -\ \$16.\ 1\ 1 \\ \hline \$3.\ 8\ 9 \end{array}$$

P = **$3.89**

21.
$$\begin{array}{r} \$0.50 \\ \$0.25 \\ +\ \$0.10 \\ \hline \mathbf{\$0.85} \end{array}$$

22. **Eight hundred ninety-four thousand, two hundred one**

23. **6**

24. The rule is "Count up by five."
5, 10, 15, 20, 25, 30, 35, 40, 45, 50
50

25. **Even**
Example:
$$\begin{array}{r} 7 \\ \times\ 2 \\ \hline 14 \end{array}$$

26. **B. 465**

27. **Associative property**

28. **One possibility: 2 × 4 × 3 = 24**

29. **10%**

Lesson 26, Warm-Up

a. **25¢; 50¢; 75¢**

b. **340**

c. **170**

d. **4**

e. **2**

f. **6**

g. **672**

h. **10**

Problem Solving

Possible answers:

$$\begin{array}{r} \mathbf{58} \\ +\ \ \mathbf{9} \\ \hline \mathbf{67} \end{array}; \quad \begin{array}{r} \mathbf{59} \\ +\ \ \mathbf{8} \\ \hline \mathbf{67} \end{array};$$

$$\begin{array}{r} \mathbf{76} \\ +\ \ \mathbf{9} \\ \hline \mathbf{85} \end{array}; \quad \begin{array}{r} \mathbf{79} \\ +\ \ \mathbf{6} \\ \hline \mathbf{85} \end{array}$$

Lesson 26, Lesson Practice

a.
$$\begin{array}{r} \mathbf{\$1.39} \\ 4\overline{)\$5.56} \\ -4 \\ \hline 1\,5 \\ -1\,2 \\ \hline 36 \\ -36 \\ \hline 0 \end{array}$$

b.
$$\begin{array}{r} \mathbf{41\ R\ 6} \\ 9\overline{)375} \\ -36 \\ \hline 15 \\ -9 \\ \hline 6 \end{array}$$

c.
$$\begin{array}{r} \mathbf{\$1.55} \\ 3\overline{)\$4.65} \\ -3 \\ \hline 1\,6 \\ -1\,5 \\ \hline 15 \\ -15 \\ \hline 0 \end{array}$$

d.
$$\begin{array}{r} \mathbf{129} \\ 5\overline{)645} \\ -5 \\ \hline 14 \\ -10 \\ \hline 45 \\ -45 \\ \hline 0 \end{array}$$

e.
$$\begin{array}{r} \mathbf{\$0.52} \\ 7\overline{)\$3.64} \\ -3\;5 \\ \hline 14 \\ -14 \\ \hline 0 \end{array}$$

f.
$$\begin{array}{r} \mathbf{52\ R\ 1} \\ 7\overline{)365} \\ -35 \\ \hline 15 \\ -14 \\ \hline 1 \end{array}$$

g.
$$\begin{array}{r} \mathbf{54\ R\ 6} \\ 10\overline{)546} \\ -50 \\ \hline 46 \\ -40 \\ \hline 6 \end{array}$$

h.
$$\begin{array}{r} \mathbf{\$1.14} \\ 4\overline{)\$4.56} \\ -4 \\ \hline 0\;5 \\ -4 \\ \hline 16 \\ -16 \\ \hline 0 \end{array}$$

i.
$$\begin{array}{r} {}^{1} \\ \mathbf{12} \\ \times\ \ \mathbf{6} \\ \hline \mathbf{72} \\ +\ \ \mathbf{3} \\ \hline \mathbf{75} \end{array}$$

j.
$$\begin{array}{r} 17 \\ 3\overline{)51} \\ -3 \\ \hline 21 \\ -21 \\ \hline 0 \end{array}$$

x = **17**

k.
$$\begin{array}{r} 23 \\ 4\overline{)92} \\ -8 \\ \hline 12 \\ -12 \\ \hline 0 \end{array}$$

y = **23**

l.
$$\begin{array}{r} 42 \\ 6\overline{)252} \\ -24 \\ \hline 12 \\ -12 \\ \hline 0 \end{array}$$

z = **42**

LESSON 26, MIXED PRACTICE

1. **$5.00 − $2.98 = *M***

$$\begin{array}{r} {}^{4\,9} \\ \$\not{5}.\not{0}{}^{1}0 \\ -\ \$2.9\;8 \\ \hline \mathbf{\$2.0\;2} \end{array}$$

2. **3 × 12 = *t***

$$\begin{array}{r} 12 \\ \times\ \ 3 \\ \hline \mathbf{36\ cupcakes} \end{array}$$

3. ***S* + 3 = 28**

$$\begin{array}{r} 28\text{ members} \\ -\ \ 3\text{ members} \\ \hline \mathbf{25\ members} \end{array}$$

4. (a) **10**

(b) **5**

(c) $\mathbf{\frac{5}{10}}$

5. Factors of 16: **1, 2, 4, 8, 16**

6.
$$\begin{array}{r} \mathbf{\$0.75} \\ 5\overline{)\$3.75} \\ -3\;5 \\ \hline 25 \\ -25 \\ \hline 0 \end{array}$$

7.
$$\begin{array}{r} \mathbf{91\ R\ 1} \\ 4\overline{)365} \\ -36 \\ \hline 05 \\ -4 \\ \hline 1 \end{array}$$

8.
$$\begin{array}{r} 39 \\ 6\overline{)234} \\ -18 \\ \hline 54 \\ -54 \\ \hline 0 \end{array}$$

m = **39**

9.
$$\begin{array}{r} \mathbf{\$0.72} \\ 6\overline{)\$4.32} \\ -4\,2 \\ \hline 12 \\ -12 \\ \hline 0 \end{array}$$

10.
$$\begin{array}{r} \mathbf{41} \\ 3\overline{)123} \\ -12 \\ \hline 03 \\ -3 \\ \hline 0 \end{array}$$

11.
$$\begin{array}{r} \mathbf{96} \\ 6\overline{)576} \\ -54 \\ \hline 36 \\ -36 \\ \hline 0 \end{array}$$

12.
$$\begin{array}{r} {}^{1\ 3} \\ \$7.48 \\ \times \quad 4 \\ \hline \mathbf{\$29.92} \end{array}$$

13.
$$\begin{array}{r} {}^{7} \\ 609 \\ \times \quad 8 \\ \hline \mathbf{4872} \end{array}$$

14. $7 \times 8 \times 10$

$56 \times 10 =$ **560**

15. $7 \times 8 \times 0$

$56 \times 0 =$ **0**

16.
$$\begin{array}{r} {}^{8}\ \ \ \ {}^{6}\ \ \ \\ \not 9\,{}^{1}3\ \not 7\,{}^{1}4 \\ -\ 4\ 9\ 3\ 8 \\ \hline 4\ 4\ 3\ 6 \end{array}$$

M = **4436**

17.
$$\begin{array}{r} {}^{9}\ \ \ {}^{9}\ \ \ \\ \$\not 1\not 0.\ \not 0\,{}^{1}0 \\ -\quad \$6.\ 2\ 4 \\ \hline \mathbf{\$3.\ 7\ 6} \end{array}$$

18.
$$\begin{array}{r} L \\ 427 \\ +\ \ 85 \\ \hline 2010 \end{array} \quad 427 + 85 = 512 \qquad \begin{array}{r} {}^{19}\ {}^{1}0\ \ \\ \not 2\not 0\ \not 1\,{}^{1}0 \\ -\quad 5\ 1\ 2 \\ \hline 1\ 4\ 9\ 8 \end{array}$$

L = **1498**

19.
$$\begin{array}{r} {}^{1\ 1}\ \ \ \\ \$12.43 \\ \$0.68 \\ +\ \$10.00 \\ \hline \mathbf{\$23.11} \end{array}$$

20. $3 \times 40 \ \textcircled{=}\ 3 \times 4 \times 10$

21. $8 \times 90 = 8 \times 9 \times N$

$720 = 72 \times N$

$720 = 72 \times 10$

$N =$ **10**

22. **8 × 9 = 72**

9 × 8 = 72

72 ÷ 8 = 9

72 ÷ 9 = 8

23. $8 \times N = 64$

$64 \div 8 =$ **8 squares**

24.
$$\begin{array}{r} {}^{1}\ \ \ \ \\ \$0.75 \\ +\ \$0.30 \\ \hline \mathbf{\$1.05} \end{array}$$

25. **500**

26. **13**

27.
$$\begin{array}{r} {}^{1}\ \ \\ 347 \\ +\ 809 \\ \hline \mathbf{1156} \end{array}$$

28. $\begin{array}{r} \overset{2}{1}6 \\ \times\ 4 \\ \hline 64 \\ +\ 3 \\ \hline 67 \end{array}$ **Todd's answer is not correct, because the product plus the remainder does not equal the dividend.**

29. B. 2

Lesson 27, Warm-Up

365 days; 366 days

a. **14 days; 21 days; 28 days**

b. **240**

c. **144**

d. **50**

e. **25**

f. **75**

g. **10**

h. **70**

i. **7**

Problem Solving

Possible answers:

$\begin{array}{r} 67 \\ -\ 8 \\ \hline 59 \end{array}$; $\begin{array}{r} 67 \\ -\ 9 \\ \hline 58 \end{array}$;

$\begin{array}{r} 85 \\ -\ 6 \\ \hline 79 \end{array}$; $\begin{array}{r} 85 \\ -\ 9 \\ \hline 76 \end{array}$

Lesson 27, Lesson Practice

a. 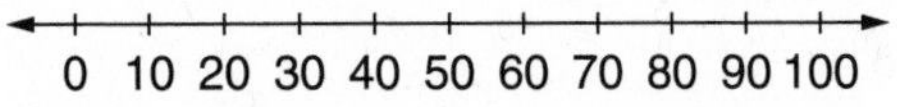

b. **−5°C**

c. **30 < 80 or 80 > 30**

Lesson 27, Mixed Practice

1. **408 + 347 + 419 = *t***

$\begin{array}{r} 4\overset{2}{0}8 \text{ miles} \\ 347 \text{ miles} \\ +\ 419 \text{ miles} \\ \hline \textbf{1174 miles} \end{array}$

2. **5 × 12 = *t***

$\begin{array}{r} \overset{1}{1}2 \\ \times\ \ 5 \\ \hline \textbf{60 inches} \end{array}$

3. ***F* − 27 = 7**

$\begin{array}{r} 27 \text{ autographed footballs} \\ +\ \ 7 \text{ autographed footballs} \\ \hline \textbf{34 autographed footballs} \end{array}$

4. **9 × $0.15 = *M***

$\begin{array}{r} \$\overset{1}{0}.\overset{4}{1}5 \\ \times\ \ \ 9 \\ \hline \textbf{\$1.35} \end{array}$

5. 12 ÷ 2 = **6 years old**

6. $\begin{array}{r} \textbf{172 R 4} \\ 5\overline{)864} \\ -5 \\ \hline 36 \\ -35 \\ \hline 14 \\ -10 \\ \hline 4 \end{array}$

7. $\begin{array}{r} \textbf{\$0.68} \\ 4\overline{)\$2.72} \\ -2\,4 \\ \hline 32 \\ -32 \\ \hline 0 \end{array}$

8. $\begin{array}{r} \textbf{67 R 5} \\ 9\overline{)608} \\ -54 \\ \hline 68 \\ -63 \\ \hline 5 \end{array}$

9. 378 ÷ (18 ÷ 3)
378 ÷ 6

$$\begin{array}{r} \mathbf{63} \\ 6\overline{)378} \\ -36 \\ \hline 18 \\ -18 \\ \hline 0 \end{array}$$

10. **82°F**

11.
$$\begin{array}{r} {\scriptstyle 1\,4} \\ \$52.60 \\ \times \quad 7 \\ \hline \mathbf{\$368.20} \end{array}$$

12.
$$\begin{array}{r} {\scriptstyle 5\,4\,2} \\ 3874 \\ \times \quad 6 \\ \hline \mathbf{23{,}244} \end{array}$$

13.
$$\begin{array}{r} {\scriptstyle 5\,2} \\ 9063 \\ \times \quad 8 \\ \hline \mathbf{72{,}504} \end{array}$$

14. **350**

15.
$$\begin{array}{r} 386 \\ 4287 \\ 672 \\ + \quad M \\ \hline 5350 \end{array} \quad 386 + 4287 + 672 \rightarrow 5345 \qquad \begin{array}{r} {\scriptstyle 4} \\ 5\,3\,\not{5}\,{}^{1}0 \\ -\ 5\,3\,4\,5 \\ \hline 5 \end{array}$$

M = **5**

16.

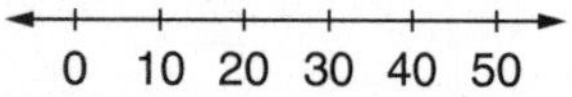

17. **B. 70 and 80**

18. Factors of 30: **1, 2, 3, 5, 6, 10, 15, 30**

19.
$$\begin{array}{r} {\scriptstyle 39} \\ \not{4}\not{0}\,{}^{1}5 \\ -\ 39\,7 \\ \hline \mathbf{8} \end{array}$$

20. **C. 28**

21. **−10°C**

22. The rule is "Count up by tens."
190, 200, 210

23. **5**

24. **Three hundred twenty-seven thousand, forty**

25. **50**

26. $\mathbf{24 \div 3;\ \frac{24}{3};\ 3\overline{)24}}$

27.
$$\begin{array}{r} {\scriptstyle 2} \\ 14 \\ \times \ 7 \\ \hline 98 \end{array} \qquad \mathbf{98 + 2 = 100}$$

Madeline's answer is correct. The product plus the remainder equals the dividend.

28. 12 ÷ (6 ÷ 2) ⓢ (12 ÷ 6) ÷ 2

12 ÷ 3   2 ÷ 2

4   1

No, the associative property does not apply to division.

29. **30%**

Lesson 28, Warm-Up

a. **14 days; 21 days; 28 days**

b. **2500**

c. **175**

d. **20**

e. **10**

f. **30**

g. **4**

h. **36**

i. **2**

Problem Solving

4 × 2 × 2 × 2 = **32 children**

LESSON 28, LESSON PRACTICE

a. 4×100 years = **400 years**

b. **June 19, 2014**

c. **366 days**

d. **Decade**

e. **8:02 p.m.**

f. **8:45 a.m.**

g. **12:20 p.m.**

h. **12:30 a.m.**

i. **9:15 a.m.**

j. **11:25 a.m.**

k. **1:25 p.m.**

LESSON 28, MIXED PRACTICE

1. $M - \$600 = \1267

$$\begin{array}{r} \$1267 \\ + \quad \$600 \\ \hline \mathbf{\$1867} \end{array}$$

2. $\$1873 + \$200 = t$

$$\begin{array}{r} \overset{1}{\$1873} \\ + \quad \$200 \\ \hline \mathbf{\$2073} \end{array}$$

3. $4C = 52$

$$\begin{array}{r} \textbf{13 cards} \\ 4\overline{)52} \\ -4 \\ \hline 12 \\ -12 \\ \hline 0 \end{array}$$

4. $10 \div 2 =$ **5 years**

5. Factors of 18:
 ①,②,③,⑥, 9, 18
 Factors of 24:
 ①,②,③, 4,⑥, 8, 12, 24
 1, 2, 3, 6

6.
$$\begin{array}{r} \mathbf{\$1.81} \\ 3\overline{)\$5.43} \\ -3 \\ \hline 2\,4 \\ -2\,4 \\ \hline 03 \\ -3 \\ \hline 0 \end{array}$$

7.
$$\begin{array}{r} \mathbf{\$0.75} \\ 8\overline{)\$6.00} \\ -5\,6 \\ \hline 40 \\ -40 \\ \hline 0 \end{array}$$

8. $528 \div (28 \div 7)$
 $528 \div 4$

$$\begin{array}{r} \mathbf{132} \\ 4\overline{)528} \\ -4 \\ \hline 12 \\ -12 \\ \hline 08 \\ -8 \\ \hline 0 \end{array}$$

9.
$$\begin{array}{r} 116 \\ 6\overline{)696} \\ -6 \\ \hline 09 \\ -6 \\ \hline 36 \\ -36 \\ \hline 0 \end{array}$$

$w =$ **116**

10. **9:05 p.m.; 12:05 a.m.**

11. **12:30 p.m.**

12.
$$\begin{array}{r} \overset{1}{\$0.50} \\ + \quad \$0.50 \\ \hline \mathbf{\$1.00} \end{array}$$

13. **Saturday**

14. **756**

15. $$\begin{array}{r} {}^{2\,1\,1} \\ 4387 \\ 2965 \\ +\ 4943 \\ \hline \mathbf{12{,}295} \end{array}$$

16. $$\begin{array}{r} \$\overset{5}{\not{6}}\,\overset{12}{\not{3}}.\,\overset{16}{\not{7}}\,{}^{1}5 \\ -\ \$4\,6.\,8\,8 \\ \hline \mathbf{\$1\,6.\,8\,7} \end{array}$$

17. $$\begin{array}{r} \overset{3}{\not{4}}\,\overset{9}{\not{0}}\,\overset{10}{\not{1}}\,{}^{1}0 \\ -\quad 5\,6\,3 \\ \hline 3\,4\,4\,7 \end{array}$$
$F = \mathbf{3447}$

18. $$\begin{array}{r} {}^{2\,5\,\ } \\ 3408 \\ \times\quad 7 \\ \hline \mathbf{23{,}856} \end{array}$$

19. $$\begin{array}{r} {}^{4\,4\,\ } \\ \$3.56 \\ \times\quad 8 \\ \hline \mathbf{\$28.48} \end{array}$$

20. $$\begin{array}{r} {}^{7\,6\,\ } \\ 487 \\ \times\quad 9 \\ \hline \mathbf{4383} \end{array}$$

21. **8:55 a.m.**

22. $\mathbf{10 \times 2 = 20}$
$\mathbf{2 \times 10 = 20}$
$\mathbf{20 \div 10 = 2}$
$\mathbf{20 \div 2 = 10}$

23. $$\begin{array}{r} {}^{1\ } \\ \mathbf{22} \\ \times\ \mathbf{9} \\ \hline \mathbf{198} \end{array} \qquad \mathbf{198 + 2 = 200}$$

The answer is correct.

24. The rule is "Count up by hundreds."
800, 900, 1000

25. **50**

26. $\mathbf{4 \times 7 = 28}$ **or** $\mathbf{7 \times 4 = 28}$

27. **Ten centuries**

28. **4 quarter circles**

29. (a) **60 minutes**
(b) **30 minutes**
(c) $\mathbf{\dfrac{30}{60}}$

Lesson 29, Warm-Up

12 months; 52 weeks; January, February, March, April, May, June, July, August, September, October, November, December

a. **24 months; 36 months; 48 months**

b. **500**

c. **258**

d. **25**

e. **5**

f. **25**

g. **500**

h. **7**

Problem Solving

The bottom factor must be either 2 or 7. Because $36 \times 2 = 72$, and because the product has three digits, we know that the bottom factor must be 7.

$$\begin{array}{r} \mathbf{36} \\ \times\ \underline{\mathbf{7}} \\ \underline{\mathbf{25}}\mathbf{2} \end{array}$$

Lesson 29, Lesson Practice

a. $$\begin{array}{r} 34 \\ \times\ 20 \\ \hline \mathbf{680} \end{array}$$

b. $$\begin{array}{r} {}^{4\ } \\ 48 \\ \times\ 50 \\ \hline \mathbf{2400} \end{array}$$

c. $$\begin{array}{r} 34 \\ \times\ 200 \\ \hline \mathbf{6800} \end{array}$$

d.
$$\begin{array}{r} \overset{3}{3}6 \\ \times\ \ 500 \\ \hline \mathbf{18{,}000} \end{array}$$

e.
$$\begin{array}{r} \overset{1}{5}5 \\ \times\ \ 30 \\ \hline \mathbf{1650} \end{array}$$

f.
$$\begin{array}{r} \$1.\overset{1}{2}5 \\ \times\ \ \ \ 30 \\ \hline \mathbf{\$37.50} \end{array}$$

g.
$$\begin{array}{r} \overset{1}{5}5 \\ \times\ \ 300 \\ \hline \mathbf{16{,}500} \end{array}$$

h.
$$\begin{array}{r} \$1.\overset{1}{2}5 \\ \times\ \ \ \ 300 \\ \hline \mathbf{\$375.00} \end{array}$$

i.
$$\begin{array}{r} \overset{3}{4}5 \\ \times\ \ 60 \\ \hline \mathbf{2700} \end{array}$$

j.
$$\begin{array}{r} \$\overset{1}{2}.\overset{2}{3}5 \\ \times\ \ \ \ 40 \\ \hline \mathbf{\$94.00} \end{array}$$

k.
$$\begin{array}{r} \overset{2}{3}7 \\ \times\ \ 400 \\ \hline \mathbf{14{,}800} \end{array}$$

l.
$$\begin{array}{r} \$1.43 \\ \times\ \ \ \ 200 \\ \hline \mathbf{\$286.00} \end{array}$$

Lesson 29, Mixed Practice

1. **3*C* = 12**

12 ÷ 3 = **4 cookies**

2. **\$841 − \$75 = *M***

$$\begin{array}{r} \$\overset{7}{\cancel{8}}\,\overset{13}{\cancel{4}}\,{}^{1}1 \\ -\ \ \ \$7\ 5 \\ \hline \mathbf{\$7\ 6\ 6} \end{array}$$

3. **10 × 10 = *t***

$$\begin{array}{r} 10 \\ \times\ \ 10 \\ \hline \mathbf{100} \end{array}$$ **stamps**

4. 1776 + 100 = **1876**

5. Factors of 10: **1, 2, 5, 10**

6.
$$\begin{array}{r} \overset{4}{3}7 \\ \times\ \ 60 \\ \hline \mathbf{2220} \end{array}$$

7. 37 × 6 × 10

222 × 10 = **2220**

8.
$$\begin{array}{r} \overset{3}{4}6 \\ \times\ \ 50 \\ \hline \mathbf{2300} \end{array}$$

9.
$$\begin{array}{r} \$0.\overset{4}{7}\overset{1}{3} \\ \times\ \ \ \ 60 \\ \hline \mathbf{\$43.80} \end{array}$$

10. 50 × (1000 − 200)

50 × 800

$$\begin{array}{r} 50 \\ \times\ \ 800 \\ \hline \mathbf{40{,}000} \end{array}$$

11. **Tens**

12. **11:30 a.m.**

13.
$$\begin{array}{r} \$\overset{1}{0}.50 \\ \$0.75 \\ +\ \$0.30 \\ \hline \mathbf{\$1.55} \end{array}$$

14.
$$\begin{array}{r} \overset{3}{3}8 \\ \times\ \ 40 \\ \hline \mathbf{1520} \end{array}$$

15. **Nine hundred forty-four thousand**

16.
$$\begin{array}{r} {}^{111}\\ 4637 \\ 2843 \\ +\ 6464 \\ \hline \mathbf{13,944} \end{array}$$

17.
$$\begin{array}{r} {}^{3\ {}^{1}5}\\ \not{4}\,\not{6}\,{}^{1}1\,8 \\ -\ 2\,7\,2\,8 \\ \hline \mathbf{1\,8\,9\,0} \end{array}$$

18.
$$\begin{array}{r} {}^{59\ \ 9}\\ \$\not{6}\not{0}.\,\not{0}\,{}^{1}0 \\ -\quad \$7.\,6\,3 \\ \hline \mathbf{\$5\,2.\,3\,7} \end{array}$$

19.
$$\begin{array}{r} \mathbf{36\ R\ 4} \\ 10\overline{)364} \\ -30 \\ \hline 64 \\ -60 \\ \hline 4 \end{array}$$

20.
$$\begin{array}{r} 52 \\ 7\overline{)364} \\ -35 \\ \hline 14 \\ -14 \\ \hline 0 \end{array}$$

w = **52**

21.
$$\begin{array}{r} \mathbf{52} \\ 7\overline{)364} \\ -35 \\ \hline 14 \\ -14 \\ \hline 0 \end{array}$$

22. **Odd**

Example:

$2 \times 2 = 4$

$4 + 1 = 5$

23. **May 19, 1957**

24. **B. 350 and 360**

25. The rule is "Count up by hundreds."
900, 1000, 1100

26. **C. 250**

27.
$$\begin{array}{r} {}^{2}\\ \mathbf{43} \\ \times\ \ \mathbf{7} \\ \hline \mathbf{301} \\ +\quad \mathbf{1} \\ \hline \mathbf{302} \end{array}$$

The answer is not correct.

28. $12 - (6 - 2)$ ⓘ $(12 - 6) - 2$

$12 - 4 \qquad\qquad 6 - 2$

$8 \qquad\qquad\qquad 4$

The associative property does not apply to subtraction.

29. **50%**

Lesson 30, Warm-Up

100 years; 10 years

a. **12 months; 24 months; 36 months**

b. **82**

c. **420**

d. **4000**

e. **144**

f. **6**

g. **3**

h. **100**

Problem Solving

HHT, HTH, THH

Lesson 30, Lesson Practice

a. $\mathbf{\frac{1}{4}}$

b. **25%**

c. $\mathbf{\frac{3}{4}}$

d. **75%**

e. $\mathbf{\frac{2}{4}, \frac{1}{2}}$

f. **50%**

g. $\mathbf{\frac{1}{10}}$

h. **10%**

i. \$0.25 + \$0.05 = \$0.30
30%

j. **100%**

k. **75%**

l. **50%**

m. **25%**

n. **100%**

o. **90%**

p. **80%**

q. **70%**

r. **60%**

s. **50%**

t. **40%**

u. **30%**

v. **20%**

w. **10%**

Lesson 30, Mixed Practice

1. **100 − 36 = *S***

$$\begin{array}{r} \overset{9}{\cancel{10}}{}^{1}0 \\ - \quad 3\,6 \\ \hline \mathbf{6\,4} \end{array} \textbf{ stamps}$$

2. **365 − 31 = *D***

$$\begin{array}{r} 365 \\ - \quad 31 \\ \hline \mathbf{334} \end{array} \textbf{ days}$$

3. **4*q* = 28**
28 ÷ 4 = **7 quarts**

4. **5 × \$0.45 = *m***

$$\begin{array}{r} {}^{2\,2} \\ \$0.45 \\ \times \quad\;\; 5 \\ \hline \mathbf{\$2.25} \end{array}$$

5. 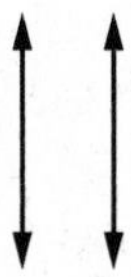

6. Factors of 25: **1, 5, 25**

7. (a) $\mathbf{\frac{2}{9}}$

(b) $\mathbf{\frac{7}{9}}$

8. **3**

9. **7:45 a.m.**

10.
$$\begin{array}{r} {}^{1\,1} \\ \$28.93 \\ + \;\$19.46 \\ \hline \$48.39 \end{array}$$
W = **\$48.39**

11.
$$\begin{array}{r} \overset{29}{\cancel{30}}\,\overset{0}{\cancel{1}}{}^{1}0 \\ - \;13\,4\,2 \\ \hline \mathbf{16\,6\,8} \end{array}$$

12.
$$\overset{2}{28}$$
$$54$$
$$75$$
$$91$$
$$\underline{+\ 26}$$
$$\mathbf{274}$$

13.
$$\overset{1\ 1}{764}$$
$$\underline{\times\qquad 30}$$
$$\mathbf{22{,}920}$$

14.
$$\overset{4}{\$9.08}$$
$$\underline{\times\qquad 60}$$
$$\mathbf{\$544.80}$$

15.
$$\begin{array}{r} \mathbf{\$1.24} \\ 6\overline{)\$7.44} \\ \underline{-6} \\ 1\ 4 \\ \underline{-1\ 2} \\ 24 \\ \underline{-24} \\ 0 \end{array}$$

16.
$$\begin{array}{r} \mathbf{36\ R\ 2} \\ 10\overline{)362} \\ \underline{-30} \\ 62 \\ \underline{-60} \\ 2 \end{array}$$

17.
$$\begin{array}{r} \mathbf{224\ R\ 2} \\ 4\overline{)898} \\ \underline{-8} \\ 09 \\ \underline{-8} \\ 18 \\ \underline{-16} \\ 2 \end{array}$$

18.
$$\overset{1\ 1\ 2}{\$42.37}$$
$$\$7.58$$
$$\$0.68$$
$$\underline{+\ \$15.00}$$
$$\mathbf{\$65.63}$$

19. $(48 \times 6) - 9$

$$\overset{4}{4\,8}$$
$$\underline{\times\qquad 6}$$
$$2\,\overset{7}{8}{}^{1}8$$
$$\underline{-\qquad 9}$$
$$\mathbf{2\,7\,9}$$

20. $6 \times 30 \times 12$

$$72 \times 30$$

$$72$$
$$\underline{\times\quad 30}$$
$$\mathbf{2160}$$

21. **7 months**

22.
$$\overset{1\ 1}{605}$$
$$\underline{+\ 597}$$
$$\mathbf{1202}$$

23. **B. 367**

24. The rule is "Count up by tens."
290, 300, 310

25. **34°C**

26. $1802 + 10 = \mathbf{1812}$

27. $\$0.25 + \$0.10 = \$0.35$
35%

28.
$$\overset{2}{14} \qquad \textbf{The answer is correct.}$$
$$\underline{\times\quad 7}$$
$$98$$
$$\underline{+\quad 2}$$
$$\mathbf{100}$$

29. $100 \div 4 \;\bigcirc\!\!\!> \; 100 \div 5$
$$\quad 25 \qquad\qquad 20$$

INVESTIGATION 3

1. 120 questions ÷ 2 = **60 questions**

2. 120 questions ÷ 3 = **40 questions**

3. 120 questions ÷ 5 = **24 questions**

4. 120 questions ÷ 8 = **15 questions**

5. 40 + 24 + 15, or 79, of the questions are not multiple choice. So 120 − 79, or **41 questions**, on the test are multiple choice.

6. 120 questions ÷ 3 = 40 questions
$41 > 40$
more than $\frac{1}{3}$

7. 40 questions + 24 questions = 64 questions
$64 > 60$
more than half

8. 40 questions + 15 questions = 55 questions
$55 < 60$
less than half

9. 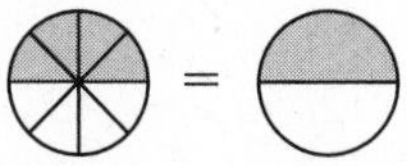

10. 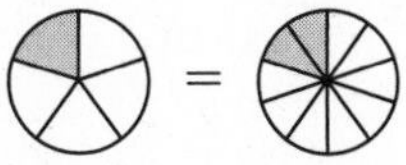

11. **2 eighths**

12. **less than one half**

13. $\mathbf{33\frac{1}{3}\%}$

14. 20% + 20% + 20% = **60%**

15. $12\frac{1}{2}\% + 12\frac{1}{2}\% + 12\frac{1}{2}\% + 12\frac{1}{2}\%$
$= \mathbf{50\%}$

16. $\mathbf{\frac{1}{8}}$ **circle**

17. **no**

18. **no**

19. **No. The denominator is an odd number. Half of an odd number is not a whole number of pieces.**

20. $\mathbf{\frac{1}{4}}$

21. $\mathbf{\frac{1}{8}}$

22. $\mathbf{\frac{1}{10}}$

23. $\mathbf{\frac{1}{6}}$

24. $\mathbf{\frac{1}{4}}$

25. $12\frac{1}{2}\% + 12\frac{1}{2}\% = \mathbf{25\%}$

26. $\mathbf{\frac{1}{5} + \frac{2}{5} = \frac{3}{5}}$

27. $\mathbf{\frac{3}{8} + \frac{5}{8} = \frac{8}{8} = 1}$

28. $\mathbf{\frac{2}{3} - \frac{1}{3} = \frac{1}{3}}$

29. $\mathbf{\frac{5}{8} - \frac{2}{8} = \frac{3}{8}}$

30. 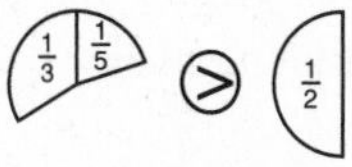

$\frac{1}{3} + \frac{1}{5} \; \textcircled{>} \; \frac{1}{2}$

31. 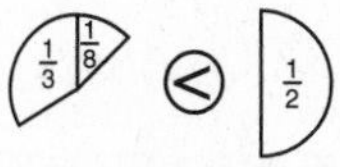

$\frac{1}{3} + \frac{1}{8} \; \textcircled{<} \; \frac{1}{2}$

32. $33\frac{1}{3}\% + 33\frac{1}{3}\% = 66\frac{2}{3}\%$

$66\frac{2}{3}\% \; \textcircled{<} \; 67\%$

33. 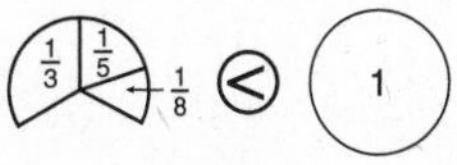

$\frac{1}{3} + \frac{1}{5} + \frac{1}{8} \; \textcircled{<} \; 1$

34. $\frac{1}{10}, \frac{1}{8}, \frac{1}{5}, \frac{1}{4}, \frac{1}{3}, \frac{1}{2}$

35. **10%, $12\frac{1}{2}$%, 20%, 25%, $33\frac{1}{3}$%, 50%**

Lesson 31, Warm-Up

366 days; 365 days

a. **24; 36; 48**

b. **73**

c. **1540**

d. **10**

e. **5**

f. **2**

g. **224**

h. **32**

Problem Solving

If the bottom factor were 3 or greater, the product would be a three-digit number. So the bottom factor must be 1 or 2.

$$\begin{array}{r} \mathbf{45} \\ \times\ \mathbf{2} \\ \hline \mathbf{90} \end{array}$$

Lesson 31, Lesson Practice

a.

b.

c.

Lesson 31, Mixed Practice

1.

2. **\$10 − \$4.19 = *M***

$$\begin{array}{r} \overset{9}{\cancel{\$10}}.\overset{9}{\cancel{0}}{}^{1}0 \\ -\ \$4.19 \\ \hline \mathbf{\$5.81} \end{array}$$

3.
$$\begin{array}{r} \overset{2}{2}4 \text{ hours per day} \\ \times\ 7 \text{ days} \\ \hline \mathbf{168\ hours} \end{array}$$

4. ***D* + 23 = 71**

$$\begin{array}{r} \overset{6}{\cancel{7}}{}^{1}1° \\ -\ 2\,3° \\ \hline 4\,8°\text{ F} \end{array}$$

D = **48°F**

5. Three of the seven circles are shaded, so $\frac{3}{7}$ of the group is shaded.

6. **1, 19**

7.
$$\begin{array}{r} \$\overset{0}{\cancel{1}}\overset{15}{\cancel{6}}.{}^{1}3\,8 \\ -\ \$9.\,4\,7 \\ \hline \mathbf{\$6.\,9\,1} \end{array}$$

8.
$$\begin{array}{r} \overset{9}{\cancel{10}}\,\overset{9}{\cancel{0}}{}^{1}0 \\ -\ 5\,7\,6 \\ \hline 4\,2\,4 \end{array}$$

Q = **424**

9.
$$\begin{array}{r} 56 \\ 5\overline{)280} \\ -25 \\ \hline 30 \\ -30 \\ \hline 0 \end{array}$$

n = **56**

10.
$$\begin{array}{r} \overset{6\,4}{476} \\ \times\ 80 \\ \hline \mathbf{38{,}080} \end{array}$$

11.
$$\begin{array}{r} \overset{4\,4}{\$9.68} \\ \times \quad 60 \\ \hline \mathbf{\$580.80} \end{array}$$

12.
$$\begin{array}{r} \mathbf{\$2.43} \\ 8\overline{)\$19.44} \\ -16 \\ \hline 3\,4 \\ -3\,2 \\ \hline 24 \\ -24 \\ \hline 0 \end{array}$$

13. Thirty minutes before midnight is "p.m." So the time is **11:30 p.m.**

14. $\frac{1}{10}$ of 100 ⊜ $\frac{1}{2}$ of 20

10 10

15. **5 × $2.87**

$$\begin{array}{r} \overset{4\,3}{\$2.87} \\ \times \quad 5 \\ \hline \mathbf{\$14.35} \end{array}$$

16.
$$\begin{array}{r} \$96.00 \\ \$128.13 \\ \$27.49 \\ + \quad W \\ \hline \$300.00 \end{array} \quad \$96.00 + \$128.13 + \$27.49 \rightarrow \$251.62$$

$$\begin{array}{r} \$\overset{29}{\cancel{30}}\,\overset{9}{\cancel{0}}.\,\overset{9}{\cancel{0}}{}^{1}0 \\ -\ \$25\,1.\,6\,2 \\ \hline \$4\,8.\,3\,8 \end{array}$$

W = **$48.38**

17. $328 \div (32 \div 8)$
$= 328 \div 4$

$$\begin{array}{r} \mathbf{82} \\ 4\overline{)328} \\ -32 \\ \hline 08 \\ -8 \\ \hline 0 \end{array}$$

18. $648 - (600 + 48)$
$= 648 - 648 =$ **0**

19. No matter what number is chosen the answer is **odd.**

20. **B. 251**

21. Afternoon is "p.m.", so the time is **3:49 p.m.**

22. **2**

23. **One hundred twenty-three thousand, four hundred**

24. $\mathbf{66\frac{2}{3}\%}$

25.

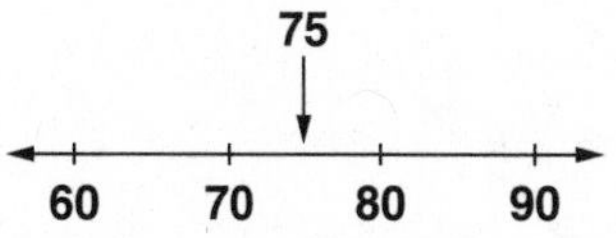

26.
$$\begin{array}{r} \overset{5}{37} \\ \times \ 8 \\ \hline 296 \end{array} \qquad \mathbf{296 + 6 = 302}$$

The answer is not correct.

27. (a) **100 years**

(b) **50 years**

(c) $\mathbf{\frac{50}{100}}$

28. **15 minutes**

29. The pattern is "Count down by nines."
0, −9, −18, −27

Lesson 32, Warm-Up

10 years; 100 years

a. **24 hours; 48 hours**

b. **84**

c. **280**

d. **600**

e. **15**

f. **3**

g. **360**

h. **3**

Problem Solving

HTT, TTH, THT

Lesson 32, Lesson Practice

a. **One possibility:**

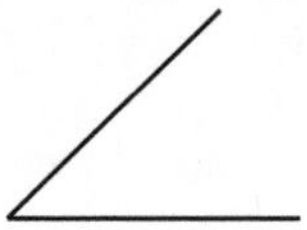

b. **One possibility:**

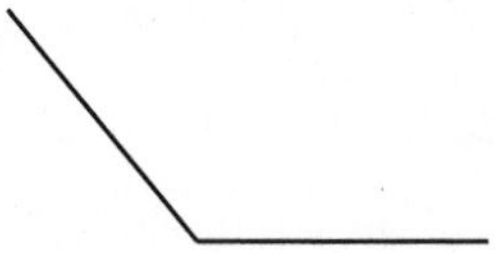

c. **One possibility:**

d. **Right**

e. **Acute**

f. **Obtuse**

g. **Straight**

h. **One possibility:**

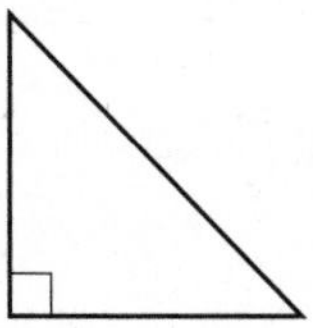

i. **4 sides**

j. **One possibility:**

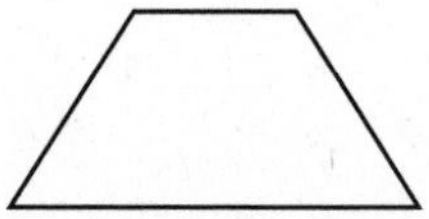

k. **One possibility:**

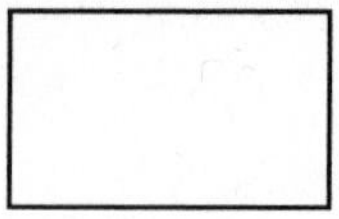

l. **One possibility:**

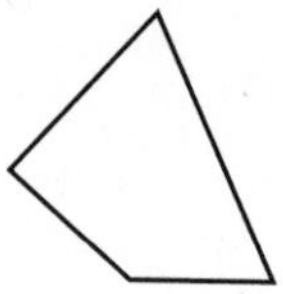

m. **Pentagon**

n. **Hexagon**

o. **Octagon**

p. 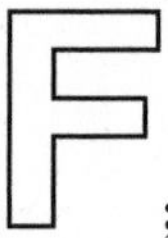 **; Decagon**

q. **The triangles should have the same shape and size. One possibility:**

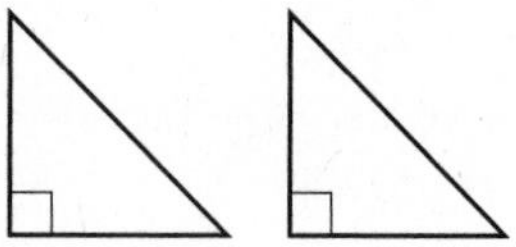

Lesson 32, Mixed Practice

1. **Rides:** $2\overline{)\$20}$ = **$10**

Food: $4\overline{)\$20}$ = **$5**

Parking: $10\overline{)\$20}$ = **$2**

2. **18 × 3 = *t***

$$\begin{array}{rl} \overset{2}{1}8 & \text{buckets} \\ \times\ \ 3 & \text{gallons in each bucket} \\ \hline 54 & \text{gallons} \end{array}$$

t = **54 gallons**

3. **4*L* = 52**

$$\begin{array}{r} 13 \text{ feet} \\ 4\overline{)52} \\ \underline{-4} \\ 12 \\ \underline{-12} \\ 0 \end{array}$$

L = **13 feet**

4. **45 − 17 = *L***

$$\begin{array}{r} \overset{3}{\not4}{}^{1}5 \text{ questions} \\ -\ 1\ 7 \text{ questions} \\ \hline 2\ 8 \text{ questions} \end{array}$$

L = **28 questions**

5.

$$\begin{array}{r} 60 \text{ minutes} \\ \times\ 60 \text{ seconds in each minute} \\ \hline \mathbf{3600\ seconds} \end{array}$$

6.

$$\begin{array}{r} \overset{2\,1\ \ 2}{\$56.37} \\ \$34.28 \\ +\ \ \$9.75 \\ \hline \mathbf{\$100.40} \end{array}$$

7.

$$\begin{array}{r} \overset{4}{\not5}{}^{1}2\ \overset{7}{\not8}{}^{1}6 \\ -\ 4\ 3\ 1\ 9 \\ \hline \mathbf{9\ 6\ 7} \end{array}$$

8.

$$\begin{array}{r} \$\overset{39}{\not4\not0}.\ \overset{9}{\not0}{}^{1}0 \\ -\ \$39.\ 5\ 6 \\ \hline \mathbf{\$0.\ 4\ 4} \end{array}$$

9.

$$\begin{array}{r} 67 \\ 72 \\ 43 \\ 91 \\ 48 \\ 19 \\ 648 \\ +\ \ M \\ \hline 996 \end{array} \Bigg\} 988 \qquad \begin{array}{r} 9\ \overset{8}{\not9}{}^{1}6 \\ -\ 9\ 8\ 8 \\ \hline 8 \end{array}$$

M = **8**

10. 936 ÷ (36 ÷ 9)

936 ÷ 4

$$\begin{array}{r} \mathbf{234} \\ 4\overline{)936} \\ \underline{-8} \\ 13 \\ \underline{-12} \\ 16 \\ \underline{-16} \\ 0 \end{array}$$

11.

$$\begin{array}{r} \overset{5\,3}{596} \\ \times\ \ \ 600 \\ \hline \mathbf{357{,}600} \end{array}$$

12.

$$\begin{array}{r} \mathbf{\$5.82} \\ 8\overline{)\$46.56} \\ \underline{-40} \\ 6\,5 \\ \underline{-6\,4} \\ 16 \\ \underline{-16} \\ 0 \end{array}$$

13.

$$\begin{array}{r} \$4.\overset{5}{0}7 \\ \times\ \ \ \ \ 80 \\ \hline \mathbf{\$325.60} \end{array}$$

14. 9 × 12 × 0 = **0**

15.

$$\begin{array}{r} \mathbf{133\ R\ 5} \\ 7\overline{)936} \\ \underline{-7} \\ 23 \\ \underline{-21} \\ 26 \\ \underline{-21} \\ 5 \end{array}$$

16. $\frac{1}{3}$ of 60 ⊜ $\frac{1}{5}$ of 100

20 20

17. **C.**

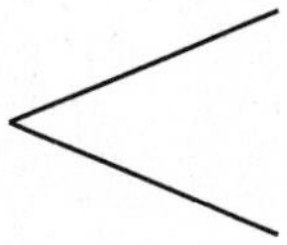

18. **1, 2, 3, 6, 9, 18**

19. Three of the four squares are shaded, so $\frac{3}{4}$ of the group is shaded; **75%**

20.

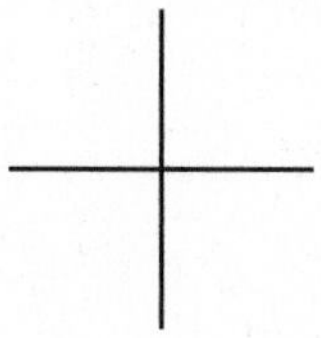

21. **Thursday**

22. **Denominator**

23.
9 × 10 = 90
10 × 9 = 90
90 ÷ 9 = 10
90 ÷ 10 = 9

24. The pattern is "Count up by tens."
690, 700, 710

25. 130

26.  **; Dodecagon**

27.
$$\begin{array}{r} \overset{4}{5}7 \\ \times\ \ 7 \\ \hline 399 \end{array} \qquad 399 + 1 = 400$$

The answer is correct.

28. 250 or 520

29. (a) **12**

(b) $2\overline{)12}$ = **6**

(c) $\frac{6}{12}$

Lesson 33, Warm-Up

6 months; 18 months

a. \$1.00; \$1.25; \$1.50

b. 378

c. 250

d. 20

e. 10

f. 4

g. 700

h. 5

Problem Solving

The ones digit of the top factor must be either 2 or 7. Since there are only two digits in the product, the top factor must be closer to 10 than to 20. So the top factor is 12.

$$\begin{array}{r} \underline{12} \\ \times\ \ 8 \\ \hline \underline{96} \end{array}$$

Lesson 33, Lesson Practice

a. 70

b. 90

c. 50

d. 100

e. 700

f. 400

g. 800

h. 400

Lesson 33, Mixed Practice

1. 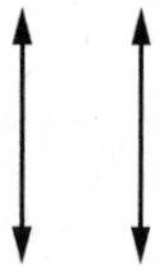

2. 500

3. 80

4. Mammals: $2\overline{)40}$ = **20 mammals**

Fish: $4\overline{)40}$ = **10 fish**

Reptiles: $10\overline{)40}$ = **4 reptiles**

Birds:
$$\begin{array}{r} 20 \\ 10 \\ 4 \\ +\ B \\ \hline 40 \end{array} \left.\begin{array}{r} \\ \\ \\ \end{array}\right\} 34 \qquad \begin{array}{r} 40 \\ -\ 34 \\ \hline 6 \end{array}$$

B = **6 birds**

5. 4 people

6. $7 \times 60 = t$

$$\begin{array}{l} \quad 7 \text{ hours} \\ \times\ 60 \text{ minutes in each hour} \\ \hline \quad 420 \text{ minute} \end{array}$$

t = **420 minutes**

7. $M - \$7.50 = \$3.75$

$$\begin{array}{r} {\scriptstyle 1} \\ \$3.75 \\ +\ \$7.50 \\ \hline \$11.25 \end{array}$$

$M = \mathbf{\$11.25}$

8. $\mathbf{427 + m = 902}$

$$\begin{array}{r} \overset{8}{\cancel{9}}\ \overset{9}{\cancel{0}}{}^{1}2 \text{ miles} \\ -\ 4\ 2\ 7 \text{ miles} \\ \hline 4\ 7\ 5 \text{ miles} \end{array}$$

$m =$ **475 miles**

9.
$$\begin{array}{r} {\scriptstyle 1\,1\ \ 1} \\ \$34.28 \\ \$9.76 \\ +\ \$20.84 \\ \hline \mathbf{\$64.88} \end{array}$$

10.
$$\begin{array}{r} \overset{2}{\cancel{3}}{}^{1}5\ \overset{1}{\cancel{2}}{}^{1}6 \\ -\ 1\ 6\ 1\ 7 \\ \hline 1\ 9\ 0\ 9 \end{array}$$

$V =$ **1909**

11.
$$\begin{array}{r} \$\overset{9}{\cancel{10}}.\ \overset{9}{\cancel{0}}{}^{1}0 \\ -\ \ \$0.\ 8\ 6 \\ \hline \mathbf{\$9.\ 1\ 4} \end{array}$$

12.
$$\begin{array}{r} {\scriptstyle 4\,3} \\ 499 \\ 25 \\ 43 \\ 756 \\ 67 \\ 94 \\ +\ \ 32 \\ \hline \mathbf{1516} \end{array}$$

13.
$$\begin{array}{r} {\scriptstyle 5\,2} \\ 563 \\ \times\ \ \ 90 \\ \hline \mathbf{50,670} \end{array}$$

14.
$$\begin{array}{r} {\scriptstyle 6\ \,4} \\ \$2.86 \\ \times\ \ \ \ 70 \\ \hline \mathbf{\$200.20} \end{array}$$

15.
$$\begin{array}{r} {\scriptstyle 6\,7} \\ 479 \\ \times\ \ 800 \\ \hline \mathbf{383,200} \end{array}$$

16.
$$\begin{array}{r} \mathbf{374} \\ 3\overline{)1122} \\ -9 \\ \hline 22 \\ -21 \\ \hline 12 \\ -12 \\ \hline 0 \end{array}$$

17.
$$\begin{array}{r} \$0.96 \\ 6\overline{)\$5.76} \\ -5\,4 \\ \hline 36 \\ -36 \\ \hline 0 \end{array}$$

$m =$ **$0.96**

18.
$$\begin{array}{r} \mathbf{273\ R\ 5} \\ 10\overline{)2735} \\ -20 \\ \hline 73 \\ -70 \\ \hline 35 \\ -30 \\ \hline 5 \end{array}$$

19.
$$\begin{array}{r} \$\overset{0}{\cancel{1}}.\ \overset{9}{\cancel{0}}{}^{1}0 \\ -\ \$0.\ 1\ 6 \\ \hline \$0.\ 8\ 4 \end{array} \qquad \begin{array}{r} {\scriptstyle 1\,2\ \ 1} \\ \$64.23 \\ \$5.96 \\ \$17.00 \\ +\ \ \$0.84 \\ \hline \mathbf{\$88.03} \end{array}$$

20. **9 months**

21. $\mathbf{\frac{1}{3}}$**; 33$\frac{1}{3}$%; less than 50%**

22. **C. Horizontal**

23. **1:15 pm**

24. 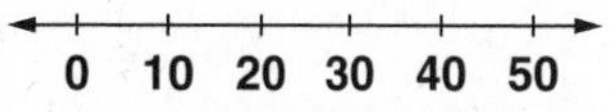

25. The pattern is "Count up by sevens."
7, 14, 21, 28, 35, 42, 49, 56, 63, 70 . . .
The tenth term is **70.**

26. **One possibility:**

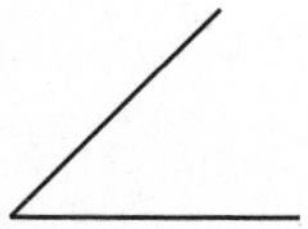

27. **1, 7**

28. **C. 9:00**

29. (a)

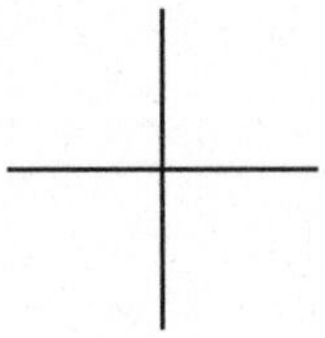

(b) **B. Perpendicular**

Lesson 34, Warm-Up

50 years; 5 years

a. **50**

b. **130**

c. **400**

d. **108**

e. **25**

f. **5**

g. **1720**

h. **1**

Problem Solving

HHH, HHT, HTH, HTT,
THH, THT, TTH, TTT

Lesson 34, Lesson Practice

a.
$$\begin{array}{r} \mathbf{20\text{ R }1} \\ 3\overline{)61} \\ -6 \\ \hline 01 \\ -0 \\ \hline 1 \end{array}$$

b.
$$\begin{array}{r} \mathbf{40\text{ R }2} \\ 6\overline{)242} \\ -24 \\ \hline 02 \\ -0 \\ \hline 2 \end{array}$$

c.
$$\begin{array}{r} \mathbf{40\text{ R }1} \\ 3\overline{)121} \\ -12 \\ \hline 01 \\ -0 \\ \hline 1 \end{array}$$

d.
$$\begin{array}{r} \mathbf{407} \\ 4\overline{)1628} \\ -16 \\ \hline 02 \\ -0 \\ \hline 28 \\ -28 \\ \hline 0 \end{array}$$

e.
$$\begin{array}{r} \mathbf{30\text{ R }2} \\ 4\overline{)122} \\ -12 \\ \hline 02 \\ -0 \\ \hline 2 \end{array}$$

f.
$$\begin{array}{r} \mathbf{\$1.05} \\ 5\overline{)\$5.25} \\ -5 \\ \hline 0\,2 \\ -0 \\ \hline 25 \\ -25 \\ \hline 0 \end{array}$$

g.
$$\begin{array}{r} \mathbf{\$3.09} \\ 2\overline{)\$6.18} \\ -6 \\ \hline 0\,1 \\ -0 \\ \hline 18 \\ -18 \\ \hline 0 \end{array}$$

h.
830 R 1
6)4981
−48
18
−18
01
−0
1

i.
30 R 1
10)301
−30
01
−0
1

j.
\$2.06
4)\$8.24
−8
0 2
−0
24
−24
0

k.
\$0.80
7)\$5.60
−5 6
00
−00
0

l.
602 R 2
8)4818
−48
01
−0
18
−16
2

m.
$\overset{4}{1}08$
× 6
648

648 + 2 = 650

The answer is correct.

Lesson 34, Mixed Practice

1.

2. Beans: 2)100 = **50 children**

Broccoli: 4)100 = **25 children**

Peas: 10)100 = **10 children**

Spinach:

50, 25, 10 } 85

50
25
10
+ S
100

100
− 85
15

S = **15 children**

3.
1849
+ 100
1949

4. **24 × 60 = *t***

$\overset{2}{2}4$ hours
× 60 minutes in one hour
1440 minutes

t = **1440 minutes**

5. **10 × 12 = *t***

12 eggs in one dozen
× 10 dozen
120 eggs

t = **120 eggs**

6. **300 − 127 = *L***

$\overset{29}{\cancel{30}}\,{}^{1}0$ pages
− 12 7 pages
17 3 pages

L = **173 pages**

7.
60 R 5
6)365
−36
05
−0
5

8.
```
    $1.06
6)$6.36
  -6
   0 3
  -0
    36
   -36
     0
```

9.
```
   107 R 1
5)536
 -5
  03
 -0
  36
 -35
   1
```

10.
```
     65 R 3
10)653
  -60
    53
   -50
     3
```

11.
```
    $1.09
4)$4.36
  -4
   0 3
  -0
    36
   -36
     0
```

12.
```
   2
   95
×  500
 47,500
```
(answer: **47,500**)

13. **80**

14.
```
   345 ⎫
    57 ⎪
   760 ⎬ 2906
   398 ⎪
   762 ⎪
   584 ⎭
+    W
  3000
```
```
  2 9 9
  3̸ 0̸ 0̸¹0
− 2 9 0 6
      9 4
```

W = **94**

15. 3004 − (3000 − 4)

3004 − 2996
```
  29 9
  3̸0̸ 0̸¹4
− 29 9 6
       8
```
(answer: **8**)

16.
```
   3 1
   $5.93
×     40
 $237.20
```
(answer: **$237.20**)

17. $\frac{1}{3}$ of 12 ⓑ> $\frac{1}{8}$ of 24

4   3

18.
```
   1 1 1
   $12.00
    $8.75
+   $0.96
   $21.71
```
(answer: **$21.71**)

19.
```
   19  9
   $2̸0̸. 0̸¹0
−  $12. 4 6
    $7. 5 4
```
(answer: **$7.54**)

20. 8 × 30 × 15

120 × 30
```
  120
×  30
 3600
```
(answer: **3600**)

21. 6 × 7 × 8 × 9

42 × 8 × 9

336 × 9

3024
```
   1
   42
×   8
  3 5
  336
×   9
 3024
```
(answer: **3024**)

22. The pattern is "Count up by tens."

490, 500, 510

23. $\mathbf{\frac{1}{4}}$**; 25%**

24. **B. Right**

25. Morning is "a.m.", so the time is **9:52 a.m.**

26. A.  **and D.** 

27.
$$\begin{array}{r} \overset{3}{8}4 \\ \times\ \ 9 \\ \hline 756 \end{array} \qquad 756 + 8 = 764$$

The answer is correct.

28. Both Abigail and Moe are correct since the associative property applies to multiplication.

29. The number 600 ends in zero, so 600 is divisible by 10. The answer is **C.** $\mathbf{\frac{600}{10}}$.

Lesson 35, Warm-Up

36 months; 48 months; 60 months

a. **300**

b. **1100**

c. **2400**

d. **210**

e. **21**

f. **$4.00**

g. **$2.00**

h. **6**

Problem Solving

AER, ARE, EAR
ERA, RAE, REA;
Three of the six arrangements spell words (ARE, EAR, and ERA). So **50% of the arrangements spell words.**

Lesson 35, Lesson Practice

a. $\mathbf{17 - S = 4}$

$$\begin{array}{r} 17 \\ -\ \ 4 \\ \hline 13 \text{ girls} \end{array}$$

$S =$ **13 girls**

b. $\mathbf{L - 3800 = 400}$

$$\begin{array}{r} \overset{1}{3}800 \text{ feet} \\ +\ \ 400 \text{ feet} \\ \hline 4200 \text{ feet} \end{array}$$

$L =$ **4200 feet**

c. $\mathbf{1448 - 1120 = D}$

$$\begin{array}{r} 1448 \text{ kilometers} \\ -\ 1120 \text{ kilometers} \\ \hline 328 \text{ kilometers} \end{array}$$

$D =$ **328 kilometers**

d. $\mathbf{1776 - 1066 = D}$

$$\begin{array}{r} 1776 \\ -\ 1066 \\ \hline \mathbf{710 \text{ years}} \end{array}$$

$D =$ **710 years**

e. $\mathbf{1787 - 1776 = D}$

$$\begin{array}{r} 1787 \\ -\ 1776 \\ \hline 11 \text{ years} \end{array}$$

$D =$ **11 years**

Lesson 35, Mixed Practice

1. One possibility:

2. $\mathbf{109 + 98 + 135 = t}$

$$\begin{array}{r} \overset{1}{1}\overset{2}{0}9 \\ 98 \\ +\ 135 \\ \hline 342 \end{array}$$

$t =$ **342**

3. **63 − S = 8**

$\overset{5}{\cancel{6}}{}^{1}3$ inches
− 8 inches
5 5 inches

S = **55 inches**

4. **1986 − 1886 = *D***

1986
− 1886
100 years

D = **100 years**

5. **40 × $1.50 = *t***

$\$\overset{2}{1}.50$
× 40
$60.00

t = **$60.00**

6. $\overset{1\,8}{919}$
× 90
82,710

7. **A.**  **and C.** 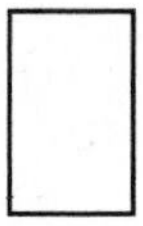

8. Factors of 28: **1, 2, 4, 7, 14, 28**

9.
108
4)432
−4
03
−0
32
−32
0

m = **108**

10.
70 R 3
6)423
−42
03
−0
3

11.
30 R 3
8)243
−24
03
−0
3

12.
500 R 1
4)2001
−20
00
−0
01
−0
1

13.
102
10)1020
−10
02
−0
20
−20
0

14. 420 ÷ (42 ÷ 6)
= 420 ÷ 7 = **60**

15. **500**

16. $\overset{1\,1\,1}{4657}$
285
+ 1223
6165

17. $\overset{2}{\cancel{3}}{}^{1}1\,6\,5$
− 1 6 3 5
1 5 3 0

18. $\$\overset{9}{\cancel{10}}.\overset{9}{\cancel{0}}{}^{1}0$
− $8. 9 3
$1. 0 7

19. $\overset{2\,4}{4\,36}$
× 70
30,520

20. $\$\overset{3\,4}{8.57}$
× 7
$59.99

21.
$$\begin{array}{r} 600 \\ \times \quad 900 \\ \hline \mathbf{540{,}000} \end{array}$$

22. $\mathbf{\frac{2}{5}}$**; 40%; less than 50%**

23. Afternoon is "p.m.", so the time is **2:45 p.m.**

24. **4 months**

25. The pattern is "Count up by one hundreds."
2200, 2300, 2400

26.
$$\begin{array}{r} 72 \\ \times \quad 6 \\ \hline 432 \end{array}$$
The answer is correct.

27. $2\overline{)14}$ = **7 girls**

28. **Sixty-eight thousand, two hundred**

29. (a)

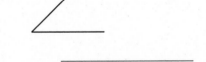

(b)

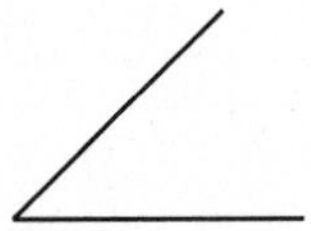

LESSON 36, WARM-UP

6; 18; 30

a. **70**

b. **150**

c. **560**

d. **584**

e. **12**

f. **$6**

g. **$3**

h. **560**

i. **6**

Problem Solving

walk, walk, walk;
walk, walk, wait;
walk, wait, walk;
walk, wait, wait;
wait, walk, walk;
wait, walk, wait;
wait, wait, walk;
wait, wait, wait

LESSON 36, LESSON PRACTICE

a. **Acute**

b. **Obtuse**

c. **Right**

d. **Isosceles**

e. **Scalene**

f. **Equilateral**

g. **One possibility:**

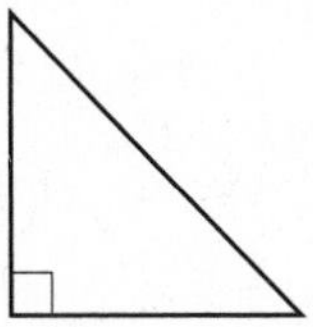

h.

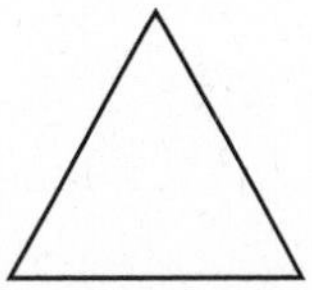

i. **Isosceles right triangle**
One possibility:

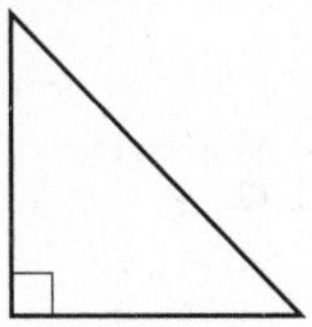

j. **Two scalene right triangles would be formed.**

LESSON 36, MIXED PRACTICE

1. ______
______

2. **402 − 336 = *p***

$$\begin{array}{r} \overset{39}{\cancel{40}}{}^{1}2 \text{ pages} \\ -\ 33\ 6 \text{ pages} \\ \hline 6\ 6 \text{ pages} \end{array}$$

p = **66 more pages**

3. **7*p* = 336**

$$\begin{array}{r} 48 \text{ pages} \\ 7\overline{)336} \\ -28 \\ \hline 56 \\ -56 \\ \hline 0 \end{array}$$

p = **48 pages**

4. **2 × 7 = *t***
t = **14 days**

5. **800**

6. **B.**

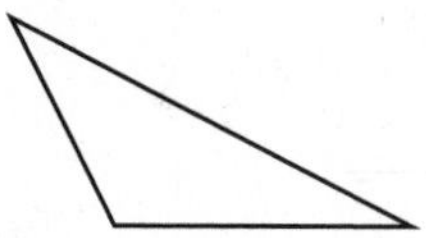

7.
$$\begin{array}{r} 1976 \\ -\ 1776 \\ \hline \textbf{200 years} \end{array}$$

8.
$$\begin{array}{r} 1\ \overset{3}{\cancel{4}}{}^{1}2 \text{ votes} \\ -\ 1\ 1\ 9 \text{ votes} \\ \hline \textbf{2 3 votes} \end{array}$$

9. **Numerator**

10. **A.**  **; The figure is not closed.**

11. $\mathbf{\frac{2}{10}\left(\text{or } \frac{1}{5}\right)}$ **of a circle**

12.
$$\begin{array}{r} \overset{5\,2}{763} \\ \times\ \ 800 \\ \hline \textbf{610,400} \end{array}$$

13.
$$\begin{array}{r} \$\overset{2}{2}\overset{4}{4}.08 \\ \times\ \ \ \ 6 \\ \hline \textbf{\$144.48} \end{array}$$

14.
$$\begin{array}{r} \overset{3\,2}{976} \\ \times\ \ 40 \\ \hline \textbf{39,040} \end{array}$$

15.
$$\begin{array}{r} 400 \\ \times\ \ 50 \\ \hline \textbf{20,000} \end{array}$$

16.
$$\begin{array}{r} 5\ \overset{7}{\cancel{8}}{}^{1}1\ 8 \\ -\ 4\ 7\ 4\ 7 \\ \hline 1\ 0\ 7\ 1 \end{array}$$

M = **1071**

17.
$$\begin{array}{r} \$\overset{2\,2}{98}.\overset{2}{9}8 \\ \$36.25 \\ \$4.97 \\ +\ \$87.64 \\ \hline \textbf{\$227.84} \end{array}$$

18.
$$\begin{array}{r} \overset{9}{\cancel{10}}\ \overset{0}{\cancel{1}}{}^{1}0 \\ -\ \ 9\ 1\ 8 \\ \hline \textbf{9 2} \end{array}$$

19.
$$\begin{array}{r} \$1.09 \\ 7\overline{)\$7.63} \\ -7 \\ \hline 0\,6 \\ -0 \\ \hline 63 \\ -63 \\ \hline 0 \end{array}$$

w = **$1.09**

20.
$$\begin{array}{r} \textbf{40 R 8} \\ 9\overline{)368} \\ -36 \\ \hline 08 \\ -0 \\ \hline 8 \end{array}$$

21.
$$\begin{array}{r} \textbf{708} \\ 6\overline{)4248} \\ -42 \\ \hline 04 \\ -0 \\ \hline 48 \\ -48 \\ \hline 0 \end{array}$$

22.
```
      $1.25
  8)$10.00
     −8
      2 0
     −1 6
        40
       −40
         0
```

23. The pattern is "Count up by one hundreds."
3000, 3100, 3200

24. $\frac{1}{6}$**; less than 25%; more than 10%**

25. **860**

26.
```
   1
   13        78 + 4 = 82
 ×  6
   78
```
The answer is not correct.

27. **6 × 6 = 36**

28. **70%**

29. **Check for a right angle and two sides of equal length.**
One possibility:

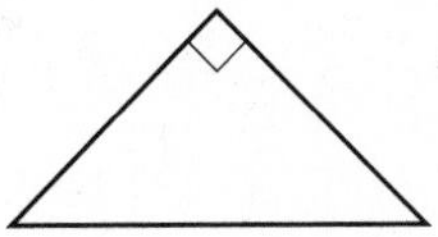

LESSON 37, WARM-UP

60 months; 72 months

a. **900**

b. **1800**

c. **3600**

d. **356**

e. **356**

f. **8**

g. **30¢**

h. **15¢**

i. **6¢**

Problem Solving
6 different license plates:
CAR 123
CAR 132
CAR 213
CAR 231
CAR 312
CAR 321

LESSON 37, LESSON PRACTICE

a. **One possibility:**

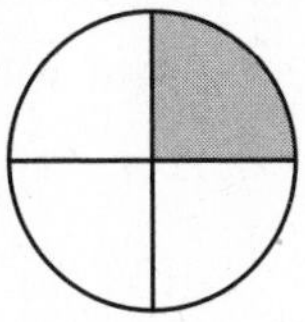

b. **One possibility:**

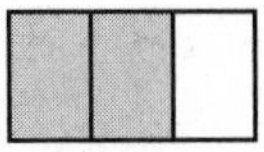

c. **One possibility:**

d. **One possibility:**

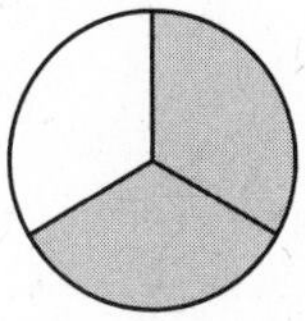

LESSON 37, MIXED PRACTICE

1.

2. **Possibilities include:**

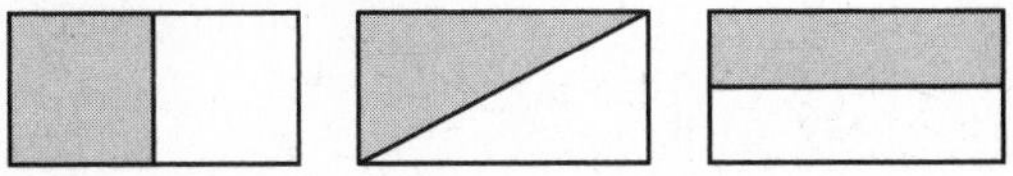

50%

3. **39 + 20 + 1 + 4 + 1 =** t
t **= 65 items**

4. $3 \times 12 = t$

$$\begin{array}{r} 12 \text{ inches in each foot} \\ \times \quad 3 \text{ feet} \\ \hline 36 \text{ inches} \end{array}$$

t = **36 inches**

5. $1620 - 1517 = y$

$$\begin{array}{r} 1\,6\,\overset{1}{\cancel{2}}{}^{1}0 \\ -\ 1\,5\,1\,7 \\ \hline 1\,0\,3 \text{ years} \end{array}$$

y = **103 years**

6. **1, 2, 4, 5, 8, 10, 20, 40**

7. Five of the eight triangles are not shaded, so $\frac{5}{8}$ of the group is not shaded. **More than 50%; $37\frac{1}{2}$%**

8. **9 months**

9. **50**

10. **One possibility:**

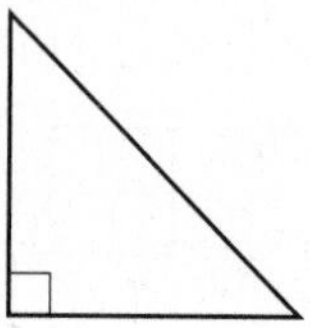

11.

$$\begin{array}{r} \scriptstyle 1\,1\ \ 1 \\ \$36.51 \\ \$74.15 \\ +\ \$25.94 \\ \hline \mathbf{\$136.60} \end{array}$$

12.

$$\begin{array}{r} \overset{29}{\cancel{30}}{}^{1}4\,0 \\ -\ 29\,5\,0 \\ \hline 9\,0 \end{array}$$

W = **90**

13.

$$\begin{array}{r} \$\overset{89}{\cancel{90}}.{}^{1}0\,0 \\ -\ \$20.\,3\,0 \\ \hline \mathbf{\$69.\,7\,0} \end{array}$$

14.

$$\begin{array}{r} \scriptstyle 8\ 1 \\ 5\,92 \\ \times \quad 90 \\ \hline \mathbf{53,280} \end{array}$$

15.

$$\begin{array}{r} \scriptstyle 6\ 4 \\ \$4.75 \\ \times \quad 80 \\ \hline \mathbf{\$380.00} \end{array}$$

16.

$$\begin{array}{r} 43 \\ C \\ 29 \\ 467 \\ +\ 94 \\ \hline 700 \end{array} \quad 633 \qquad \begin{array}{r} \overset{6}{\cancel{7}}\,\overset{9}{\cancel{0}}{}^{1}0 \\ -\ 6\,3\,3 \\ \hline 6\,7 \end{array}$$

C = **67**

17. $\frac{840}{8}$ ⓒ< $\frac{460}{4}$

$$\begin{array}{r} 105 \\ 8\overline{)840} \\ -8 \\ \hline 04 \\ -0 \\ \hline 40 \\ -40 \\ \hline 0 \end{array} \qquad \begin{array}{r} 115 \\ 4\overline{)460} \\ -4 \\ \hline 06 \\ -4 \\ \hline 20 \\ -20 \\ \hline 0 \end{array}$$

18.

$$\begin{array}{r} 720 \\ \times \quad 400 \\ \hline \mathbf{288,000} \end{array}$$

19.

$$\begin{array}{r} \$2.04 \\ 6\overline{)\$12.24} \\ -12 \\ \hline 0\,2 \\ -0 \\ \hline 24 \\ -24 \\ \hline 0 \end{array}$$

w = **$2.04**

20. $1000 \div (100 \div 10) =$
$1000 \div 10 =$ **100**

$$\begin{array}{r} 100 \\ 10\overline{)1000} \\ -10 \\ \hline 00 \\ -0 \\ \hline 00 \\ -0 \\ \hline 0 \end{array}$$

21. $60 \times (235 \div 5) =$
$60 \times 47 =$ **2820**

$$\begin{array}{r} \scriptstyle 4 \\ 47 \\ \times \quad 60 \\ \hline 2820 \end{array}$$

22. 42 × 30 × 7

1260 × 7 = **8820**

$$\begin{array}{r} 42 \\ \times\ \ 30 \\ \hline \overset{1\,4}{1260} \\ \times\ \ \ \ 7 \\ \hline 8820 \end{array}$$

23. \$20 − (\$3.48 + \$12 + \$4.39)

$$\begin{array}{r} \$3.\overset{1}{4}8 \\ \$12.00 \\ +\ \ \$4.39 \\ \hline \$19.87 \end{array} \qquad \begin{array}{r} \$\overset{19}{\not{2}\not{0}}.\ \overset{9}{\not{0}}{}^{1}0 \\ -\ \$19.\ 8\ 7 \\ \hline \mathbf{\$0.\ 1\ 3} \end{array}$$

24.

$$\begin{array}{r} 50\% \\ 25\% \\ +\ 10\% \\ \hline 85\% \end{array} \qquad \begin{array}{r} \overset{9}{\not{1}\not{0}}{}^{1}0\% \\ -\ \ 8\ 5\% \\ \hline \mathbf{1\ 5\%} \end{array}$$

25. One possibility: B.

26. One possibility:

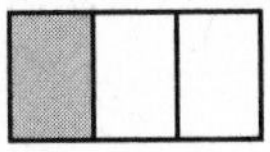

27.

$$\begin{array}{r} 1932 \\ -\ \ \ 10 \\ \hline \mathbf{1922} \end{array}$$

28. (a) **8 angles**

(b) **4 angles**

(c) $\mathbf{\frac{4}{8}}$

29. D.

Lesson 38, Warm-Up

grandfather; 60; 72; 84

a. 600

b. 1500

c. 700

d. 120

e. 22

f. 40¢

g. 20¢

h. 8¢

i. 0

Problem Solving

Since the top addend has only two digits, and adding 1 to that addend results in a number with 3 digits, the top addend must be 99 and the sum must be 100.

$$\begin{array}{r} \mathbf{99} \\ \mathbf{+\ \ 1} \\ \hline \mathbf{100} \end{array}$$

Lesson 38, Lesson Practice

a. $\mathbf{\frac{3}{4}}$

b. $\mathbf{1\frac{1}{4}}$

c. $\mathbf{6\frac{2}{5}}$

d. $\mathbf{7\frac{3}{5}}$

e. $\frac{1}{3} < \frac{1}{2}$

f. $\frac{1}{2} < \frac{3}{4}$

g. $\frac{3}{4} > \frac{1}{3}$

LESSON 38, MIXED PRACTICE

1.

2. $4\overline{)28}$ = **7 points**

3. **4 × \$4.75 = *t***

```
  3 2
 $4.75
 ×   4
 $19.00
```

t = **\$19.00**

4. **\$1020 − \$725 = *m***

```
   9 ¹1
 $1̸0̸ 2̸¹0
 −  $7 2 5
    $2 9 5
```

m = **\$295**

5. **3*C* = 1347**

```
   449 cherries
 3)1347
  −12
    14
   −12
     27
    −27
      0
```

C = **449 cherries**

6. **1776 − 1620 = *y***

```
   1776
 − 1620
   156 years
```

y = **156 years**

7. **A.**

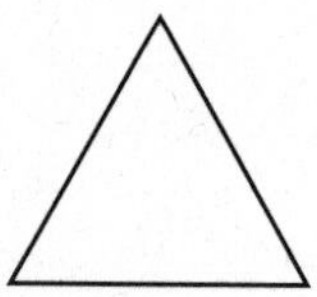

8. 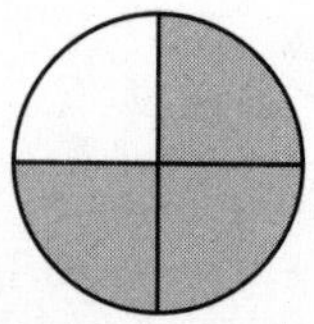 **; 75%**

9. **366 days**

10. **8 sides**

11.
```
   111
   3647
     92
 +  429
   4168
```

12.
```
   2 ¹4
   3̸ 5̸¹1 8
 − 1 8 5 3
   1 6 6 5
```

13. 4 × 6 × 8 × 0 = **0**

14.
```
     351 R 8
 10)3518
    −30
      51
     −50
      18
     −10
       8
```

15.
```
    $4.76 ⟩
   $12.00 ⟩ 17.73
    $0.97 ⟩
 +      W
   $20.00
```

```
     19 9
   $2̸0̸.0̸¹0
 −  $17.7 3
     $2.2 7
```

W = **\$2.27**

16.
```
   9  9  9
  $1̸0̸ 0̸. 0̸¹0
 −  $8 7. 2 3
    $1 2. 7 7
```

17.
```
   7 5
    786
 ×  900
 707,400
```

18.
```
    $7.02
 9)$63.18
   −63
     0 1
     −0
      18
     −18
       0
```

19. $375 \times (640 \div 8) =$

$375 \times 80 =$ **30,000**

$$\begin{array}{r} \scriptstyle 6\,4 \\ 375 \\ \times \quad 80 \\ \hline 30{,}000 \end{array}$$

20. $(3 \times 5) \times 7$ ⊜ $3 \times (5 \times 7)$

$15 \times 7 \qquad 3 \times 35$

$105 \qquad 105$

21. **C. Quadrilateral**

22. The pattern is "Count up by one hundred."
2100, 2200, 2300

23. $\mathbf{6\frac{2}{3}}$

24. $\mathbf{\frac{3}{4}}$

25. In 4 hours the time will be afternoon, so the "a.m." will switch to "p.m." The time will be **1:45 p.m.**

26. **600**

27. $6 \times 3 =$ **18**

28. (a) **No**

(b) **Yes**

29. $\frac{1}{2}$ ⧁ $\frac{1}{6}$

Lesson 39, Warm-Up

aunt; 21 days; 28 days; 42 days

a. **80**

b. **630**

c. **2400**

d. **144**

e. **1000**

f. $\mathbf{10\frac{1}{2}}$

g. **$1.00**

h. **$0.50**

i. **1**

Problem Solving

Counting by 6's to 36: 6, 12, 18, 24, 30, 36

Counting by 8's to 48: 8, 16, 24, 32, 40, 48

24

Lesson 39, Lesson Practice

a. **One possibility:**

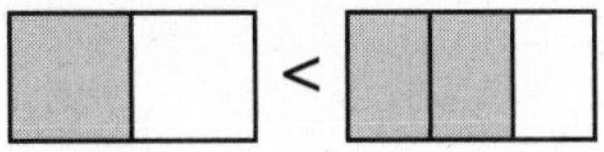

$\frac{1}{2}$ ⧀ $\frac{2}{3}$

b. **One possibility:**

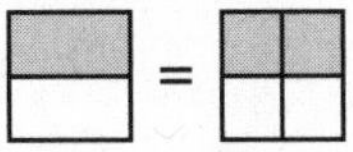

$\frac{1}{2}$ ⊜ $\frac{2}{4}$

c. **One possibility:**

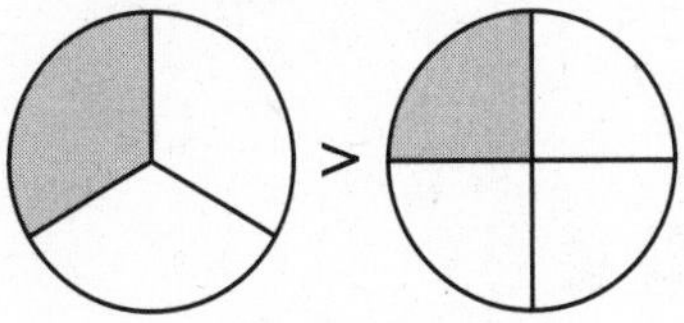

$\frac{1}{3}$ ⧁ $\frac{1}{4}$

d. **One possibility:**

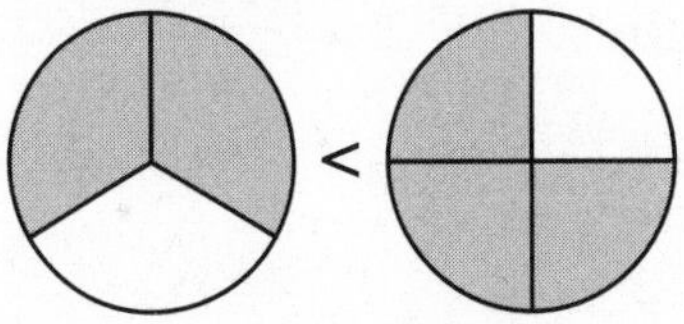

$\frac{2}{3}$ ⧀ $\frac{3}{4}$

Lesson 39, Mixed Practice

1. **Possibilities include:**

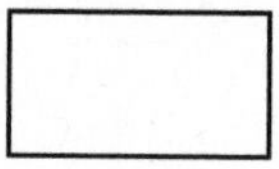 **and** 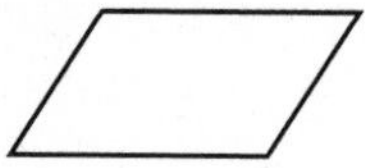

2. $\mathbf{5 \times 100 = t}$

$$\begin{array}{r} 5 \\ \times\ 100 \\ \hline 500 \text{ years} \end{array}$$

t = **500 years**

3. $\mathbf{13 - s = 6}$

$$\begin{array}{r} 13 \\ -\ \ 6 \\ \hline 7 \text{ years} \end{array}$$

s = **7 years old**

4. $\mathbf{2f = 488}$

$$\begin{array}{r} 244 \text{ feet} \\ 2\overline{)488} \\ -4 \\ \hline 08 \\ -8 \\ \hline 08 \\ -8 \\ \hline 0 \end{array}$$

f = **244 feet**

5. **One possibility:**

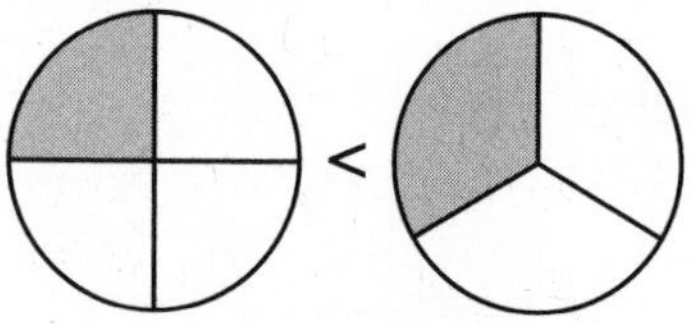

$\frac{1}{4} \textcircled{<} \frac{1}{3}$

6. $\mathbf{11\frac{1}{2}}$

7. $4\overline{)12}$ **3 fish filets**

8. **80**

9. **1, 5, 7, 35**

10.
$$\begin{array}{r} {}^{1\,2\,2} \\ \$93.18 \\ \$42.87 \\ +\ \$67.95 \\ \hline \mathbf{\$204.00} \end{array}$$

11.
$$\begin{array}{r} \$\overset{2\,9}{\cancel{30}}.\overset{9}{\cancel{0}}{}^{1}0 \\ -\ \ \$8.\ 7\ 5 \\ \hline \mathbf{\$21.\ 2\ 5} \end{array}$$

12.
$$\begin{array}{r} \overset{3}{4}6 \\ 23 \\ 97 \\ 15 \\ 24 \\ 55 \\ +\ 55 \\ \hline \mathbf{315} \end{array}$$

13.
$$\begin{array}{r} \overset{3}{\cancel{4}}\ \overset{{}^{1}2}{\cancel{3}}{}^{1}0\ 4 \\ -\ 3\ 4\ 5\ 2 \\ \hline 8\ 5\ 2 \end{array}$$

B = **852**

14.
$$\begin{array}{r} {}^{2\ 4} \\ \$6.38 \\ \times\ \ \ \ 60 \\ \hline \mathbf{\$382.80} \end{array}$$

15.
$$\begin{array}{r} \overset{2}{6}40 \\ \times\ \ 700 \\ \hline \mathbf{448{,}000} \end{array}$$

16.
$$\begin{array}{r} \mathbf{80} \\ 8\overline{)640} \\ -64 \\ \hline 00 \\ -0 \\ \hline 0 \end{array}$$

17.
$$\begin{array}{r} \mathbf{72} \\ 10\overline{)720} \\ -70 \\ \hline 20 \\ -20 \\ \hline 0 \end{array}$$

18.
$$\begin{array}{r} \mathbf{\$1.04} \\ 6\overline{)\$6.24} \\ -6 \\ \hline 0\,2 \\ -0 \\ \hline 24 \\ -24 \\ \hline 0 \end{array}$$

19.
$$\begin{array}{r} \mathbf{309} \\ 4\overline{)1236} \\ -12 \\ \hline 03 \\ -0 \\ \hline 36 \\ -36 \\ \hline 0 \end{array}$$

20.
$$\begin{array}{r} \mathbf{80\ R\ 3} \\ 7\overline{)563} \\ -56 \\ \hline 03 \\ -0 \\ \hline 3 \end{array}$$

21.
$$\begin{array}{r} \mathbf{524\ R\ 2} \\ 9\overline{)4718} \\ -45 \\ \hline 21 \\ -18 \\ \hline 38 \\ -36 \\ \hline 2 \end{array}$$

22.
$$\begin{array}{r} 375 \\ 8\overline{)3000} \\ -24 \\ \hline 60 \\ -56 \\ \hline 40 \\ -40 \\ \hline 0 \end{array}$$

$m = \mathbf{375}$

23. The hours after noon and before midnight are the "p.m." hours. The time will be **11:40 p.m.**

24. $\mathbf{37\frac{1}{2}\%}$

25. **April 20, 1901**

26. (a) $\mathbf{8\frac{1}{5}}$

(b) $\mathbf{8\frac{4}{5}}$

(c) $\mathbf{8\frac{1}{5} < 8\frac{4}{5}}$ **or** $\mathbf{8\frac{4}{5} > 8\frac{1}{5}}$

27.
$$\begin{array}{r} 4\overset{3}{1}6 \\ \times\quad 60 \\ \hline \mathbf{24{,}960} \end{array}$$

28. (a) **24 hr**

(b) $\begin{array}{r} \mathbf{12\ hours} \\ 2\overline{)24} \end{array}$

(c) $\mathbf{\frac{12}{24}}$

29. $\begin{array}{r} \mathbf{180\ degrees} \\ 2\overline{)360} \end{array}$

Lesson 40, Warm-Up

Problem Solving

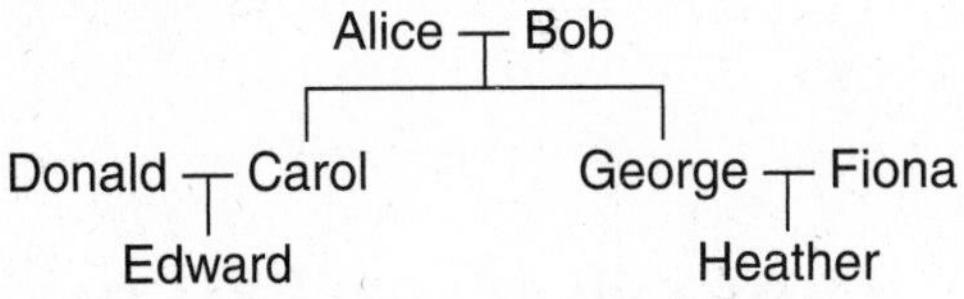

a. **grandmother**

b. **granddaughter**

c. **uncle**

d. **niece**

e. **son-in-law**

f. **mother-in-law**

g. **cousin**

Lesson 40, Lesson Practice

a. $\mathbf{1\frac{1}{3}}$

b. $\mathbf{2\frac{1}{4}}$

c. 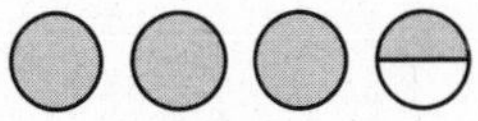

d. 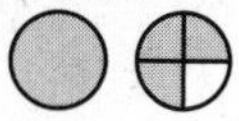

e. **Sample answer:**

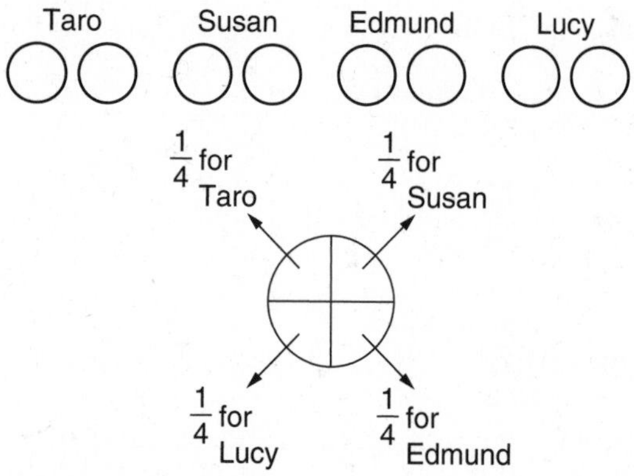

$$\begin{array}{r} \mathbf{2\tfrac{1}{4}} \text{ **pies**} \\ 4\overline{)9} \\ \underline{-8} \\ 1 \end{array}$$

Lesson 40, Mixed Practice

1. **One possibility:**

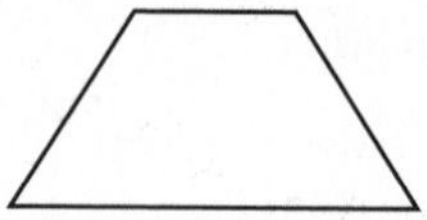

2. $\mathbf{\frac{1}{6}}$

3. $$\begin{array}{r} \mathbf{1\tfrac{1}{3}} \text{ **oranges**} \\ 3\overline{)4} \\ \underline{-3} \\ 1 \end{array}$$

4. $\mathbf{5c = 140}$

$$\begin{array}{r} 28 \text{ jelly beans} \\ 5\overline{)140} \\ \underline{-10} \\ 40 \\ \underline{-40} \\ 0 \end{array}$$

c = **28 jelly beans**

5. $\mathbf{105 - 87 = p}$

$$\begin{array}{r} \overset{9}{\not{1}\not{0}}{}^{1}5 \text{ pounds} \\ \underline{-\quad 8\ 7 \text{ pounds}} \\ 1\ 8 \text{ pounds} \end{array}$$

p = **18 pounds**

6. $\mathbf{13 + M = 50}$ **or** $\mathbf{50 - 13 = d}$

$$\begin{array}{r} \overset{4}{\not{5}}{}^{1}0 \\ \underline{-\ 1\ 3} \\ 3\ 7 \text{ stars} \end{array}$$

M = **37 more stars**

7. $\mathbf{12\frac{1}{2}\%}$

8.
$$\begin{array}{r} 6 \text{ sides} \\ \underline{-\ 5 \text{ sides}} \\ \textbf{1 more side} \end{array}$$

9.
$$\begin{array}{r} 50\% \\ \underline{+\ 25\%} \\ \mathbf{75\%} \end{array}$$

10. (a) $\mathbf{2\frac{3}{4}}$

(b) $\mathbf{3\frac{1}{4}}$

(c) $\mathbf{2\frac{3}{4} < 3\frac{1}{4}}$ **or** $\mathbf{3\frac{1}{4} > 2\frac{3}{4}}$

11.
$$\begin{array}{r} 534 \\ \underline{+\ 345} \\ 879 \end{array}$$

M = **879**

12.
$$\begin{array}{r} \overset{2\,2}{785} \\ 964 \\ 287 \\ \underline{+\ 846} \\ \mathbf{2882} \end{array}$$

13.
$$\begin{array}{r} \overset{6}{\not{7}}\ \overset{{}^{1}0}{\not{1}}{}^{1}0\ 6 \\ \underline{-\ 3\ 7\ 5\ 4} \\ \mathbf{3\ 3\ 5\ 2} \end{array}$$

14.
$$\begin{array}{r} \overset{5\ 2}{\$3.84} \\ \underline{\times \quad 60} \\ \mathbf{\$230.40} \end{array}$$

15.
$$\begin{array}{r} \overset{5\,7}{769} \\ \underline{\times \quad 800} \\ \mathbf{615{,}200} \end{array}$$

16.
$$\begin{array}{r} \$3.06 \\ 8\overline{)\$24.48} \\ -24 \\ \hline 0\,4 \\ -0 \\ \hline 48 \\ -48 \\ \hline 0 \end{array}$$

17.
$$\begin{array}{r} 480 \\ 9\overline{)4320} \\ -36 \\ \hline 72 \\ -72 \\ \hline 00 \\ -0 \\ \hline 0 \end{array}$$

18. $20 - (\$1.45 + \$6.23 + \$8)$

$$\begin{array}{r} \$1.45 \\ \$6.23 \\ + \$8.00 \\ \hline \$15.68 \end{array} \qquad \begin{array}{r} {}^{19}\ {}^{9} \\ \$2\cancel{0}.\,\cancel{\emptyset}{}^{1}0 \\ - \$15.\,6\,8 \\ \hline \$4.\,3\,2 \end{array}$$

19.
$$\begin{array}{r} {}^{3\,2\,1} \\ 3742 \\ \times \quad 5 \\ \hline 18{,}710 \end{array}$$

20. $2\frac{2}{3}$

21. 700

22. $\frac{1}{10}$; 10%

23. C.

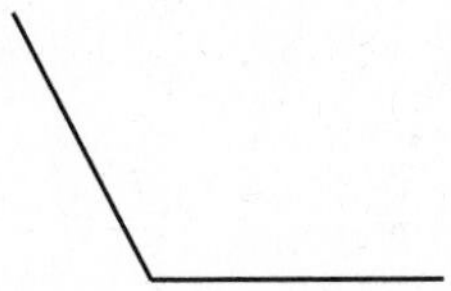

24. The pattern is "Count up by tens."

 90, 100, 110

25. 150

26.$$\frac{1}{4} \,\textcircled{<}\, \frac{1}{3}$$

27. B. $\frac{3}{5}$

28.
$$\begin{array}{r} 90 \text{ degrees} \\ 4\overline{)360} \end{array}$$

29.
$$\begin{array}{r} 1789 \\ - \ 1776 \\ \hline 13 \text{ years} \end{array}$$

INVESTIGATION 4

1. Observe student actions.

2. Observe student actions.

3. Observe student actions.

4. a. **45°**
 b. **90°**
 c. **120°**
 d. **60°**

5.

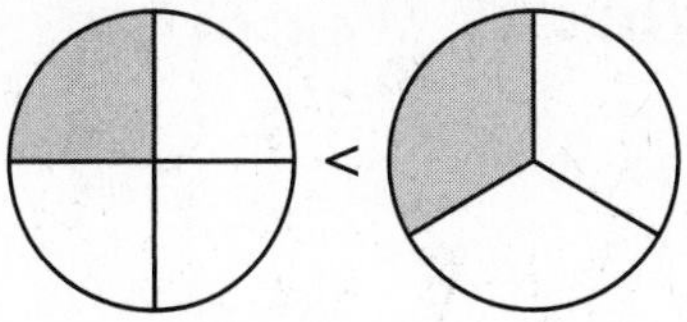

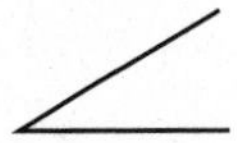

6.

7.

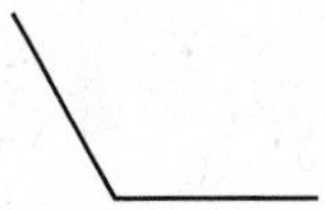

8.

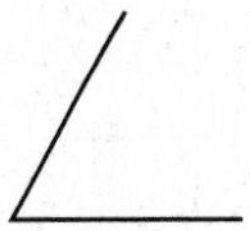

Lesson 41, Warm-Up

cousin; 6; 24; 30

a. **400**

b. **620**

c. **168**

d. **6400**

e. **250**

f. $\mathbf{7\frac{1}{2}}$

g. **5¢**

h. **1¢**

i. **0**

Problem Solving

In this division problem, we are given the quotient and the divisor. To find the dividend, we multiply: $24 \times 4 = 96$.

$$\begin{array}{r} \mathbf{24} \\ 4\overline{)\mathbf{96}} \end{array}$$

Lesson 41, Lesson Practice

a. **See student work;**

$$\frac{1}{10} + \frac{2}{10} = \frac{3}{10}$$

b. **See student work;**

$$\frac{3}{4} - \frac{2}{4} = \frac{1}{4}$$

c. **See student work;**

$$1\frac{1}{2} + 1\frac{1}{2} = 2\frac{2}{2} = 3$$

d. **See student work;**

$$3\frac{4}{10} - 1\frac{1}{10} = 2\frac{3}{10}$$

Lesson 41, Mixed Practice

1. **One possibility:**

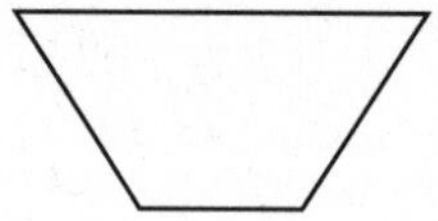

2. $\mathbf{\frac{1}{10}}$**; 10%**

3.
$$\begin{array}{r} 1492 \\ +\ \ 200 \\ \hline \mathbf{1692} \end{array}$$

4. $\mathbf{S - 4360 = 340}$

$$\begin{array}{r} \overset{1}{4}360 \\ +\ \ 340 \\ \hline 4700 \end{array}$$

$S =$ **4700**

5. $\mathbf{3 \times 6 = t}$

$3 \times 6 = 18$ plants

$t =$ **18 plants**

6. $\mathbf{243 - 167 = P}$

$$\begin{array}{r} \overset{1}{\cancel{2}}\,\overset{13}{\cancel{4}}\,{}^{1}3 \text{ pages} \\ -\ 1\ 6\ 7 \text{ pages} \\ \hline 7\ 6 \text{ pages} \end{array}$$

$P =$ **76 pages**

7. $3\frac{3}{10} - 1\frac{2}{10} = \mathbf{2\frac{1}{10}}$

8. $\frac{5}{10} + \frac{4}{10} = \mathbf{\frac{9}{10}}$

9. $\frac{1}{2} - \frac{1}{2} = \frac{0}{2} = \mathbf{0}$

10. $2\frac{1}{4} + 3\frac{2}{4} = \mathbf{5\frac{3}{4}}$

11. $\mathbf{10\frac{2}{3}}$

12.
$$\begin{array}{r} \scriptstyle 333 \\ 3784 \\ 2693 \\ 429 \\ 97 \\ 856 \\ +\ \ 907 \\ \hline \mathbf{8766} \end{array}$$

13.
$$\begin{array}{r} \scriptstyle 2\ {}^{1}0\ \ 9\ \ \ \\ \not{3}\ \not{1}\ \not{0}\,{}^{1}6 \\ -\quad 5\ 2\ 8 \\ \hline \mathbf{2\ 5\ 7\ 8} \end{array}$$

14.
$$\begin{array}{r} \scriptstyle 79\ \ \ 9\ \ \ \\ \$\not{8}\not{0}.\ \not{0}\,{}^{1}0 \\ -\ \$77.\ 5\ 6 \\ \hline \mathbf{\$2.\ 4\ 4} \end{array}$$

15.
$$\begin{array}{r} \scriptstyle 2\ \ \\ 804 \\ \times\quad 700 \\ \hline \mathbf{562{,}800} \end{array}$$

16.
$$\begin{array}{r} \scriptstyle 1\ \ \\ 43 \\ \times\quad 60 \\ \hline \scriptstyle 46\ \ \ \\ 2580 \\ \times\quad\ \ 8 \\ \hline \mathbf{20{,}640} \end{array}$$

17.
$$\begin{array}{r} \mathbf{1002} \\ 4\overline{)4008} \\ -4\ \ \ \ \\ \hline 00\ \ \\ -0\ \ \\ \hline 00\ \\ -00\ \\ \hline 08 \\ -8 \\ \hline 0 \end{array}$$

18.
$$\begin{array}{r} \mathbf{604} \\ 7\overline{)4228} \\ -42\ \ \\ \hline 02\ \\ -0\ \\ \hline 28 \\ -28 \\ \hline 0 \end{array}$$

19.
$$\begin{array}{r} \mathbf{1204\ R\ 3} \\ 8\overline{)9635} \\ -8\ \ \ \ \\ \hline 16\ \ \ \\ -16\ \ \ \\ \hline 03\ \ \\ -0\ \ \\ \hline 35\ \\ -32\ \\ \hline 3 \end{array}$$

20.
$$\begin{array}{r} \mathbf{\$1.33} \\ 6\overline{)\$7.98} \\ -6\ \ \ \ \\ \hline 1\,9\ \ \\ -1\,8\ \ \\ \hline 18 \\ -18 \\ \hline 0 \end{array}$$

21.
$$\begin{array}{r} \scriptstyle 1\ \ \ \\ \$4.56 \\ \$3.00 \\ +\ \$1.29 \\ \hline \$8.85 \end{array} \qquad \begin{array}{r} \scriptstyle 9\ \ \ 9\ \ \ \\ \$\not{1}\not{0}.\ \not{0}\,{}^{1}0 \\ -\quad \$8.\ 8\ 5 \\ \hline \mathbf{\$1.\ 1\ 5} \end{array}$$

22. **100**

23. **One possibility:**

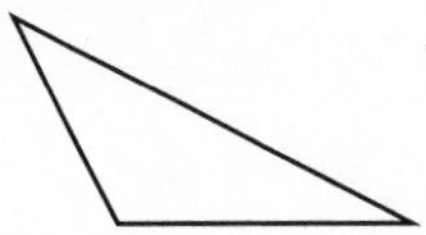

24. $30 \div 5 =$ **6 children**

25. Evening is "p.m." so the time is **6:06 p.m.**

26.
$$\begin{array}{r} \mathbf{2\frac{1}{4}\ miles} \\ 4\overline{)9} \\ -8 \\ \hline 1 \end{array}$$

27.

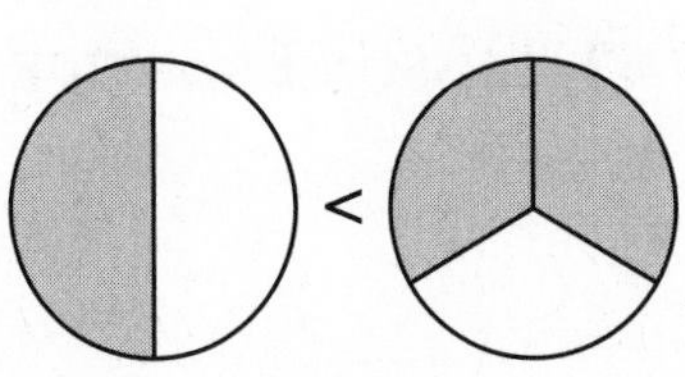

$\frac{1}{2}$ (<) $\frac{2}{3}$

28. B.

29. $\frac{1}{4}$ of 100 $\circled{=}$ 100 ÷ 4

25 25

Lesson 42, Warm-Up

granddaughter

a. **40**

b. **400**

c. **1**

d. **0**

e. **192**

f. **$50.00**

g. **$25.00**

h. **150**

i. $\mathbf{12\frac{1}{2}}$

j. **1**

Problem Solving

3456, 3465, 3546, 3564, 3645, 3654
4356, 4365, 4536, 4563, 4635, 4653
5346, 5364, 5436, 5463, 5634, 5643
6345, 6354, 6435, 6453, 6534, 6543
24 different phone numbers

Lesson 42, Lesson Practice

a. $\begin{array}{r} \mathbf{145} \\ 3\overline{)4^{1}3^{1}5} \end{array}$

b. $\begin{array}{r} \mathbf{89} \\ 6\overline{)5\ 3^{5}4} \end{array}$

c. $\begin{array}{r} \mathbf{63} \\ 9\overline{)5\ 6^{2}7} \end{array}$

d. $\begin{array}{r} \mathbf{125} \\ 4\overline{)5^{1}0^{2}0} \end{array}$

e. $\begin{array}{r} \mathbf{114\ R\ 2} \\ 7\overline{)8^{1}0^{3}0} \end{array}$

f. $\begin{array}{r} \mathbf{83\ R\ 6} \\ 10\overline{)8\ 3^{3}6} \end{array}$

g. $\begin{array}{r} \mathbf{120} \\ 5\overline{)6^{1}0\ 0} \end{array}$

h. $\begin{array}{r} \mathbf{205\ R\ 1} \\ 3\overline{)6\ 1^{1}6} \end{array}$

i. $\begin{array}{r} \mathbf{143} \\ 6\overline{)8^{2}5^{1}8} \end{array}$

j. **3, 6, 9**

k. **3**

l. **3, 6**

m. **None**

n. **3, 6, 9**

o. **3, 9**

Lesson 42, Mixed Practice

1. **Polygon should have five straight sides. One possibility:**

2. **60 × 50 = t**

$$\begin{array}{r} 60 \text{ seconds} \\ \times\ 50 \text{ times per second} \\ \hline 3000 \text{ times} \end{array}$$

t = **3000 times**

3. $\mathbf{101 - 98 = \mathit{p}}$

$$\begin{array}{r} \overset{9}{\cancel{10}}{}^{1}1 \text{ pounds} \\ -\quad 9\ 8 \text{ pounds} \\ \hline 3 \text{ pounds} \end{array}$$

p = **3 pounds**

4. $\mathbf{\mathit{P} - 27 = 194}$

$$\begin{array}{r} \overset{1\,1}{194} \text{ pounds} \\ +\quad 27 \text{ pounds} \\ \hline 221 \text{ pounds} \end{array}$$

P = **221 pounds**

5.
$$\begin{array}{r} \mathbf{5\tfrac{1}{4}\ \text{feet}} \\ 4\overline{)21} \\ -20 \\ \hline 1 \end{array}$$

6. **One possibility:**

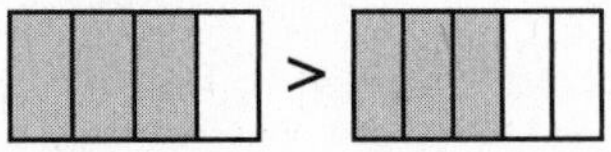

$$\frac{3}{4} \,\textcircled{>}\, \frac{3}{5}$$

7. $\frac{3}{10} + \frac{4}{10} = \mathbf{\frac{7}{10}}$

8. $1\frac{1}{3} + 2\frac{1}{3} = \mathbf{3\frac{2}{3}}$

9. $\frac{7}{10} - \frac{4}{10} = \mathbf{\frac{3}{10}}$

10. $5\frac{1}{4} - 2\frac{1}{4} = 3\frac{0}{4} = \mathbf{3}$

11. $\mathbf{2\frac{1}{2}}$

12. **200**

13. $\mathbf{\frac{2}{5}}$

14. $\frac{1}{3}$ of 30 $\textcircled{=}$ $\frac{1}{5}$ of 50

10   10

15. **50%**

16.
$$\begin{array}{r} \$18.73 \\ \$34.26 \\ +\quad M \\ \hline \$79.33 \end{array} \Big\rangle\ \$52.99$$

$$\begin{array}{r} \$7\,\overset{8}{\cancel{9}}.\,\overset{12}{\cancel{3}}{}^{1}3 \\ -\ \$5\,2.\,9\,9 \\ \hline \$2\,6.\,3\,4 \end{array}$$

M = **$26.34**

17.
$$\begin{array}{r} \overset{59}{\cancel{60}}\,\overset{10}{\cancel{1}}{}^{1}0 \\ -\quad 5\,4\,3 \\ \hline 54\,6\,7 \end{array}$$

R = **5467**

18.
$$\begin{array}{r} \overset{3\,2}{936} \\ 47 \\ 18 \\ 493 \\ 71 \\ +\quad 82 \\ \hline \mathbf{1647} \end{array}$$

19.
$$\begin{array}{r} \overset{3\,4}{346} \\ \times\quad 80 \\ \hline \mathbf{27{,}680} \end{array}$$

20.
$$\begin{array}{r} \$\overset{2\,4}{7.25} \\ \times\quad 90 \\ \hline \mathbf{\$652.50} \end{array}$$

21.
$$\begin{array}{r} \overset{4}{670} \\ \times\quad 700 \\ \hline \mathbf{469{,}000} \end{array}$$

22.
$$\begin{array}{r} \mathbf{4\,3\,4} \\ 4\overline{)1\,7^{1}3^{1}6} \end{array}$$

23.
$$\begin{array}{r} \mathbf{\$2.\,2\,0} \\ 8\overline{)\$17.^{1}6\,0} \end{array}$$

24.
$$\begin{array}{r} \mathbf{3\,3\ R\,1} \\ 3\overline{)10^{1}0} \end{array}$$

25. **A. Acute**

26. C.

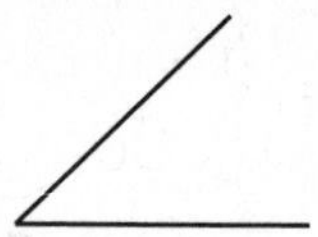

27. **C. 468**

28.
$$\begin{aligned} 3 + (4 + 5) &= (3 + 4) + M \\ 3 + 9 &= 7 + M \\ 12 &= 7 + M \\ M &= 12 - 7 \\ M &= \mathbf{5} \end{aligned}$$

29. $\frac{2}{12} + \frac{3}{12} = \mathbf{\frac{5}{12}}$

Lesson 43, Warm-Up

nephew; dime; quarter

a. **160**

b. **500**

c. **1**

d. $\mathbf{\frac{1}{3}}$

e. **280**

f. **240**

g. $\mathbf{4\frac{1}{2}}$

h. **4**

i. **7**

j. **25**

Problem Solving

8 ways (3 across, 3 down, and 2 diagonally)

Lesson 43, Lesson Practice

a.
$$\begin{array}{r} 3 \\ 4\overline{)15} \\ \underline{12} \\ 3 \end{array}$$

3 ← There are three whole pizzas for each team.

3 ← Three pizzas remain to be divided.

"Remainder" pizzas to be divided:

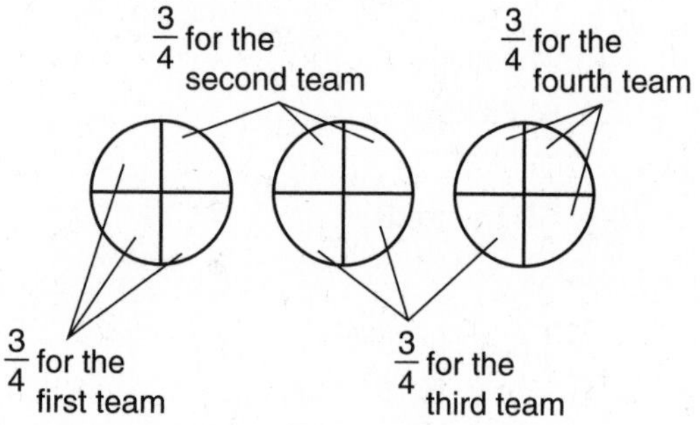

$$\begin{array}{r} \mathbf{3\frac{3}{4}} \\ 4\overline{)15} \\ \underline{12} \\ 3 \end{array}$$ **pizzas**

b.
$$\begin{array}{r} \mathbf{14\frac{2}{7}\%} \\ 7\overline{)100} \\ \underline{7} \\ 30 \\ \underline{28} \\ 2 \end{array}$$

c. $3\frac{1}{2} + 2 = \mathbf{5\frac{1}{2}}$

d. $3\frac{2}{4} - \frac{1}{4} = \mathbf{3\frac{1}{4}}$

e. $6\frac{2}{3} - 3 = \mathbf{3\frac{2}{3}}$

f. $3\frac{2}{4} + \frac{1}{4} = \mathbf{3\frac{3}{4}}$

g. $2\frac{1}{2} - \frac{1}{2} = \mathbf{2}$

h. $\frac{3}{4} + 2 = \mathbf{2\frac{3}{4}}$

Lesson 43, Mixed Practice

1. **One possibility:**

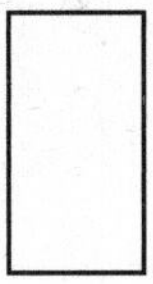

2.
$$4c = 32$$
$$32 \div 4 = 8 \text{ ounces}$$
$$c = \mathbf{8\ ounces}$$

3. $\mathbf{100 - 48 = L}$

$$\begin{array}{r} \overset{9}{\cancel{10}}{}^{1}0 \text{ centimeters} \\ -\ \ 4\,8 \text{ centimeters} \\ \hline 5\,2 \text{ centimeters} \end{array}$$

$L = \mathbf{52\ centimeters}$

4. $\mathbf{\$28.75 + M = \$34.18}$

$$\begin{array}{r} \$\overset{2}{\cancel{3}}\,\overset{13}{\cancel{4}}.{}^{1}1\,8 \\ -\ \$2\,8.\,7\,5 \\ \hline \mathbf{\$5.\,4\,3} \end{array}$$

5. **One possibility:**

 **; 25%**

6. **160**

7. $5 + 2\frac{1}{2} = \mathbf{7\frac{1}{2}}$

8. $12\frac{1}{2} + 12\frac{1}{2} = 24\frac{2}{2} = \mathbf{25}$

9. $1 + \frac{1}{3} = \mathbf{1\frac{1}{3}}$

10. $3\frac{1}{2} - \frac{1}{2} = \mathbf{3}$

11. $4\frac{3}{5} - 3\frac{1}{5} = \mathbf{1\frac{2}{5}}$

12. $5\frac{3}{8} - 1 = \mathbf{4\frac{3}{8}}$

13.
$$\begin{array}{r} \mathbf{12\frac{4}{8}\%} \text{ (or } \mathbf{12\frac{1}{2}\%}) \\ 8\overline{)100} \\ \underline{8} \\ 20 \\ \underline{16} \\ 4 \end{array}$$

14. **4 eighths**

15.
$$\begin{array}{r} \overset{5}{4}08 \\ \times\ \ 70 \\ \hline \mathbf{28,560} \end{array}$$

16.
$$\begin{array}{r} \$9.\overset{4}{6}\overset{4}{7} \\ \times\ \ \ 60 \\ \hline \mathbf{\$580.20} \end{array}$$

17.
$$\begin{array}{r} \overset{6}{9}70 \\ \times\ 900 \\ \hline \mathbf{873,000} \end{array}$$

18.
$$\begin{array}{r} \$\overset{1}{3}.\overset{2}{4}7 \\ \$5.23 \\ \$7.68 \\ +\ \$2.42 \\ \hline \mathbf{\$18.80} \end{array}$$

19.
$$\begin{array}{r} \overset{1}{3}\overset{1}{9}77 \\ +\ \ 309 \\ \hline 4286 \end{array}$$

$R = \mathbf{4286}$

20.
$$\begin{array}{r} \overset{8}{\cancel{9}}{}^{1}0\,\overset{0}{\cancel{1}}{}^{1}3 \\ -\ 3\,6\,0\,8 \\ \hline 5\,4\,0\,5 \end{array}$$

$W = \mathbf{5405}$

21.
$$\begin{array}{r} \mathbf{1\,2\,7\,R\,1} \\ 7\overline{)8^{1}9^{5}0} \end{array}$$

22.
$$\begin{array}{r} \mathbf{1\,6\,R\,4} \\ 6\overline{)1\,0^{4}0} \end{array}$$

23.
$$\begin{array}{r} \mathbf{2\,0\,0\,8\,R\,3} \\ 4\overline{)8\,0\,3^{3}5} \end{array}$$

24. $\begin{array}{r} \textbf{6 minutes} \\ 10\overline{)60 \text{ minutes}} \\ \underline{60} \\ 0 \end{array}$

25. Friday

26. $\mathbf{2\frac{1}{2}}$

27. (a) **90°**

(b) $\begin{array}{r} 45° \\ 2\overline{)90} \\ \underline{8} \\ 10 \\ \underline{10} \\ 0 \end{array}$ **45°**

(c) $\mathbf{\frac{45}{90}}$

28. B. 510

29. $\begin{array}{r} \mathbf{1\,2\,5} \\ 8\overline{)1\,0^{2}0^{4}0} \end{array}$

Lesson 44, Warm-Up

Problem Solving

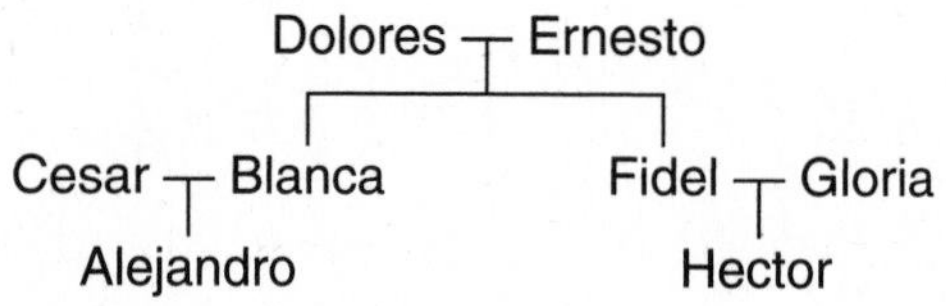

a. **cousin**

b. **uncle**

c. **nephew**

d. **brother-in-law**

e. **sister-in-law**

f. **grandmother**

g. **grandson**

Lesson 44, Lesson Practice

a. **3 cm, 30 mm**

b. **6 cm, 60 mm**

c. **9 cm, 90 mm**

d. **28 cm**

e. **10 millimeters**

f. **50 millimeters**

g. **cm, mm**

h. **15 mm**

i. **3 cm**

j. $\mathbf{\frac{1}{2}}$ **in.**

k. $\mathbf{1\frac{1}{4}}$ **in.**

l. $\mathbf{1\frac{7}{8}}$ **in.**

m. $\mathbf{2\frac{3}{8}}$ **in.**

n. $\mathbf{3\frac{1}{8}}$ **in.**

o. $\mathbf{3\frac{3}{4}}$ **in.**

p. $\mathbf{3\frac{1}{4}}$ **in.**

q. $\mathbf{2\frac{1}{2}}$ **in.**

r. $\mathbf{1\frac{5}{8}}$ **in.**

s. **6 in.**

t. **6 in.**

Lesson 44, Mixed Practice

1. **One possibility:**

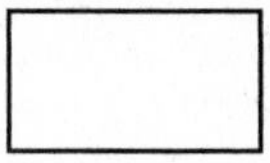

2.
$$\begin{array}{r} \overset{1}{\$0.03} \\ \$0.10 \\ \$0.10 \\ \$0.75 \\ +\ \$0.50 \\ \hline \mathbf{\$1.48} \end{array}$$

3. **3*B* = 138**

$$\begin{array}{r} 4\ 6 \text{ children} \\ 3\overline{)1\ 3^{1}8} \end{array}$$

B = **46 children**

4. **2 × 10 = *t***

t = **20 mm**

5. **1976 − 1776 = *d***

$$\begin{array}{r} 1976 \\ -\ 1776 \\ \hline 200 \text{ years} \end{array}$$

d = **200 years**

6.
$$\begin{array}{l} \mathbf{1\frac{2}{3}} \textbf{ oranges} \\ 3\overline{)5} \\ \ \ 3 \\ \ \ 2 \end{array}$$

7. $\mathbf{2\frac{3}{5}}$

8. $3\frac{3}{4} - 1\frac{2}{4} = \mathbf{2\frac{1}{4}}$

9. $4\frac{1}{2} - \frac{1}{2} = \mathbf{4}$

10. $5\frac{1}{4} - 4 = \mathbf{1\frac{1}{4}}$

11. $33\frac{1}{3} + 33\frac{1}{3} = \mathbf{66\frac{2}{3}}$

12. $5\frac{1}{3} + 3 = \mathbf{8\frac{1}{3}}$

13. $8\frac{3}{8} + \frac{4}{8} = \mathbf{8\frac{7}{8}}$

14. **16 mm**

15.
$$\begin{array}{r} \overset{3\,2}{{}_{2}352} \\ 4287 \\ 593 \\ 7684 \\ +\ 9856 \\ \hline \mathbf{22{,}772} \end{array}$$

16.
$$\begin{array}{r} 3\ \overset{5}{\not 6}\ \overset{1}{\not 2}{}^{1}7 \\ -\ \ 4\ 2\ 9 \\ \hline \mathbf{3\ 1\ 9\ 8} \end{array}$$

17.
$$\begin{array}{r} \overset{19}{\not 2\not 0}\ \overset{9}{\not 0}{}^{1}0 \\ -\ \ 6\ 6 \\ \hline 19\ 3\ 4 \end{array} \qquad \begin{array}{r} \overset{8}{\not 9}\ \overset{10}{\not 1}{}^{1}0\ 4 \\ -\ 1\ 9\ 3\ 4 \\ \hline \mathbf{7\ 1\ 7\ 0} \end{array}$$

18.
$$\begin{array}{r} \overset{6}{491} \\ \times\ \ 700 \\ \hline \mathbf{343{,}700} \end{array}$$

19.
$$\begin{array}{r} \overset{4}{37} \\ \times\ \ 60 \\ \hline \overset{1\,1}{2220} \\ \times\ \ \ 8 \\ \hline \mathbf{17{,}760} \end{array}$$

20.
$$\begin{array}{r} 6\ 3\ 5 \\ 5\overline{)3\ 1^{1}7^{2}5} \end{array}$$

n = **635**

21.
$$\begin{array}{r} \mathbf{2\ 9\ 6\ R\ 4} \\ 10\overline{)2\ 9^{9}6^{6}4} \end{array}$$

22. 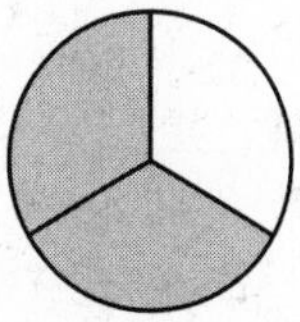 **; $66\frac{2}{3}$%**

23. **C. 260**

24. Morning is "a.m." so the time is **10:47 a.m.**

25. **1, 2, 5, 10, 25, 50**

26. 12 inches ÷ 4 = **3 inches**

27. (a) **90°**

(b) $\begin{array}{r} 90° \\ \times\ 2 \\ \hline \mathbf{180°} \end{array}$

(c) $\mathbf{\dfrac{90}{180}}$

28. **B. 315**

29. $\frac{1}{3} + \frac{1}{3} = \mathbf{\frac{2}{3}}$ **cup**

Lesson 45, Warm-Up

aunt; 75¢; 150¢

a. **280**

b. **1000**

c. **1**

d. $\mathbf{\frac{2}{4}\left(\text{or } \frac{1}{2}\right)}$

e. **294**

f. **$2.50**

g. **5**

h. **5**

i. **5**

Problem Solving

In this division problem, we are given the quotient and the divisor. To find the dividend, we multiply: 56 × 3 = 168.

$$\mathbf{3\overline{)168}} \quad \text{quotient } \mathbf{56}$$

Lesson 45, Lesson Practice

a. **Parallelogram; rectangle**

b. **Trapezoid**

c. **Parallelogram; rhombus**

d. **Trapezium**

e. **Parallelogram**

f. **Parallelogram; rectangle; rhombus; square**

g. **A parallelogram has two pairs of parallel sides. A trapezoid has only one pair of parallel sides.**

h. **One possibility:**

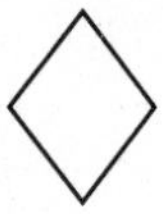

Lesson 45, Mixed Practice

1.

2. **$10 − *M* = $2.47**

$$\begin{array}{r} \overset{9}{\cancel{\$10}}.\ \overset{9}{\cancel{0}}{}^{1}0 \\ -\ \ \$2.\ 4\ 7 \\ \hline \$7.\ 5\ 3 \end{array}$$

M = **$7.53**

3. **50 × 2 = *t***

$$\begin{array}{l} \quad\ 50 \text{ states} \\ \times\ \ \ 2 \text{ senators in each state} \\ \hline \ \ 100 \text{ senators} \end{array}$$

t = **100 senators**

4. **4*a* = 2500**

$$\begin{array}{r} 6\ 2\ 5 \\ 4\overline{)2\ 5^{1}0^{2}0} \end{array}$$

a = **625**

5. $4\overline{)15}$ → $\mathbf{3\frac{3}{4}}$ **gallons**
$\underline{12}$
3

6. **1 cm**

7. $\mathbf{1\frac{1}{2}}$ **inches**

8. $3\frac{1}{3} + 1\frac{1}{3} = \mathbf{4\frac{2}{3}}$

9. $4\frac{1}{4} + 2 = \mathbf{6\frac{1}{4}}$

10. $3 + \frac{3}{4} = \mathbf{3\frac{3}{4}}$

11. $5\frac{3}{8} - 2 = \mathbf{3\frac{3}{8}}$

12. $6\frac{3}{4} - 1\frac{2}{4} = \mathbf{5\frac{1}{4}}$

13. $5\frac{1}{2} - 1\frac{1}{2} = \mathbf{4}$

14.
$$\begin{array}{r} {}^{2\,2\;\;2} \\ \$87.93 \\ \$35.16 \\ \$42.97 \\ +\ \$68.74 \\ \hline \mathbf{\$234.80} \end{array}$$

15.
$$\begin{array}{r} {}^{49\;\;\;1} \\ \$\not{5}\not{0}.\not{2}{}^{1}6 \\ -\ \$13.87 \\ \hline \mathbf{\$36.39} \end{array}$$

16.
$$\begin{array}{r} {}^{5\;\;10} \\ \not{6}\,\not{1}{}^{1}0\,9 \\ -\ 4\,9\,3\,7 \\ \hline 1\,1\,7\,2 \end{array}$$
$A = \mathbf{1172}$

17.
$$\begin{array}{r} {}^{2\;2} \\ 9314 \\ \times\quad 70 \\ \hline \mathbf{651{,}980} \end{array}$$

18.
$$\begin{array}{r} {}^{2\;2} \\ \$2.34 \\ \times\quad 600 \\ \hline \mathbf{\$1404.00} \end{array}$$

19.
$$\begin{array}{r} {}^{1\,4\,3} \\ 4287 \\ \times\quad 5 \\ \hline \mathbf{21{,}435} \end{array}$$

20. $9\overline{)9\,6^{6}3\,6}$ → **1 0 7 0 R 6**

21. $8\overline{)\$34.{}^{2}1{}^{5}6}$ → **$4. 2 7**

22. **One possibility:**

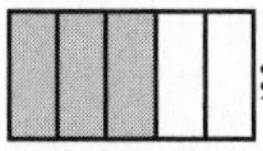 **; 60%**

23. **260**

24. **B.**

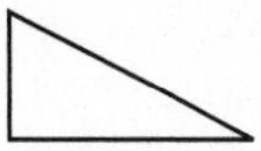

25. **150**

26. $\mathbf{2\frac{3}{4}}$ **in.**

27.
$$\begin{array}{r} {}^{1\,2} \\ \mathbf{123} \\ \times\quad \mathbf{8} \\ \hline \mathbf{984} \\ +\quad \mathbf{3} \\ \hline \mathbf{987} \end{array}$$

The answer is correct.

28.

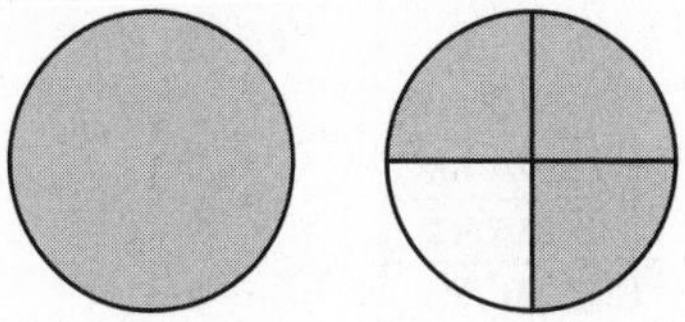

29. $1\frac{2}{8} + 1\frac{5}{8} = \mathbf{2\frac{7}{8}}$ **in.**

LESSON 46, WARM-UP

brother-in-law; quarter

a. **270**

b. **520**

c. $\mathbf{\frac{7}{10}}$

d. $\mathbf{\frac{3}{10}}$

e. **492**

f. **$7.50**

g. $\mathbf{25\frac{1}{2}}$

h. **6**

i. **0**

Problem Solving

First roll	Second roll
0	10
1	9
2	8
3	7
4	6
5	5
6	4
7	3
8	2
9	1

LESSON 46, LESSON PRACTICE

a.

30 ducks

$\frac{2}{5}$ swam in the pond.	6 ducks
	6 ducks
$\frac{3}{5}$ did not swim in the pond.	6 ducks
	6 ducks
	6 ducks

12 ducks

b.

60 minutes

Susan practiced for $\frac{3}{4}$ of the hour.	15 minutes
	15 minutes
	15 minutes
Susan did not practice for $\frac{1}{4}$ of the hour.	15 minutes

45 minutes

c.

30 students

$\frac{3}{5}$ were girls.	6 students
	6 students
	6 students
$\frac{2}{5}$ were boys.	6 students
	6 students

12 boys

LESSON 46, MIXED PRACTICE

1. $3\overline{)11}$ = $\mathbf{3\frac{2}{3}}$ **miles**

   $\underline{9}$

   2

2. **625 − *f* = 139 (or 139 + *f* = 625)**

   $$\begin{array}{r} \overset{5}{\not6}\,\overset{1}{\not2}\,{}^{1}5 \text{ seats} \\ -\ 1\ 3\ 9 \text{ seats} \\ \hline 4\ 8\ 6 \text{ seats} \end{array}$$

   f = **486 seats**

3. **4 × 10 = *t***

   t = **40 millimeters**

4. **8*p* = 7000**

   $$8\overline{)7\,0^{6}0^{4}0} = 8\,7\,5 \text{ passengers}$$

   p = **875 passengers**

5. $$\begin{array}{r} 1976 \\ -\ \ 200 \\ \hline \mathbf{1776} \end{array}$$

6.

48 points

Nia scored $\frac{2}{3}$.	16 points
	16 points
Nia did not score $\frac{1}{3}$.	16 points

32 points

7. $\frac{1}{4}$ of 60 $\bigcirc<$ $\frac{1}{3}$ of 60

 15 20

8. **300**

9. One possibility:

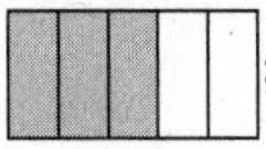

 ; 40%

10. **May**

11. $1\frac{3}{4}$ **inches**

12. $3\frac{3}{7} + 2 + \frac{2}{7} = \mathbf{5\frac{5}{7}}$

13. $2\frac{2}{5} - 1 = \mathbf{1\frac{2}{5}}$

14. $3\frac{2}{3} - \frac{1}{3} = \mathbf{3\frac{1}{3}}$

15. $4 + 1\frac{4}{12} = 5\frac{4}{12}$

 $6\frac{5}{12} - 5\frac{4}{12} = \mathbf{1\frac{1}{12}}$

16.
$$\begin{array}{r} \overset{2\ 2\ 2}{1396} \\ 727 \\ 854 \\ +\ 4685 \\ \hline \mathbf{7662} \end{array}$$

17.
$$\begin{array}{r} \overset{4\ \ \overset{1}{0}}{\cancel{5}\ \cancel{1}2} \\ -\ \ \ 97 \\ \hline 4\ 1\ 5 \end{array}$$
$W = \mathbf{415}$

18.
$$\begin{array}{r} \overset{3\ 6}{938} \\ \times\ \ \ \ 800 \\ \hline \mathbf{750,400} \end{array}$$

19.
$$\begin{array}{r} \overset{2}{54} \\ \times\ \ 60 \\ \hline \overset{1\ 2}{3240} \\ \times\ \ \ \ 7 \\ \hline \mathbf{22,680} \end{array}$$

20.
$$\begin{array}{r} 6\ 0\ 5 \\ 9\overline{)5\ 4\ 4\ ^4 5} \end{array}$$
$n = \mathbf{605}$

21.
$$\begin{array}{r} \mathbf{3\ 2\ 0\ R\ 5} \\ 10\overline{)3\ 2^2 0\ 5} \end{array}$$

22.
$$\begin{array}{r} \$15.37 \\ -\ \$12.00 \\ \hline \$3.37 \end{array} \qquad \begin{array}{r} \overset{19\ \ \ 9}{\$20.\ \cancel{0}^1 0} \\ -\ \ \$3.\ 3\ 7 \\ \hline \mathbf{\$16.\ 6\ 3} \end{array}$$

23.
$$\begin{array}{r} \overset{3\ 1\ 2}{4826} \\ \times\ \ \ \ \ 4 \\ \hline \mathbf{19,304} \end{array}$$

24.
$$\begin{array}{r} \mathbf{20\%} \\ 5\overline{)100} \\ \underline{10} \\ 00 \\ \underline{\ 0} \\ 0 \end{array}$$

25. **B**

26. **C. Obtuse**

27. **Each corner is a right angle.**

28. **A. 234**

29. **2 inches**

LESSON 47, WARM-UP

daughter-in-law; 40; 60; 80

a. **380**

b. **360**

c. **2**

d. **1**

e. $1\frac{1}{2}$

f. 1

g. 5

h. 102

Problem Solving

The ones digit of one of the factors must be 5, because the ones digit of the product is 5. Since the product is greater than 600, the bottom factor cannot be 5, because any number in the seventies times 5 is much less than 600. So the ones digit of the top factor is 5, giving us 75. The bottom factor must be odd, because the ones digit of the product is 5. Only $75 \times 9 = 675$ is greater than 600. Therefore, the bottom factor is 9, and the tens digit of the product is 7.

$$\begin{array}{r} \mathbf{75} \\ \times\ \mathbf{9} \\ \hline \mathbf{675} \end{array}$$

Lesson 47, Lesson Practice

a. 6 feet × 12 inches in each foot
= 72 inches
72 inches + 2 inches = **74 inches**

b. 3 minutes × 60 seconds in each minutes
= 180 seconds
180 seconds + 20 seconds = **200 seconds**

c. 2 hours × 60 minutes in each hour
= 120 minutes
120 minutes + 30 minutes = **150 minutes**

d. 2 pounds × 16 ounces in each pound
= 32 ounces
32 ounces + 12 ounces = **44 ounces**

Lesson 47, Mixed Practice

1. $\mathbf{36 + 29 + 73 = t}$

$$\begin{array}{r} {}^{1} \\ 36 \\ 29 \\ +\ 73 \\ \hline 138 \end{array}$$ children

t = **138 children**

2. $\mathbf{7 \times 10 = t}$

t = **70 years**

3. $\mathbf{68 - 12 = d}$

$$\begin{array}{r} 68 \\ -\ 12 \\ \hline 56 \end{array}$$ years

d = **56 years**

4. 5 feet × 12 inches in each foot = 60 inches
60 inches + 6 inches = **66 inches**

5. C. 70,000

6. 14 months

7. One possibility:

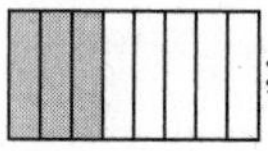 **; $37\frac{1}{2}$%**

8. $\mathbf{2\frac{1}{4}}$ **in.**

9. $4 + 3\frac{3}{4} = \mathbf{7\frac{3}{4}}$

10. $3\frac{3}{5} + 1\frac{1}{5} = \mathbf{4\frac{4}{5}}$

11. $2\frac{3}{8} + \frac{2}{8} = \mathbf{2\frac{5}{8}}$

12. $5\frac{1}{3} - \frac{1}{3} = 5$

$5\frac{1}{3} - 5 = \mathbf{\frac{1}{3}}$

13. $2\frac{1}{2} - \frac{1}{2} = \mathbf{2}$

14. $3\frac{5}{9} - 1\frac{1}{9} = \mathbf{2\frac{4}{9}}$

15.

$$\begin{array}{r} {}^{2\,1\ \,1\,2} \\ \$48{,}748 \\ \$37{,}145 \\ +\ \$26{,}498 \\ \hline \mathbf{\$112{,}391} \end{array}$$

$$\begin{array}{r} \$\overset{5}{6}\,\overset{12}{3},\overset{1}{1}\,\overset{3}{4}\overset{12}{2} \\ -\ \$1\,7,9\,3\,6 \\ \hline \$4\,5,2\,0\,6 \end{array}$$

17.
$$\begin{array}{r} \$\overset{4\ 2}{5.63} \\ \times\qquad 700 \\ \hline \$3941.00 \end{array}$$

18.
$$\begin{array}{r} \overset{5\,2\,7}{4729} \\ \times\qquad 8 \\ \hline 37,832 \end{array}$$

19.
$$\begin{array}{r} \overset{4}{9006} \\ \times\qquad 80 \\ \hline 720,480 \end{array}$$

20.
$$8\overline{)3\ 4^2 5^1 6}\ \ \begin{array}{c}\mathbf{4\ 3\ 2}\end{array}$$

21.
$$9\overline{)1\ 8\ 3^3 6}\ \ \begin{array}{c}\mathbf{2\ 0\ 4}\end{array}$$

22.
$$7\overline{)1405}\ \ \begin{array}{c}\mathbf{200\ R\ 5}\end{array}$$

23.
$$\begin{array}{r}\overset{1}{25}\\ \times\ 20\\ \hline 500\end{array} \qquad \begin{array}{r}\overset{2}{25}\\ \times\ 5\\ \hline 125\end{array}$$

$$\begin{array}{r}500\\ +\ 125\\ \hline 625\end{array}$$

24.

```
                        60 seeds
            swallowed 2/5 { 12 seeds
                            12 seeds
                            12 seeds
      did not swallow 3/5 { 12 seeds
                            12 seeds

                        24 seeds
```

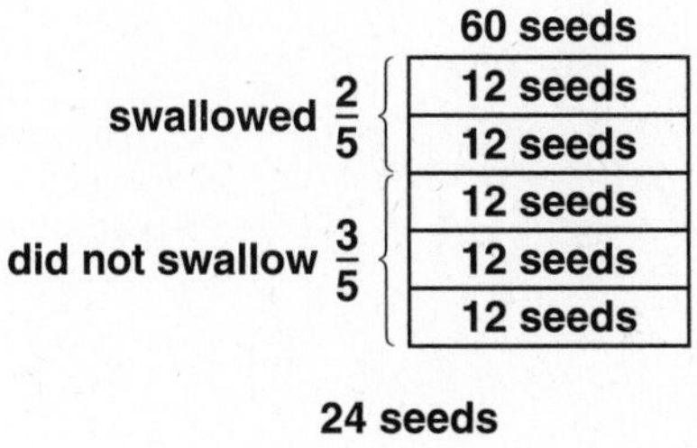

25. Evening is "p.m." so the time is **10:35 p.m.**

26.

45°

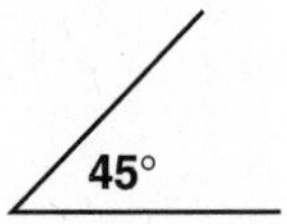

27. C. **7650**

28. $\frac{1}{5}$ of 10 \bigcirc 10 ÷ 5

 2 2

29. **2 inches**

LESSON 48, WARM-UP

Problem Solving

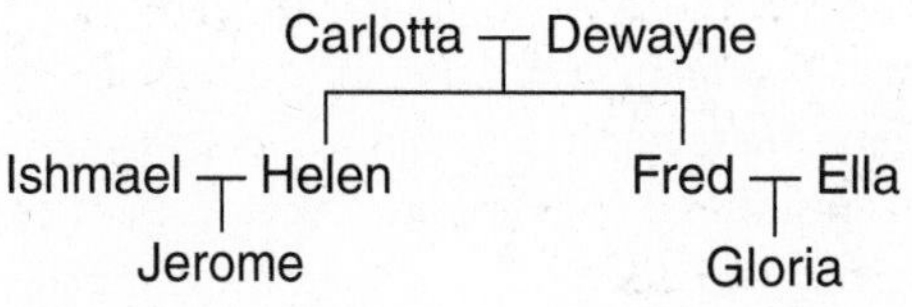

a. nephew

b. daughter-in-law

c. son-in-law

d. grandmother

e. cousin

f. brother-in-law

g. grandson

h. brother

LESSON 48, LESSON PRACTICE

a. (5 × 10) + (6 × 1)

b. (5 × 1000) + (2 × 100) + (8 × 10)

c. (2 × 100,000) + (5 × 10,000)

d. **6040**

e. **570**

f. **84,000**

g. **930,000**

LESSON 48, MIXED PRACTICE

1.

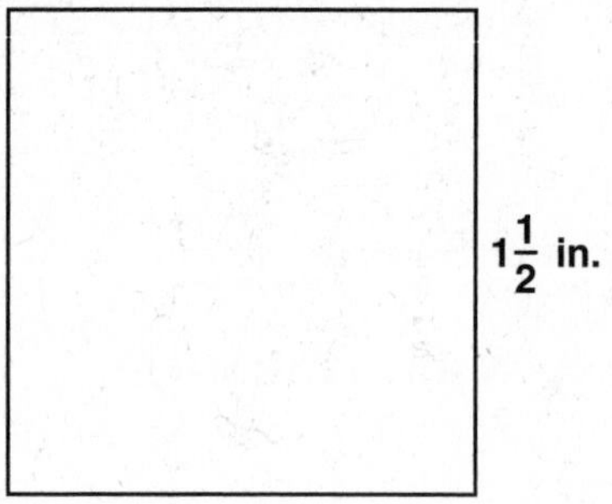

2. **21 − J = 6**

$$\begin{array}{r} \overset{1}{\cancel{2}}{}^{1}1 \\ -\ \ 6 \\ \hline 1\ 5 \end{array}$$ years old

J = **15 years old**

3. **366 − 25 = d**

$$\begin{array}{r} 366 \\ -\ \ 25 \\ \hline 341 \end{array}$$ days

d = **341 days**

4. **8 × 12 = t**

$$\begin{array}{r} \overset{1}{1}2 \\ \times\ \ 8 \\ \hline 96 \end{array}$$ eggs

t = **96 eggs**

5. **C. 6000**

6. **190 millimeters**

7.

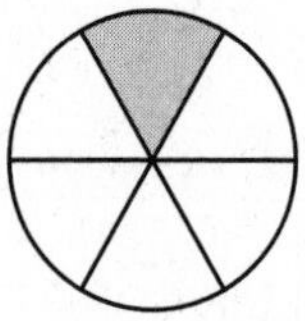

$6\overline{)10^40}$ = $16\frac{4}{6}$ or **$16\frac{2}{3}$%**

8. **290**

9. **502**

10. **(4 × 10,000) + (7 × 1000)**

11.
$$\begin{array}{r} \scriptsize{2\,2\ 2\,2} \\ 98{,}572 \\ 42{,}156 \\ 37{,}428 \\ +\ 16{,}984 \\ \hline \mathbf{195{,}140} \end{array}$$

12.
$$\begin{array}{r} \scriptsize{1\,1\ \ 1} \\ 19{,}724 \\ +\ 32{,}436 \\ \hline \mathbf{52{,}160} \end{array}$$

13.
$$\begin{array}{r} \overset{9}{\cancel{10}},\overset{9}{\cancel{0}}\,\overset{9}{\cancel{0}}{}^{1}0 \\ -\ \ 1{,}7\,4\,6 \\ \hline \mathbf{8{,}\,2\,5\,4} \end{array}$$

14.
$$\begin{array}{r} \scriptsize{2\,4\ 4} \\ \$34.78 \\ \times\ \ \ \ \ 6 \\ \hline \mathbf{\$208.68} \end{array}$$

15.
$$\begin{array}{r} \scriptsize{3\,2\,5} \\ 6549 \\ \times\ \ \ 60 \\ \hline \mathbf{392{,}940} \end{array}$$

16.
$$\begin{array}{r} \scriptsize{3\,6} \\ 8037 \\ \times\ \ \ 90 \\ \hline \mathbf{723{,}330} \end{array}$$

17. $6\overline{)364^47}$ = **60 7 R 5**

18. $9\overline{)5408}$ = **600 R 8**

19.
$$\begin{array}{r} 100 \\ 10\overline{)1000} \\ 10 \\ \hline 00 \\ 0 \\ \hline 00 \\ 0 \\ \hline 0 \end{array}$$

W = **100**

20. $4\frac{1}{3} - 2 = 2\frac{1}{3}$

$3\frac{1}{3} + 2\frac{1}{3} = \mathbf{5\frac{2}{3}}$

21. $6 \times 800 = 6 \times 8 \times 100$

H = **100**

22.
$$\begin{array}{r} \overset{1\,1}{\$6.00} \\ \$1.47 \\ +\ \$0.93 \\ \hline \$8.40 \end{array} \qquad \begin{array}{r} \$\overset{9}{\cancel{10}}.{}^{1}0\ 0 \\ -\ \ \$8.\ 4\ 0 \\ \hline \mathbf{\$1.\ 6\ 0} \end{array}$$

23.
$$\begin{array}{r} 62 \\ \times\ \ 20 \\ \hline 1240 \end{array} \qquad \begin{array}{r} 62 \\ \times\ \ 3 \\ \hline 186 \end{array} \qquad \begin{array}{r} \overset{1}{1240} \\ +\ \ 186 \\ \hline \mathbf{1426} \end{array}$$

24.

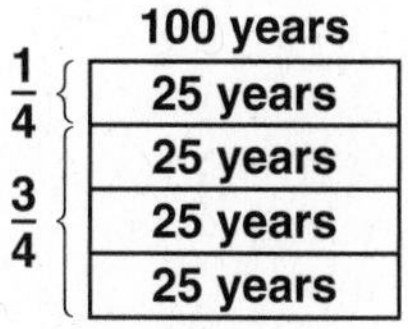

25 years

25. The hours after noon and before midnight are the "p.m." hours. The time will be **11:59 p.m.**

26. **125**

27. **C. 963**

28. 10 ÷ 2 = **5 dimes**

29. **8 sides**

LESSON 49, WARM-UP

sister-in-law

a. **500**

b. **800**

c. $\mathbf{\frac{9}{10}}$

d. **1**

e. **5**

f. **1**

g. **4**

h. **5**

Problem Solving

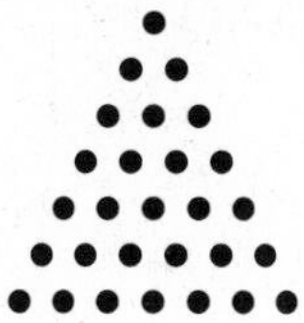

1 + 2 + 3 + 4 + 5 + 6 + 7 = **28 dots**

LESSON 49, LESSON PRACTICE

a.
$$\begin{array}{r} \overset{1}{32}\text{ eggs} \\ +\ \ 29\text{ eggs} \\ \hline 61\text{ eggs} \end{array} \qquad \begin{array}{r} \overset{7}{\cancel{8}}{}^{1}0\text{ eggs} \\ -\ \ 6\ 1\text{ eggs} \\ \hline \mathbf{1\ 9\text{ eggs}} \end{array}$$

b.
$$\begin{array}{r} 37\text{ cans} \\ +\ \ 21\text{ cans} \\ \hline 58\text{ cans} \end{array} \qquad \begin{array}{r} \mathbf{2\ 9\text{ cans}} \\ 2\overline{)5{}^{1}8} \end{array}$$

LESSON 49, MIXED PRACTICE

1. **One possibility:**

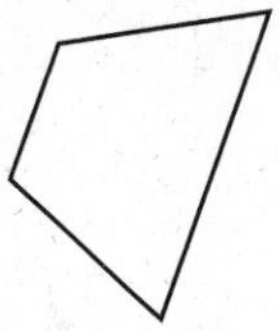

2.
$$\begin{array}{r} 15 \\ +\ \ 5 \\ \hline 20\text{ years old} \end{array} \qquad \begin{array}{r} 20 \\ +\ 10 \\ \hline \mathbf{30\text{ years old}} \end{array}$$

3.
$$\begin{array}{r} \mathbf{12\text{ years old}} \\ 3\overline{)36} \end{array}$$

4.

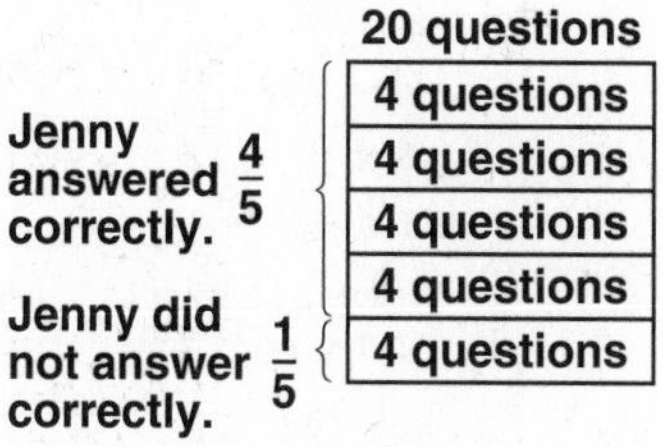

16 questions

5. **A. 700,000**

6. **506**

7. $2\overline{)\$5.{}^{1}00}$ → **$2. 50**

8. **230**

9. **25,300**

10.

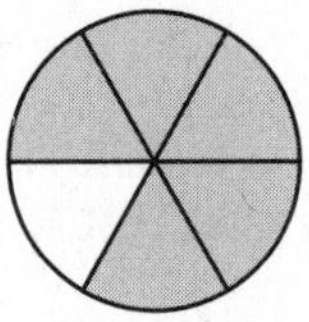

$6\overline{)10{}^{4}0}$ → $16\frac{4}{6}$ or $\mathbf{16\frac{2}{3}\%}$

11. $5\frac{2}{8} + 6 + \frac{3}{8} = \mathbf{11\frac{5}{8}}$

12. $3\frac{5}{6} - 3 = \frac{5}{6}$

$8\frac{5}{6} - \frac{5}{6} = \mathbf{8}$

13.
$$\begin{array}{r} 342 \\ 5874 \\ 63 \\ 285 \\ 8 \\ 96 \\ +\ 87 \\ \hline \mathbf{6755} \end{array}$$

14.
$$\begin{array}{r} \$4\,\overset{1}{\not2}.\,\overset{9}{\not0}{}^{1}1 \\ -\ \$2\,0.\,1\,4 \\ \hline \mathbf{\$2\,1.\,8\,7} \end{array}$$

15.
$$\begin{array}{r} \overset{9}{\not10}\,\overset{9}{\not0}{}^{1}0 \\ -\quad 1 \\ \hline 9\,9\,9 \end{array}$$
$M = \mathbf{999}$

16.
$$\begin{array}{r} 800 \\ \times\ 50 \\ \hline \mathbf{40,000} \end{array}$$

17.
$$\begin{array}{r} \overset{1}{2}5 \\ \times\ 30 \\ \hline \overset{4}{7}50 \\ \times\ 8 \\ \hline \mathbf{6000} \end{array}$$

18. $6\overline{)1205}$ → **200 R 5**

19. $8\overline{)\$76.{}^{4}3{}^{3}2}$ → **$9. 5 4**

20.
$$\begin{array}{r} \overset{1}{\$1}\overset{1}{2}.00 \\ \$4.76 \\ \$2.89 \\ +\ \$0.34 \\ \hline \$19.99 \end{array} \qquad \begin{array}{r} \$\overset{19}{\not20}.\,\overset{9}{\not0}{}^{1}0 \\ -\ \$19.\,9\,9 \\ \hline \mathbf{\$0.\,0\,1} \end{array}$$

21.
$$\begin{array}{r} \overset{1}{3}5 \\ \times\ 20 \\ \hline 700 \end{array} \qquad \begin{array}{r} \overset{2}{3}5 \\ \times\ 5 \\ \hline 175 \end{array} \qquad \begin{array}{r} 700 \\ +\ 175 \\ \hline \mathbf{875} \end{array}$$

22. **One hundred fifty thousand**

23. **(1 × 100,000) + (5 × 10,000)**

24. The pattern is "Count up by one hundreds."
1200, 1300, 1400

25. **Ten thousands**

26. **1750**

27. **December 7, 1941**

28. **5**

29. **A.**

Lesson 50, Warm-Up

6 feet; 15 feet; 30 feet

a. **600**

b. **1000**

c. **3**

d. $\mathbf{1\frac{1}{3}}$

e. 5

f. 4

g. 5

h. 272

i. 2

Problem Solving

First Cube	Second Cube
1	6
2	5
3	4
4	3
5	2
6	1

LESSON 50, LESSON PRACTICE

a.
$$
\begin{array}{r}
5 \\
6 \\
9 \\
+\ 8 \\
\hline
28
\end{array}
\qquad
\overset{\textbf{7 players}}{4\overline{)28}}
$$

b.
$$
\begin{array}{r}
11 \\
+\ 17 \\
\hline
28
\end{array}
\qquad
\overset{\textbf{14 runners}}{2\overline{)28}}
$$

c.
$$
\begin{array}{r}
5 \\
7 \\
+\ 9 \\
\hline
21
\end{array}
\qquad
\overset{\textbf{7 books}}{3\overline{)21}}
$$

d.
$$
\begin{array}{r}
8 \\
9 \\
7 \\
9 \\
_5 8 \\
10 \\
6 \\
+\ 7 \\
\hline
64
\end{array}
\qquad
\overset{\textbf{8}}{8\overline{)64}}
$$

LESSON 50, MIXED PRACTICE

1. **One possibility:**

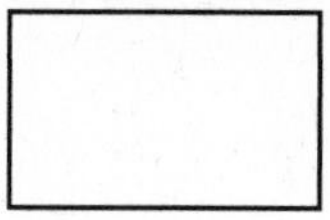

2.

Kimberly Loh

Miguel Loh

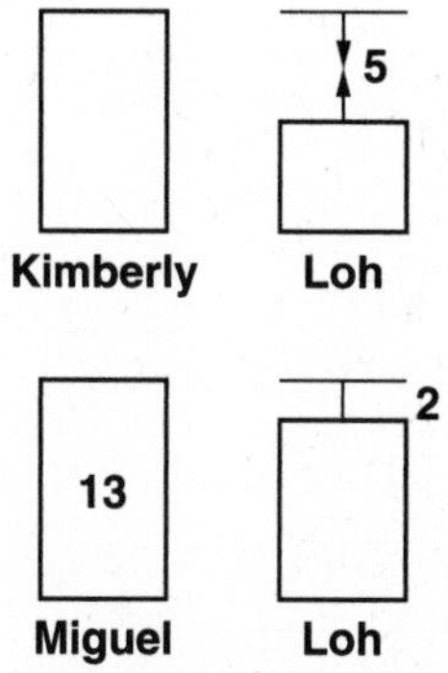

Loh is 2 years younger than Miguel, so Loh is 11 years old. Kimberly is 5 years older than Loh, so Kimberly is **16 years old.**

3.
$$
\begin{array}{r}
5 \text{ ounces} \\
8 \text{ ounces} \\
+\ 8 \text{ ounces} \\
\hline
21 \text{ ounces}
\end{array}
\qquad
\overset{\textbf{7 ounces}}{3\overline{)21}}
$$

4.
$$
\left.\begin{array}{l}
\frac{3}{5} \\[6pt]
\frac{2}{5}
\end{array}\right.
\begin{array}{|c|}
\hline
\textbf{60 minutes} \\
\hline
\textbf{12 minutes} \\
\hline
\textbf{12 minutes} \\
\hline
\textbf{12 minutes} \\
\hline
\textbf{12 minutes} \\
\hline
\textbf{12 minutes} \\
\hline
\end{array}
$$

36 minutes

5. 2 hours \times 60 minutes in each hour
 = 120 minutes
 120 minutes + 15 minutes = **135 minutes**

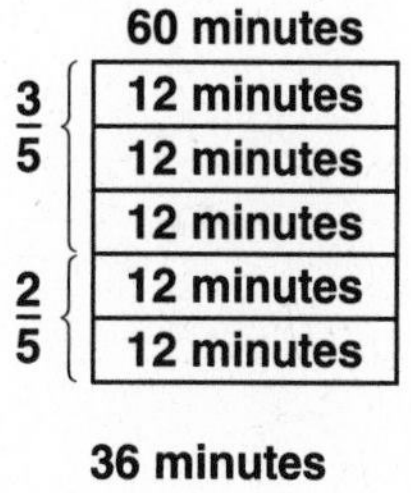

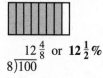

6.
$$
\overset{\textbf{4 centuries}}{100\overline{)400}}
$$
$$
\begin{array}{r}
400 \\
\hline
0
\end{array}
$$

7. **54,919**

8. **One possibility:**

$$
\overset{12\frac{4}{8} \text{ or } \mathbf{12\frac{1}{2}\%}}{8\overline{)100}}
$$

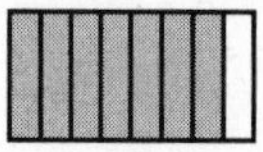

9.

$$\begin{array}{r} 15 \text{ children} \\ +\ 11 \text{ children} \\ \hline 26 \text{ children} \end{array} \qquad \begin{array}{r} \textbf{13 children} \\ 2\overline{)26} \end{array}$$

10.

$$\begin{array}{r} {}^{2\,2} \\ 342 \\ 67 \\ 918 \\ 897 \\ +\ 42 \\ \hline \mathbf{2266} \end{array}$$

11.

$$\begin{array}{r} {}^{4\quad 7} \\ \$\not{5}{}^{1}3.\ \not{8}{}^{1}7 \\ -\ \$2\,7.\,5\,9 \\ \hline \mathbf{\$2\,6.\,2\,8} \end{array}$$

12.

$$\begin{array}{r} {}^{2\,1\,4} \\ \$34.28 \\ \times\ \ 60 \\ \hline \mathbf{\$2056.80} \end{array}$$

13.

$$\begin{array}{r} 57 \\ \times\ 10 \\ \hline {}^{4} \\ 570 \\ \times\ \ 7 \\ \hline \mathbf{3990} \end{array}$$

14.

$$\begin{array}{r} 4 \\ 7 \\ +\ 7 \\ \hline 18 \end{array} \qquad \begin{array}{r} \mathbf{6} \\ 3\overline{)18} \end{array}$$

15.

$$\begin{array}{r} 5 \\ 6 \\ 9 \\ +\ 8 \\ \hline 28 \end{array} \qquad \begin{array}{r} \mathbf{7} \\ 4\overline{)28} \end{array}$$

16.

$$\begin{array}{r} \mathbf{600\ R\ 6} \\ 7\overline{)4206} \end{array}$$

17.

$$\begin{array}{r} \mathbf{\$10.0\,4} \\ 6\overline{)\$60.2^{2}4} \end{array}$$

18.

$$\begin{array}{r} \mathbf{1\ 1\ 1\ R\ 1} \\ 9\overline{)10^{1}0^{1}0} \end{array}$$

19.

$$\begin{array}{r} 30 \\ 6\overline{)180} \end{array}$$

$D = \mathbf{30}$

20. $1\frac{1}{7} + 2\frac{2}{7} + 3\frac{3}{7} = \mathbf{6\frac{6}{7}}$

21. $7\frac{7}{10} - 5\frac{5}{10} = 2\frac{2}{10}$

$9\frac{9}{10} - 2\frac{2}{10} = \mathbf{7\frac{7}{10}}$

22.

$$\begin{array}{r} 43 \\ \times\ 10 \\ \hline 430 \end{array} \qquad \begin{array}{r} 43 \\ \times\ \ 2 \\ \hline 86 \end{array} \qquad \begin{array}{r} {}^{1} \\ 430 \\ +\ \ 86 \\ \hline \mathbf{516} \end{array}$$

23. **May**

24. **B. 1200**

25.

$$\begin{array}{r} 32 \text{ miles} \\ +\ 32 \text{ miles} \\ \hline \textbf{64 miles} \end{array}$$

26.

$$\begin{array}{r} 4 \text{ miles} \\ +\ 5 \text{ miles} \\ \hline \textbf{9 miles} \end{array}$$

27.

$$\begin{array}{r} 5 \text{ miles} \\ 8 \text{ miles} \\ 8 \text{ miles} \\ 6 \text{ miles} \\ +\ 8 \text{ miles} \\ \hline 35 \text{ miles} \end{array} \qquad \begin{array}{r} \textbf{7 miles} \\ 5\overline{)35} \end{array}$$

28. $\mathbf{(3 \times 100{,}000) + (2 \times 10{,}000)}$

29. 321, the largest number is not divisible by 6.
312, the next largest is divisible by 6.

INVESTIGATION 5

1. **Frequency Table**

Number of Brothers and Sisters	Tally	Freq.
0	||||	4
1	~~||||~~ ||	7
2	~~||||~~	5
3	|||	3
4	|	1

2. **Frequency Table**

Weight	Tally	Freq.
11–15 lb	𝍸	5
16–20 lb	𝍸 𝍸	10
21–25 lb	𝍸 ||	7

3. **Frequency Table**

Rainfall	Tally	Freq.
15–19 in.	𝍸	5
20–24 in.	||||	4
25–29 in.	||	2
30–34 in.	||||	4

4. **12**

5. **4 people**

6. **9, 11, 13**

7. **19 years**

8. **21 iguanas**

9. **15 iguanas**

10. **10, 12, 28**

11. **18–20, 22–23 (Also accept 15–16.)**

12.

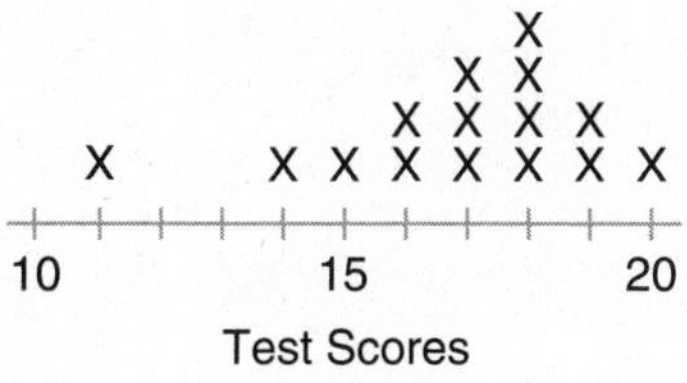

13. **11**

14. **18**

15. **12 tests**

16. **8 games; After the team won 5 games, it still had 3 games left to play.**

17. **No. The team won five games, which is half of the ten games in the season. However, it is not certain that the team won any of the last three games, so it is not certain that the team won more than half its games.**

18. $9 + 6 + 5 + 8 =$ **28 children**

19. **No. Five children named spaghetti as their favorite. Ten is twice as many as 5, but the number who named pizza was not 10—it was 9.**

20. $18 + 16 + 17 = 51$
$51 \div 3 =$ **17 answers**

21. $\mathbf{\frac{19}{20}}$

Lesson 51, Warm-Up

24 in.; 36 in.; 48 in.

a. **26 in.**

b. **400**

c. **1**

d. **2**

e. $\mathbf{12\frac{1}{2}}$

f. **6**

g. **2**

h. **272**

i. **3**

Problem Solving

+3 +4 +5 +6 +7
3, 6, 10, **15, 21, 28,** …

Lesson 51, Lesson Practice

a.
$$\begin{array}{r} 32 \\ \times\ 12 \\ \hline 64 \\ 320 \\ \hline \mathbf{384} \end{array}$$

b.
$$\begin{array}{r} \$0.62 \\ \times\ \ \ 23 \\ \hline 1\ 86 \\ 12\ 40 \\ \hline \mathbf{\$14.26} \end{array}$$

c.
$$\begin{array}{r} 48 \\ \times\ 64 \\ \hline 192 \\ 2880 \\ \hline \mathbf{3072} \end{array}$$

d.
$$\begin{array}{r} 246 \\ \times\ \ 22 \\ \hline 492 \\ 4920 \\ \hline \mathbf{5412} \end{array}$$

e.
$$\begin{array}{r} \$1.47 \\ \times\ \ \ 34 \\ \hline 5\ 88 \\ 44\ 10 \\ \hline \mathbf{\$49.98} \end{array}$$

f.
$$\begin{array}{r} 87 \\ \times\ 63 \\ \hline 261 \\ 5220 \\ \hline \mathbf{5481} \end{array}$$

g. **12 (20 + 3)**

$$\begin{array}{r} 23 \\ \times\ 12 \\ \hline 46 \\ 230 \\ \hline 276 \end{array} \quad \text{or} \quad \begin{array}{r} 12 \\ \times\ 20 \\ \hline 240 \end{array} \quad \begin{array}{r} 12 \\ \times\ 3 \\ \hline 36 \end{array} \quad \begin{array}{r} 240 \\ +\ 36 \\ \hline 276 \end{array}$$

Lesson 51, Mixed Practice

1.
$$\begin{array}{r} \overset{2}{4}7 \text{ pages} \\ 76 \text{ pages} \\ +\ 68 \text{ pages} \\ \hline 191 \text{ pages} \end{array} \qquad \begin{array}{r} \overset{2}{\cancel{3}}\,\overset{1}{\cancel{2}}{}^{1}0 \text{ pages} \\ -\ 1\ 9\ 1 \text{ pages} \\ \hline \mathbf{1\ 2\ 9 \text{ pages}} \end{array}$$

2.
$$\begin{array}{r} \overset{1}{2}3 \text{ cents} \\ 23 \text{ cents} \\ 23 \text{ cents} \\ +\ 37 \text{ cents} \\ \hline \mathbf{106 \text{ cents}} \end{array}$$

3.

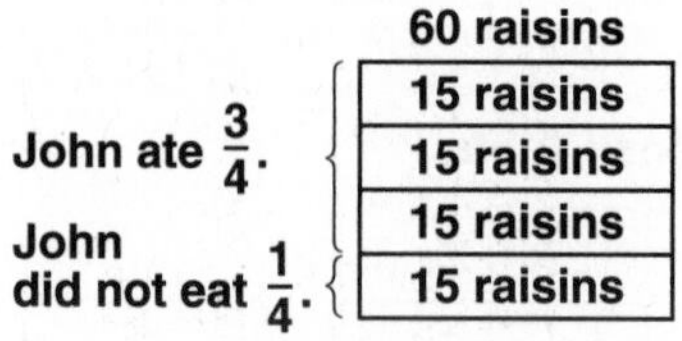

45 raisins; 75%

4. **1001**

5. $\frac{1}{2}$ of 10 $\bigcirc\!\!\!\!>$ $\frac{1}{3}$ of 12
5 4

6. **One thousand, seven hundred sixty**

7.

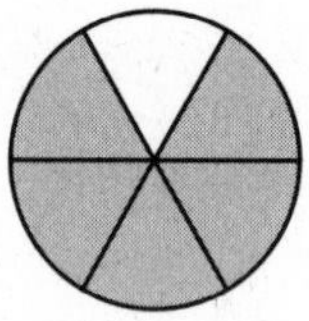

$$6\overline{)10^{4}0} = 16\tfrac{4}{6}$$ or $\mathbf{16\frac{2}{3}\%}$

8. **62,490**

9. **C. 2400**

10. $\mathbf{2\frac{1}{2}}$ **inches**

11.
$$\begin{array}{r} 4 \text{ coins} \\ +\ 8 \text{ coins} \\ \hline 12 \text{ coins} \end{array} \qquad \begin{array}{r} \mathbf{6 \text{ coins}} \\ 2\overline{)12} \end{array}$$

12.
$$\begin{array}{r} 43 \\ \times\ 12 \\ \hline 86 \\ 430 \\ \hline \mathbf{516} \end{array}$$

13.
$$\begin{array}{r} \$0.72 \\ \times\ \ \ 31 \\ \hline 72 \\ 21\ 60 \\ \hline \mathbf{\$22.32} \end{array}$$

14.
$$\begin{array}{r} 248 \\ \times\ \ 24 \\ \hline 992 \\ 4960 \\ \hline \mathbf{5952} \end{array}$$

15.
$$\begin{array}{r} \$1.96 \\ \times\ \ \ \ 53 \\ \hline 5\ 88 \\ 98\ 00 \\ \hline \mathbf{\$103.88} \end{array}$$

16.
$$\begin{array}{r} {}^{2\ 2\ 1} \\ 8{,}762 \\ 3{,}624 \\ 4{,}795 \\ +\ 8{,}473 \\ \hline \mathbf{25{,}654} \end{array}$$

17.
$$\begin{array}{r} \$\overset{9}{\cancel{10}}.\overset{9}{\cancel{0}}{}^{1}0 \\ -\ \ \$9.9\,2 \\ \hline \mathbf{\$0.0\,8} \end{array}$$

18.
$$\begin{array}{r} 600 \\ \times\ \ \ 50 \\ \hline \mathbf{30{,}000} \end{array}$$

19.
$$\begin{array}{r} \mathbf{\$0.\ 7\ 5} \\ 8\overline{)\$6.{}^{6}0{}^{4}0} \end{array}$$

20.
$$\begin{array}{r} \mathbf{\$10.\ 3\ 4} \\ 4\overline{)\$41.{}^{1}3{}^{1}6} \end{array}$$

21.
$$\begin{array}{r} 4\ 7\ 5 \\ 9\overline{)42{}^{6}7{}^{4}5} \end{array}$$

$x = \mathbf{475}$

22. $3 + \frac{1}{4} + 2\frac{2}{4} = \mathbf{5\frac{3}{4}}$

23. $5\frac{5}{8} - 3\frac{3}{8} = 2\frac{2}{8}$

$2\frac{2}{8} - 1\frac{1}{8} = \mathbf{1\frac{1}{8}}$

24. $(1 + 2 + 3 + 4 + 5) \div 5 =$

$15 \div 5 = \mathbf{3}$

25. 16 feet × 12 inches in each foot

= 192 inches

192 inches + 9 inches = **201 inches**

26. **15 × 24 or (15 × 20) + (15 × 4)**
$$\begin{array}{r} 24 \\ \times\ 15 \\ \hline 120 \\ 240 \\ \hline \mathbf{360} \end{array}$$

27. **Frequency Table**

Score	Tally	Frequency
1	||	2
2	||	2
3	~~||||~~	5
4	~~||||~~ |||	8
5	|||	3

28. **(2 × 100,000) + (5 × 1000)**

29. **11 in.**

Lesson 52, Warm-Up

120 seconds; 180 seconds

a. **130 seconds**

b. **$0.75**

c. **30 seconds**

d. **2**

e. **15 seconds**

f. **6 seconds**

g. **5**

Problem Solving

They are father and son. ("I" is the father.)

Lesson 52, Lesson Practice

a. **Hundreds**

b. **Millions**

c. **Hundred thousands**

d. **Ten millions**

e. **7**

f. **B. 1,372,486**

g. **One million**

h. **Twenty-one million, four hundred sixty-two thousand, three hundred**

i. **Nineteen billion, six hundred fifty million**

j. **19,225,500**

k. **750,300,000,000**

l. **206,712,934**

m. **(7 × 1,000,000) + (5 × 100,000)**

Lesson 52, Mixed Practice

1. 5 dozen × 12 cookies in each dozen = 60 cookies

$$\begin{array}{r} \overset{5}{\cancel{6}}\,{}^{1}0 \\ -\ 2\ 4 \\ \hline \textbf{3 6 cookies} \end{array}$$

2. 120 ÷ 2 = **60 pounds**

3.

$$\begin{array}{r} \$3.60 \\ +\ \$4.00 \\ \hline \$7.60 \end{array} \qquad \begin{array}{r} \$\overset{9}{\cancel{10}}.{}^{1}0\ 0 \\ -\ \ \$7.\ 6\ 0 \\ \hline \mathbf{\$2.\ 4\ 0} \end{array}$$

4.

$$\begin{array}{r} 1992 \\ -\ 1492 \\ \hline 500 \text{ years} \end{array} \qquad \begin{array}{r} \textbf{5 centuries} \\ 100\overline{)500} \end{array}$$

5. **148**

6. **One possibility:**

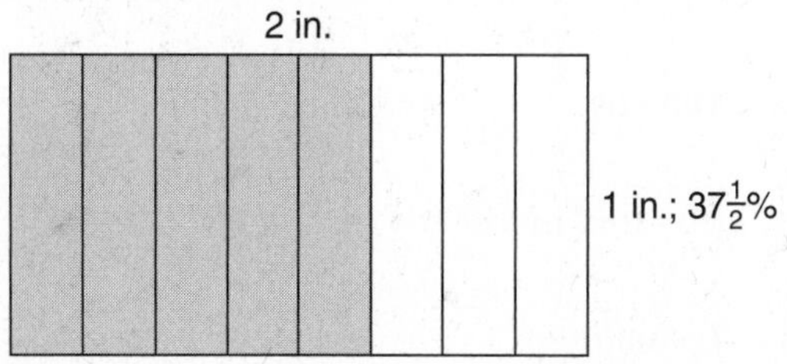

7. **Two hundred fifty thousand**

8.

$$\begin{array}{r} 4 \text{ books} \\ 9 \text{ books} \\ +\ 2 \text{ books} \\ \hline 15 \text{ books} \end{array} \qquad \begin{array}{r} \textbf{5 books} \\ 3\overline{)15} \end{array}$$

9. **7**

10. **1200**

11. **Ten millions**

12.

$$\begin{array}{r} 57 \\ \times\ 22 \\ \hline 114 \\ 1140 \\ \hline \mathbf{1254} \end{array}$$

13.

$$\begin{array}{r} \$0.83 \\ \times\ \ \ 47 \\ \hline 5\ 81 \\ 33\ 20 \\ \hline \mathbf{\$39.01} \end{array}$$

14.

$$\begin{array}{r} 167 \\ \times\ 89 \\ \hline 1\ 503 \\ 13\ 360 \\ \hline \mathbf{14,863} \end{array}$$

15.

$$\begin{array}{r} \$1.96 \\ \times\ \ \ 46 \\ \hline 11\ 76 \\ 78\ 40 \\ \hline \mathbf{\$90.16} \end{array}$$

16.

$$\begin{array}{r} {}^{2\ \ 1\,2} \\ 8\ 437 \\ 3\ 429 \\ 5\ 765 \\ +\ 9\ 841 \\ \hline \mathbf{27,472} \end{array}$$

17.

$$\begin{array}{r} \$\overset{1}{\cancel{2}}\ \overset{{}^{1}5}{\cancel{6}}.{}^{1}38 \\ -\ \$1\ 9.\ 57 \\ \hline \mathbf{\$6.\ 81} \end{array}$$

18.

$$\begin{array}{r} \overset{29}{\cancel{30}}\ \overset{{}^{1}3}{\cancel{4}}{}^{1}1 \\ -\ 29\ 7\ 5 \\ \hline 6\ 6 \end{array}$$

W = **66**

19. $\begin{array}{r}\mathbf{1082}\\ 4\overline{)43^{3}28}\end{array}$

20. $\begin{array}{r}\mathbf{5\ 6\ 7}\\ 10\overline{)56^{6}7^{7}0}\end{array}$

21. $\begin{array}{r}\mathbf{\$1\ 9.\ 60}\\ 4\overline{)\$7^{3}8.^{2}40}\end{array}$

22. $\frac{3}{10} + 2 + 1\frac{4}{10} = \mathbf{3\frac{7}{10}}$

23. $2\frac{3}{4} - 2 = \frac{3}{4}$

$5\frac{3}{4} - \frac{3}{4} = \mathbf{5}$

24. $\begin{array}{r} \scriptstyle{2\ 1}\ \ \\ \$1.43 \\ \$2.00 \\ \$2.85 \\ +\ \$0.79 \\ \hline \$7.07 \end{array}$ $\qquad$ $\begin{array}{r} \scriptstyle{9\ \ 9}\ \ \ \\ \$\not{1}\not{0}.\not{0}{}^{1}0 \\ -\ \ \$7.0\ 7 \\ \hline \mathbf{\$2.9\ 3} \end{array}$

25. ***D***

26. **25 × 24 or (25 × 20) + (25 × 4)**

27.

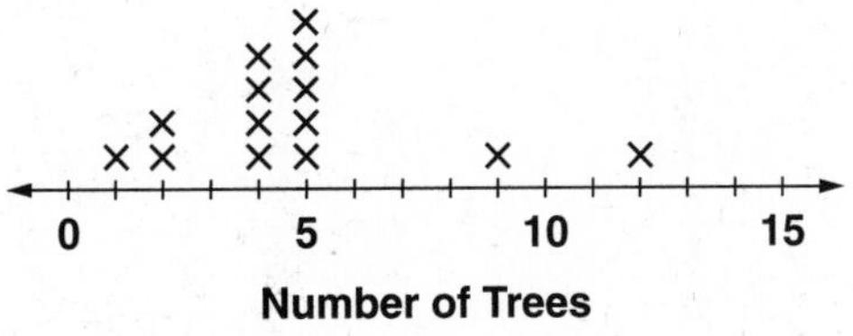

28. (a) **9 and 12**

(b) **4 and 5 (or 1 and 2)**

29. **(3 × 1,000,000) + (2 × 100,000)**

Lesson 53, Warm-Up

120 minutes; 180 minutes; 240 minutes; 600 minutes

a. **135 minutes**

b. **1500**

c. **5**

d. **0**

e. **90 minutes; 150 minutes**

f. **2**

Problem Solving

Triangular numbers:

+3 +4 +5 +6 +7 +8 +9 +10

3, 6, 10, 15, 21, 28, 36, 45, 55, ...

Square numbers include numbers of objects that can be arranged in a square pattern:

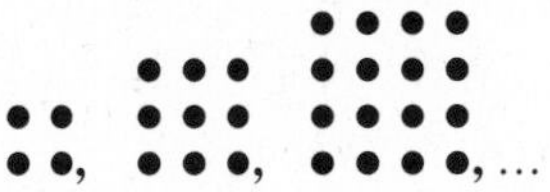

4, 9, 16, 25, 36, 49, 64, 81, ...

The number **36** is both a triangular number and a square number.

Lesson 53, Lesson Practice

a. **5 in.**

b. **3 in.**

c. 5 in. + 5 in. + 3 in. + 3 in. = **16 in.**

d. 5 cm + 4 cm + 3 cm = **12 cm**

e. 4 ft × 4 = **16 ft**

f. **Circumference**

g. **Diameter**

h. 6 in. × 2 = **12 inches**

Lesson 53, Mixed Practice

1. 3 × 12 = 36

$\begin{array}{r} \text{36 golden eggs} \\ -\ \text{15 golden eggs} \\ \hline \textbf{21 golden eggs} \end{array}$

2.
$$\begin{array}{r} \scriptstyle 1 \\ 13 \text{ players} \\ +\ 9 \text{ players} \\ \hline 22 \text{ players} \end{array} \qquad \begin{array}{r} \mathbf{11\ players} \\ 2\overline{)22} \end{array}$$

3. $\frac{1}{3}$ had blue eyes. $\frac{2}{3}$ did not have blue eyes.

30 children
10 children
10 children
10 children

10 children; $33\frac{1}{3}$%

4.
$$\begin{array}{r} 3 \text{ ounces} \\ 5 \text{ ounces} \\ +\ 7 \text{ ounces} \\ \hline 15 \text{ ounces} \end{array} \qquad \begin{array}{r} \mathbf{5\ ounces} \\ 3\overline{)15} \end{array}$$

5. **B. 20 billion**

6. Factors of 8: 1, 2, 4, 8
Factors of 12: 1, 2, 3, 4, 6, 12
Common factors: **1, 2, 4**

7.
$$\begin{array}{r} 1890 \\ -\ 1820 \\ \hline 70 \text{ years} \end{array} \qquad \begin{array}{r} \mathbf{7\ decades} \\ 10\overline{)70} \end{array}$$

8. **19,490,000**

9. $4\frac{2}{3} - 2 = 2\frac{2}{3}$

$6 + 2\frac{2}{3} = \mathbf{8\frac{2}{3}}$

10. $2\frac{2}{3} + 2 = 4\frac{2}{3}$

$4\frac{2}{3} - 4\frac{2}{3} = \mathbf{0}$

11.
$$\begin{array}{r} 300 \\ \times\ 200 \\ \hline \mathbf{60{,}000} \end{array}$$

12.
$$\begin{array}{r} 800 \\ \times\ 70 \\ \hline \mathbf{56{,}000} \end{array}$$

13.
$$\begin{array}{r} 100 \\ 5\overline{)500} \end{array}$$
$T = \mathbf{100}$

14.
$$\begin{array}{r} \$5.64 \\ \times\ 78 \\ \hline 45\ 12 \\ 394\ 80 \\ \hline \mathbf{\$439.92} \end{array}$$

15.
$$\begin{array}{r} 865 \\ \times\ 74 \\ \hline 3\ 460 \\ 60\ 550 \\ \hline \mathbf{64{,}010} \end{array}$$

16.
$$\begin{array}{r} 983 \\ \times\ 76 \\ \hline 5\ 898 \\ 68\ 810 \\ \hline \mathbf{74{,}708} \end{array}$$

17.
$$\begin{array}{r} \scriptstyle 2\ \ {}^{1}0 \\ \$6\ \not{3}.\ \not{1}{}^{1}4 \\ -\ \$4\ 2.\ 8\ 7 \\ \hline \mathbf{\$2\ 0.\ 2\ 7} \end{array}$$

18.
$$\begin{array}{r} \scriptstyle 2\ {}^{1}0 \\ \not{3}\ \not{1}{}^{1}0\ 6 \\ -\ \ 8\ 7\ 5 \\ \hline \mathbf{2\ 2\ 3\ 1} \end{array}$$

19.
$$\begin{array}{r} \scriptstyle 2\,1\ \ 3 \\ \$68.09 \\ \$43.56 \\ \$27.18 \\ +\ \$14.97 \\ \hline \mathbf{\$153.80} \end{array}$$

20.
$$\begin{array}{r} \mathbf{\$6.\ 3\ 3} \\ 5\overline{)\$31.{}^{1}6{}^{1}5} \end{array}$$

21.
$$\begin{array}{r} \mathbf{703} \\ 6\overline{)421{}^{1}8} \end{array}$$

22.
$$\begin{array}{r} \mathbf{5\ 3\ 6\ R\ 1} \\ 10\overline{)53{}^{3}6{}^{6}1} \end{array}$$

23. **B. 1240**

24. **3 cm**

25. 3 cm + 3 cm + 2 cm + 2 cm = **10 cm**

26. **35 × 21 or (35 × 20) + (35 × 1)**

27. **(2 × 1,000,000) + (5 × 10,000)**

28.

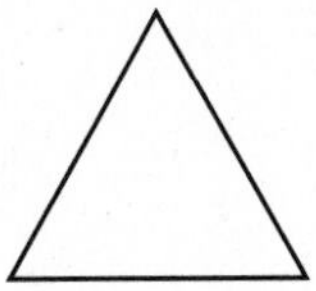

29. **9 in.**

Lesson 54, Warm-Up

a. **$0.67**

b. **9**

c. **25**

d. **50 in.**

e. **10**

Problem Solving

There are **6 dots** in each row.

Lesson 54, Lesson Practice

a.
```
     $0.14
30)$4.20
    3 0
    1 20
    1 20
       0
```

b.
```
     12 R 5
60)725
    60
    125
    120
      5
```

c.
```
     $0.12
40)$4.80
    4 0
     80
     80
      0
```

d.
```
     $0.16
20)$3.20
    2 0
    1 20
    1 20
       0
```

e.
```
     12 R 10
50)610
    50
    110
    100
     10
```

f.
```
     34 R 5
10)345
    30
     45
     40
      5
```

g.
```
    23
  × 40
   920
  +  5
   925
```
The answer is correct.

Lesson 54, Mixed Practice

1.
```
     1
   $3.18      $5.25
 + $1.02    − $4.20
   $4.20      $1.05
```
($1.05 in bold)

2.

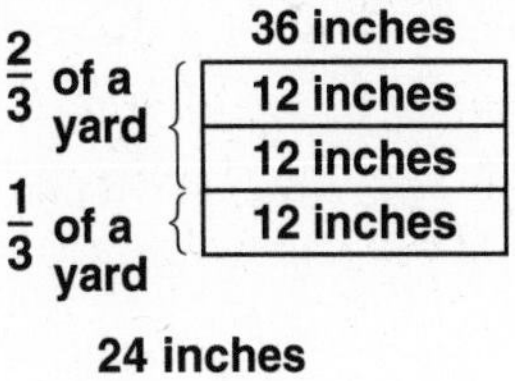

3. **1240**

4. **B. 7,000,000**

5. 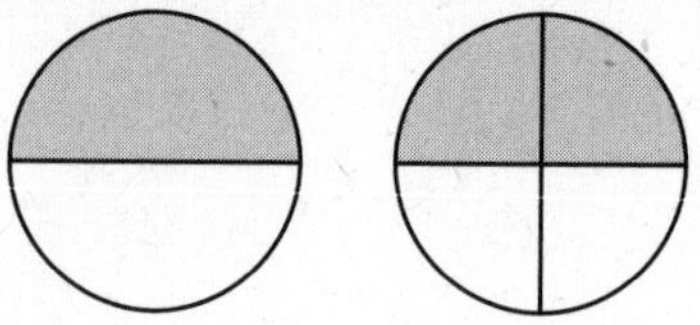 ; **50%**

6. (a) **25¢**

(b) **50¢**

7. **Three billion, one hundred fifty million**

8. The factors of 9: 1, 3, 9
The factors of 12: 1, 2, 3, 4, 6, 12
The factors of both: **1, 3**

9.
```
      15 R 4
30)454
   30
   154
   150
     4
```

10.
```
     $0.14
40)$5.60
    4 0
    1 60
    1 60
       0
```

11.
```
      15 R 10
50)760
   50
   260
   250
    10
```

12.
```
     500
×    400
 200,000
```

13.
```
     563
×     46
   3 378
  22 520
  25,898
```

14.
```
    $4.32
×      68
    34 56
   259 20
  $293.76
```

15. $25\frac{1}{4} + 8\frac{2}{4} = \mathbf{33\frac{3}{4}}$

16. $36\frac{2}{3} - 17\frac{2}{3} = \mathbf{19}$

17.
```
   368 R 3
8)2947
  24
   54
   48
    67
    64
     3
```

18.
```
     9
10)90
```

```
   840 R 4
9)7564
  72
   36
   36
    04
     0
     4
```

19.
```
   0 1 1  1 2 1 3
   1 2, 3 4 1 5
−     6, 7 8 9
      5, 5 5 6
```

20.
```
   2 2
  $3.65
  $2.47
  $4.83
+ $2.79
 $13.74
```

21.
```
   9 tables
4)36
```

22.
```
  1
  10 cm
   8 cm
+  6 cm
  24 cm
```

23. $\mathbf{1\frac{1}{2}}$ **inches**

24.
```
   1
  1896
+   50
  1946
```

25. $\begin{array}{r} \mathbf{15\ mm} \\ 2\overline{)30}\text{ mm} \\ \underline{2} \\ 10 \\ \underline{10} \\ 0 \end{array}$

26. **150 × 12 or (150 × 10) + (150 × 2)**

$\begin{array}{r} 150 \\ \times\ \ 12 \\ \hline 300 \\ \underline{1500} \\ \mathbf{1800} \end{array}$

27. **C. 645**

28. $\begin{array}{r} \mathbf{2\ 5}\text{ } \mathbf{pennies} \\ 4\overline{)10^{2}0} \end{array}$

29. **No, the triangle could not be scalene, because two of the sides have the same length. The sides of a scalene triangle all have different lengths.**

LESSON 55, WARM-UP

a. **12 hours**

b. **6 hours**

c. **64 inches**

d. **3**

e. **10**

f. **2**

Problem Solving

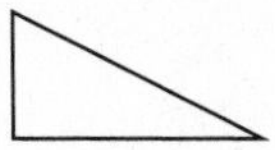

LESSON 55, LESSON PRACTICE

a. $\begin{array}{r} 346 \\ \times\ 354 \\ \hline 1\ 384 \\ 17\ 300 \\ \underline{103\ 800} \\ \mathbf{122{,}484} \end{array}$

b. $\begin{array}{r} 487 \\ \times\ 634 \\ \hline 1\ 948 \\ 14\ 610 \\ \underline{292\ 200} \\ \mathbf{308{,}758} \end{array}$

c. $\begin{array}{r} 403 \\ \times\ 768 \\ \hline 3\ 224 \\ 24\ 180 \\ \underline{282\ 100} \\ \mathbf{309{,}504} \end{array}$

d. $\begin{array}{r} 705 \\ \times\ 678 \\ \hline 5\ 640 \\ 49\ 350 \\ \underline{423\ 000} \\ \mathbf{477{,}990} \end{array}$

LESSON 55, MIXED PRACTICE

1. $\begin{array}{r} \scriptstyle 1 \\ \$1.65 \\ +\ \$0.90 \\ \hline \$2.55 \end{array}$ $\begin{array}{r} \scriptstyle 4\ \ 9 \\ \$\not{3}.\ \not{0}^{1}0 \\ -\ \$2.\ 5\ 5 \\ \hline \mathbf{\$2.\ 4\ 5} \end{array}$

2.

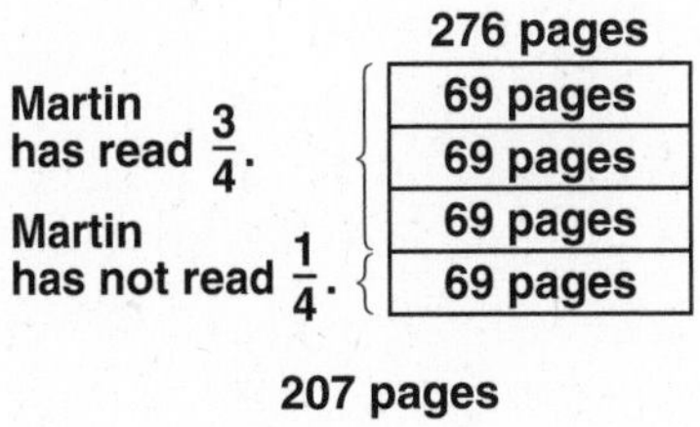

3. **J − 276 = 26**

$\begin{array}{r} \scriptstyle 1\,1 \\ 276 \\ +\ \ 26 \\ \hline 302\text{ pages} \end{array}$

J = **302 pages**

4. **9**

5. 2 yards × 36 inches in each yard = 72 inches
72 inches + 3 = **75 inches**

6.

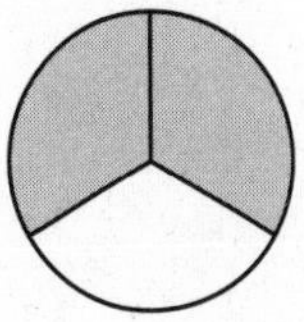

; $66\frac{2}{3}$%

7. 679,542,500

8.
$$60\overline{)7.20} \quad \mathbf{\$0.12}$$
$$\underline{6\,0}$$
$$1\,20$$
$$\underline{1\,20}$$
$$0$$

9.
$$70\overline{)850} \quad \mathbf{12\ R\ 10}$$
$$\underline{70}$$
$$150$$
$$\underline{140}$$
$$10$$

10.
$$80\overline{)980} \quad \mathbf{12\ R\ 20}$$
$$\underline{80}$$
$$180$$
$$\underline{160}$$
$$20$$

11.
$$\begin{array}{r} 234 \\ \times\ 123 \\ \hline 702 \\ 4\,680 \\ 23\,400 \\ \hline \mathbf{28{,}782} \end{array}$$

12.
$$\begin{array}{r} \$3.75 \\ \times\quad 26 \\ \hline 22\,50 \\ 75\,00 \\ \hline \mathbf{\$97.50} \end{array}$$

13.
$$\begin{array}{r} 604 \\ \times\ 789 \\ \hline 5\,436 \\ 48\,320 \\ 422\,800 \\ \hline \mathbf{476{,}556} \end{array}$$

14.
$$\begin{array}{r} 10\text{ mm} \\ \times\quad 4 \\ \hline \mathbf{40\text{ mm}} \end{array}$$

15. 320,000

16. 30,000

17. 81,000

18. 1200

19. 4000

20. 20

21. $6\frac{5}{11} + 5\frac{4}{11} = 11\frac{9}{11}$

22. $3\frac{2}{3} - 3 = \frac{2}{3}$

23. $3\frac{1}{3} - 3 = \frac{1}{3}$

$7\frac{2}{3} - \frac{1}{3} = 7\frac{1}{3}$

24.
$$\begin{array}{r} 2150 \\ \times\quad \$2 \\ \hline \mathbf{\$4300} \end{array}$$

25.
$$\begin{array}{r} \overset{29}{\cancel{3}}\overset{9}{\cancel{0}}{}^{1}0 \\ -\quad 2\,2\,7 \\ \hline 27\,7\ \mathbf{3\ fans} \end{array}$$

26. $(1 \times 1{,}000{,}000) + (2 \times 100{,}000)$

27. Check figure for at least two sides of equal length. One possibility:

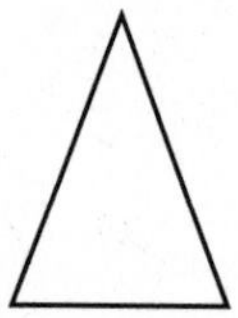

28.
$$5\overline{)10} \quad \mathbf{2\ dimes}$$

29. 6 in.

LESSON 56, WARM-UP

a. 6 in.

b. 12 in.

c. 10 in.

d. **60; 30; 20**

e. **294**

f. **21**

g. **7**

Problem Solving

The ones digits of the factors must be either 1 and 3 or 9 and 7. The products of 31×3 and of 33×1 are much less than 333. Since 333 is divisible by 9, it is reasonable to guess that the ones digit of the top factor is 7 and that the bottom factor is 9.

$$\begin{array}{r} \mathbf{3\underline{7}} \\ \times\ \ \mathbf{\underline{9}} \\ \hline \mathbf{333} \end{array}$$

Lesson 56, Lesson Practice

a. $$\begin{array}{r} 234 \\ \times\ 240 \\ \hline 9\,360 \\ 46\,800 \\ \hline \mathbf{56{,}160} \end{array}$$

b. $$\begin{array}{r} \$1.25 \\ \times\ \ \ 240 \\ \hline 5\,000 \\ 25\,000 \\ \hline \mathbf{\$300.00} \end{array}$$

c. $$\begin{array}{r} 230 \\ \times\ 120 \\ \hline 4\,600 \\ 23\,000 \\ \hline \mathbf{27{,}600} \end{array}$$

d. $$\begin{array}{r} 304 \\ \times\ 120 \\ \hline 6\,080 \\ 30\,400 \\ \hline \mathbf{36{,}480} \end{array}$$

e. $$\begin{array}{r} 234 \\ \times\ 204 \\ \hline 936 \\ 46\,80\ \ \\ \hline \mathbf{47{,}736} \end{array}$$

f. $$\begin{array}{r} \$1.25 \\ \times\ \ 2\,04 \\ \hline 5\,00 \\ 25\,00\ \ \\ \hline \mathbf{\$255.00} \end{array}$$

g. $$\begin{array}{r} 230 \\ \times\ 102 \\ \hline 460 \\ 23\,00\ \ \\ \hline \mathbf{23{,}460} \end{array}$$

h. $$\begin{array}{r} 304 \\ \times\ 102 \\ \hline 608 \\ 30\,40\ \ \\ \hline \mathbf{31{,}008} \end{array}$$

Lesson 56, Mixed Practice

1. $$\begin{array}{r} \$12 \\ +\ \ \$7 \\ \hline \$19 \end{array} \qquad \begin{array}{r} \$\overset{2}{\not{3}}{}^{1}0 \\ -\ \$1\ 9 \\ \hline \mathbf{\$1\ 1} \end{array}$$

2.

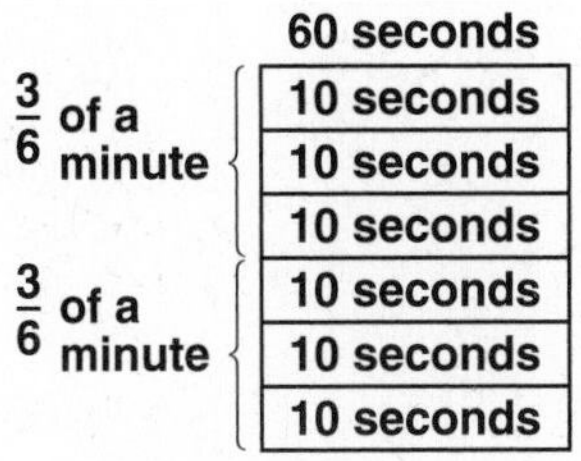

30 seconds

3. $8 \times 2 = 16$ blocks
$16 \times 5 =$ **80 blocks**

4. $$\begin{array}{r} \overset{1}{3}6 \text{ children} \\ 29 \text{ children} \\ +\ 73 \text{ children} \\ \hline 138 \text{ children} \end{array} \qquad \begin{array}{r} \mathbf{46} \textbf{ children} \\ 3\overline{)138} \\ \underline{12}\ \\ 18 \\ \underline{18} \\ 0 \end{array}$$

5. **5**

6. 5 in. $\times$ 2 = **10 in.**

7. **345,614,784**

8. 20 mm + 20 mm + 10 mm + 10 mm = **60 mm**

9. $\begin{array}{r} 900 \\ \times \quad 40 \\ \hline \mathbf{36{,}000} \end{array}$

10. $\begin{array}{r} 700 \\ \times \quad 400 \\ \hline \mathbf{280{,}000} \end{array}$

11. $\begin{array}{r} 234 \\ \times\ 320 \\ \hline 4\ 680 \\ 70\ 200 \\ \hline \mathbf{74{,}880} \end{array}$

12. $\begin{array}{r} \$3.45 \\ \times \quad 203 \\ \hline 10\ 35 \\ 690\ 0 \\ \hline \mathbf{\$700.35} \end{array}$

13. $\begin{array}{r} 468 \\ \times\ 386 \\ \hline 2\ 808 \\ 37\ 440 \\ 140\ 400 \\ \hline \mathbf{180{,}648} \end{array}$

14. $5 \times 6 = 30$
$w = \mathbf{30}$

15. $\begin{array}{r} \mathbf{7\ 1\ 9\ R\ 3} \\ 6\overline{)43^{1}1^{5}7} \end{array}$

16. $\begin{array}{r} \mathbf{300\ R\ 3} \\ 9\overline{)2703} \end{array}$

17. $\begin{array}{r} \$10.\ 7\ 6 \\ 8\overline{)\$86.^{6}0^{4}8} \end{array}$
$m = \mathbf{\$10.76}$

18. $\begin{array}{r} {}^{3\,1\ \,2\,2} \\ 79{,}089 \\ 37{,}865 \\ 29{,}453 \\ +\ 16{,}257 \\ \hline \mathbf{162{,}664} \end{array}$

19. $\begin{array}{r} 4\ \overset{2}{\cancel{3}},\ \overset{1}{\cancel{2}}{}^{1}1\ 8 \\ -\ 3\ 2,\ 4\ 6\ 1 \\ \hline \mathbf{1\ 0,\ 7\ 5\ 7} \end{array}$

20. $\begin{array}{r} \$\overset{9}{\cancel{10}}\ \overset{9}{\cancel{0}}.\ \overset{9}{\cancel{0}}{}^{1}0 \\ -\quad \$4.\ 5\ 6 \\ \hline \mathbf{\$9\ 5.\ 4\ 4} \end{array}$

21. $3\frac{5}{6} - 1\frac{5}{6} = \mathbf{2}$

22. $4\frac{1}{8} + 6 = \mathbf{10\frac{1}{8}}$

23. 3 weeks × 7 days in each week = 21 days
21 days + 3 days = **24 days**

24. ***C***

25. **Seven and one tenth**

26. **150 × 203 or (150 × 200) + (150 × 3)**

$\begin{array}{r} 203 \\ \times \quad 150 \\ \hline 10\ 150 \\ 20\ 300 \\ \hline \mathbf{30{,}450} \end{array}$

27. **C. 543 ÷ 3**

28. $\begin{array}{r} \mathbf{2\ 5} \\ 4\overline{)10^{2}0} \end{array}$ **small squares**

29. **38 in.**

LESSON 57, WARM-UP

a. **7 in.**

b. **5 in.**

c. **9 in.**

d. **dime**

e. **25**

f. **20**

g. **1**

Problem Solving

8 in. + 8 in. + 4 in. + 4 in. = 24 in.
We divide this length by 4 to find the length of each side of the square: 24 in. ÷ 4 = **6 in.**

Lesson 57, Lesson Practice

a. $\frac{1}{5}$

b. $\frac{2}{5}$

c. $\frac{2}{5}$

d. $\frac{3}{5}$

e. **Not to rain**

f.
$$\begin{array}{r} 100\% \\ -\ 20\% \\ \hline \mathbf{80\%} \end{array}$$

g. **Zero**

h. **Agree. Five of the ten marbles are red, so the probability of picking a red marble is $\frac{5}{10}$, which is a fraction equal to $\frac{1}{2}$.**

Lesson 57, Mixed Practice

1. 5 feet × 12 inches in each foot = 60 inches
   60 inches + 4 inches = **64 inches**

2.
$$\begin{array}{r} 100 \\ \times\ 10 \\ \hline \mathbf{1000} \end{array}$$ **years**

3. **Circumference**

4. **Ten and seven tenths**

5. 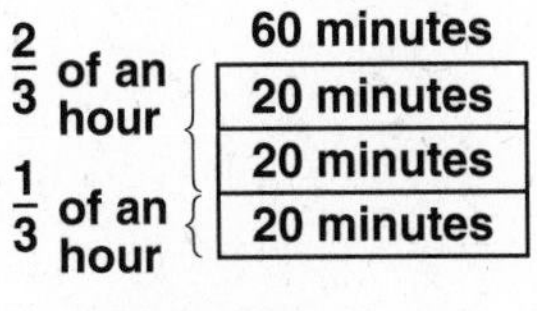

   40 minutes

6. From 11 p.m. to midnight is 1 hour and from midnight to 6 a.m. is 6 hours.
   1 hour + 6 hours = **7 hours**

7.
$$\begin{array}{r} 12 \\ \times\ 4 \\ \hline \mathbf{48} \end{array}$$

8. **Hundred thousands**

9. Factors of 15: ①, 3, ⑤,15
   Factors of 20: ①, 2, 4, ⑤,10, 20
   1, 5

10. 3 cm × 6 = **18 cm**

11. $2\frac{1}{3} + 1\frac{1}{3} = 3\frac{2}{3}$
    $3\frac{2}{3} - 3\frac{2}{3} = \mathbf{0}$

12. $2\frac{2}{3} - 1\frac{1}{3} = 1\frac{1}{3}$
    $3\frac{1}{3} + 1\frac{1}{3} = \mathbf{4\frac{2}{3}}$

13.
$$\begin{array}{r} \mathbf{\$0.13} \\ 40\overline{)\$5.20} \\ 4\,0 \\ \hline 1\,20 \\ 1\,20 \\ \hline 0 \end{array}$$

14.
$$\begin{array}{r} \mathbf{3\,9\,5\ R\,1} \\ 8\overline{)3\,1^{7}6^{4}1} \end{array}$$

15. **3**

16.
$$\begin{array}{r} \overset{3}{\cancel{\$4}}\,\overset{12}{\cancel{3}}.\,\overset{10}{\cancel{1}}{}^{1}5 \\ -\ \$2\,8.\,7\,9 \\ \hline \mathbf{\$1\,4.\,3\,6} \end{array}$$

17.
$$\begin{array}{r} 423 \\ \times\ 302 \\ \hline 846 \\ 12\,690 \\ \hline \mathbf{127{,}746} \end{array}$$

18.
$$\begin{array}{r} \overset{4}{99} \\ 36 \\ 42 \\ 75 \\ 64 \\ 98 \\ +\ 17 \\ \hline \mathbf{431} \end{array}$$

19.
$$\begin{array}{r} \$3.45 \\ \times \quad 3\,60 \\ \hline 207\,00 \\ 1035\,00 \\ \hline \mathbf{\$1242.00} \end{array}$$

20.
$$\begin{array}{r} 604 \\ \times\ 598 \\ \hline 4\,832 \\ 54\,360 \\ 302\,000 \\ \hline \mathbf{361{,}192} \end{array}$$

21. $\frac{10}{10} - \frac{9}{10} = \mathbf{\frac{1}{10}}$

22. $4\frac{2}{3} - \frac{1}{3} = \mathbf{4\frac{1}{3}}$

23. $5\frac{2}{2} - 1\frac{1}{2} = \mathbf{4\frac{1}{2}}$

24. **15 months**

25. In two hours and 20 minutes the time will be afternoon, so the "a.m." will switch to "p.m." The time will be **1:35 p.m.**

26. (a) **1000 years**

(b) 1000 years ÷ 2 = **500 years**

(c) $\mathbf{\frac{500}{1000}}$

27. $\mathbf{\frac{5}{6}}$

28.
$$\begin{array}{r} 80 \\ 80 \\ +\ 95 \\ \hline 255 \end{array} \qquad 3\overline{)25^{1}5} \text{ quotient } \mathbf{85}$$

29.
$$\begin{array}{r} \overset{9}{\cancel{10}}{}^{1}0\% \\ -\ \ 2\,5\% \\ \hline \mathbf{7\,5\%} \end{array}$$

LESSON 58, WARM-UP

a. **19 in.**

b. **12 in.**

c. **4 in.**

d. **80; 40; 20**

e. **400**

f. **7**

g. **192**

h. **8**

Problem Solving

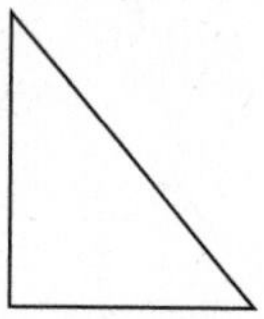

LESSON 58, LESSON PRACTICE

a. $4\overline{)17}$ = $\mathbf{4\frac{1}{4}}$
$$\begin{array}{r} 16 \\ \hline 1 \end{array}$$

b. $3\overline{)20}$ = $\mathbf{6\frac{2}{3}}$
$$\begin{array}{r} 18 \\ \hline 2 \end{array}$$

c. $5\overline{)16}$ = $\mathbf{3\frac{1}{5}}$
$$\begin{array}{r} 15 \\ \hline 1 \end{array}$$

d. $5\overline{)49}$ = $\mathbf{9\frac{4}{5}}$
$$\begin{array}{r} 45 \\ \hline 4 \end{array}$$

e. $4\overline{)21}$ = $\mathbf{5\frac{1}{4}}$
$$\begin{array}{r} 20 \\ \hline 1 \end{array}$$

f. $10\overline{)49}$ = $\mathbf{4\frac{9}{10}}$
$$\begin{array}{r} 40 \\ \hline 9 \end{array}$$

g. $12\frac{5}{6}$

$6\overline{)77}$

$\underline{6}$

17

$\underline{12}$

5

h. $4\frac{3}{10}$

$10\overline{)43}$

$\underline{40}$

3

i. $3\frac{7}{8}$

$8\overline{)31}$

$\underline{24}$

7

LESSON 58, MIXED PRACTICE

1.
$$\overset{1\ 4}{\$0.15}$$
$$\times\qquad 8$$
$$\overline{\$1.20}$$

$$\overset{1}{\$2.}\overset{1}{0}0$$
$$-\ \$1.\ 2\ 0$$
$$\overline{\$0.\ 8\ 0}$$

2. $5\frac{1}{4}$ inches

$4\overline{)21}$

$\underline{20}$

1

3.

100 stamps

Sarah used $\frac{3}{5}$ { 20 stamps / 20 stamps / 20 stamps }

Sarah did not use $\frac{2}{5}$ { 20 stamps / 20 stamps }

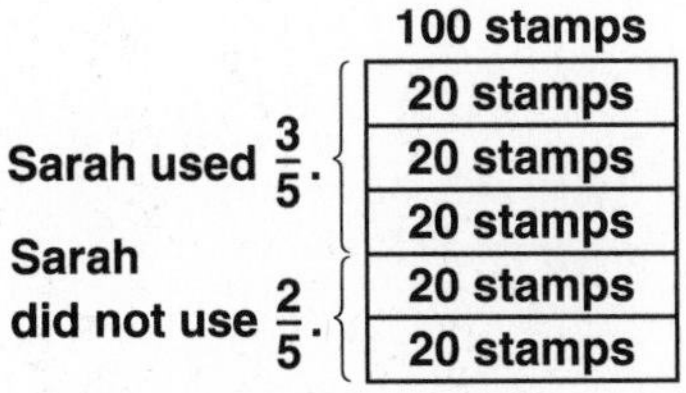

60 stamps; 60%

4. **1800**

5. A. **186,542,039**

6. 12 mm + 12 mm + 8 mm + 8 mm = **40 mm**

7. **21 R 10**

$30\overline{)640}$

$\underline{60}$

40

$\underline{30}$

10

8. **23 R 2**

$40\overline{)922}$

$\underline{80}$

122

$\underline{120}$

2

9. 16

$50\overline{)800}$

$\underline{50}$

300

$\underline{300}$

0

$w =$ **16**

10. $\overset{6}{7}\overset{1}{2}\,0\,0$

$-\ 1\,4\,0\,0$

$\overline{5\,8\,0\,0}$

$m =$ **5800**

11. $\$1.25$

$\times\qquad 80$

$\overline{\$100.00}$

12. 70

$10\overline{)700}$

$\underline{70}$

00

$\underline{0}$

0

13. 679

$\times\ 489$

$\overline{6\,111}$

$54\,320$

$\underline{271\,600}$

$\textbf{332,031}$

14. $\overset{7\ 1}{\cancel{8}}\overset{0\ 9}{\cancel{1}\cancel{0}}4$

$-\ 5\,6\,4\,7$

$\overline{\textbf{2\,4\,5\,7}}$

15. $\overset{3\ 2}{\$2.86}$

$\$6.35$

$\$1.78$

$\$0.46$

$+\ \$0.62$

$\overline{\textbf{\$12.07}}$

16. $60\ 4$

$7\overline{)422\,^2 8}$

17. $\begin{array}{r} \mathbf{5\ 1\ 5} \\ 9\overline{)46^{1}3^{4}5} \end{array}$

18. $\frac{5}{5} - \frac{1}{5} = \mathbf{\frac{4}{5}}$

19. $3\frac{1}{3} - \frac{1}{3} = \mathbf{3}$

20. $4\frac{6}{6} - 2\frac{5}{6} = \mathbf{2\frac{1}{6}}$

21. $\begin{array}{r} \mathbf{20\frac{2}{3}} \\ 3\overline{)62} \\ \underline{6} \\ 02 \\ \underline{0} \\ 2 \end{array}$

22. **4**

23. $3 \times 9 = \mathbf{27}$

24. $\begin{array}{r} 1500 \\ -\ \ 500 \\ \hline \mathbf{1000} \end{array}$

25. 12 mm × 2 = **24 mm**

26. $\mathbf{\frac{2}{11}}$

27. **C.**

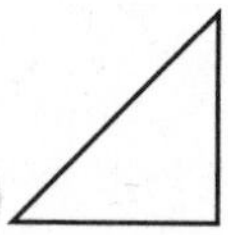

28. **75 small squares**

29. **1,284,204,000 people**

Lesson 59, Warm-Up

a. **5 in.**

b. **12 in.**

c. **9 in.**

d. **24 days**

e. $\mathbf{50\frac{1}{2}}$

f. **5**

g. **3**

Problem Solving

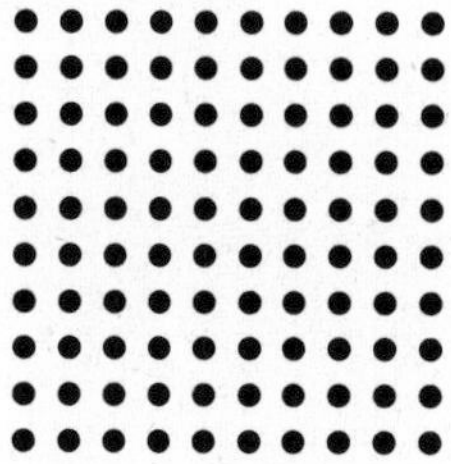

There are **10 dots** in each column.

Lesson 59, Lesson Practice

a. $\mathbf{\frac{3}{3}}$

b. $\frac{4}{4} \text{ (=) } 1$

c. $5\frac{4}{4} \text{ (=) } 6$

d. $\frac{3}{10} + \frac{7}{10} = \frac{10}{10} = \mathbf{1}$

e. $3\frac{3}{5} + 2\frac{2}{5} = 5\frac{5}{5} = \mathbf{6}$

f. $1 - \frac{1}{4}$

$\frac{4}{4} - \frac{1}{4} = \mathbf{\frac{3}{4}}$

g. $1 - \frac{2}{3}$

$\frac{3}{3} - \frac{2}{3} = \mathbf{\frac{1}{3}}$

h. **An infinite number (more names than you could count if you counted forever).**

LESSON 59, MIXED PRACTICE

1. 3 minutes \times 60 seconds in each minute
\qquad = 180 seconds
180 seconds + 24 seconds = **204 seconds**

2. $5 \times 12 = 60$ cookies
$60 \div 10 =$ **6 cookies**

3. **One possibility:**

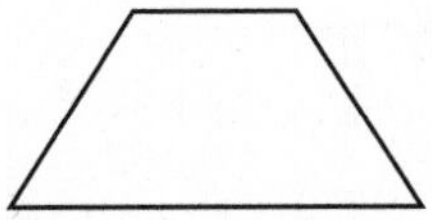

4. Factors of 8: ①, ②, ④, 8
Factors of 20: ①, ②, ④, 5, 10, 20
1, 2, 4

5.
$$\begin{array}{r} 12 \\ 5\overline{)60} \\ \underline{5} \\ 10 \\ \underline{10} \\ 0 \end{array} \qquad \begin{array}{r} 12 \\ \times\ 2 \\ \hline \textbf{24 seconds} \end{array}$$

$$\begin{array}{r} 20 \\ 5\overline{)100} \\ \underline{10} \\ 00 \\ \underline{00} \\ 0 \end{array} \qquad 20 \times 2 = \textbf{40\%}$$

6.
$$\begin{array}{r} 46 \\ +\ 60 \\ \hline 106 \end{array} \qquad \begin{array}{r} \textbf{53 pounds} \\ 2\overline{)106} \\ \underline{10} \\ 06 \\ \underline{6} \\ 0 \end{array}$$

7. $\dfrac{1}{4} + \dfrac{3}{4} = \dfrac{4}{4} = \textbf{1}$

8. $1\dfrac{1}{3} + 2\dfrac{2}{3} = 3\dfrac{3}{3} = \textbf{4}$

9. $2\dfrac{5}{8} + \dfrac{3}{8} = 2\dfrac{8}{8} = \textbf{3}$

10. $1 - \dfrac{1}{4}$

$\dfrac{4}{4} - \dfrac{1}{4} = \dfrac{\textbf{3}}{\textbf{4}}$

11. $1 - \dfrac{3}{8}$

$\dfrac{8}{8} - \dfrac{3}{8} = \dfrac{\textbf{5}}{\textbf{8}}$

12. $2\dfrac{8}{8} - \dfrac{3}{8} = \textbf{2}\dfrac{\textbf{5}}{\textbf{8}}$

13.
$$\begin{array}{r} {}^{1\ 1}\ {}^{2\ 2} \\ 98,789 \\ 41,286 \\ +\ 18,175 \\ \hline \textbf{158,250} \end{array}$$

14.
$$\begin{array}{r} {}^{6}4{}^{1}7,{}^{1}1\ {}^{4}5{}^{1}0 \\ -\ 36,2\ 4\ 7 \\ \hline \textbf{10, 9 0 3} \end{array}$$

15.
$$\begin{array}{r} 368 \\ \times\ 479 \\ \hline 3\ 312 \\ 25\ 760 \\ 147\ 200 \\ \hline \textbf{176,272} \end{array}$$

16. **Eight and nine tenths**

17.
$$\begin{array}{r} \textbf{3}\dfrac{\textbf{3}}{\textbf{4}} \\ 4\overline{)15} \\ \underline{12} \\ 3 \end{array}$$

18.
$$\begin{array}{r} \textbf{17 R 7} \\ 40\overline{)687} \\ \underline{40} \\ 287 \\ \underline{280} \\ 7 \end{array}$$

19.
$$\begin{array}{r} \textbf{14 R 10} \\ 60\overline{)850} \\ \underline{60} \\ 250 \\ \underline{240} \\ 10 \end{array}$$

20.
$$\begin{array}{r} \textbf{\$0.18} \\ 30\overline{)\$5.40} \\ \underline{3\ 0} \\ 2\ 40 \\ \underline{2\ 40} \\ 0 \end{array}$$

21.
$$\begin{array}{r} 507 \\ \times\ \$3.60 \\ \hline 304\ 20 \\ +\ 1521\ 00 \\ \hline \mathbf{\$1825.20} \end{array}$$

22.
$$\begin{array}{r} 900 \\ -\ 300 \\ \hline 600 \end{array} \qquad \begin{array}{r} \mathbf{20} \\ 30\overline{)600} \\ 60 \\ \hline 00 \\ 00 \\ \hline 0 \end{array}$$

23. **B.** $\mathbf{3\frac{2}{2}}$

24. $\mathbf{\frac{5}{5}}$

25. 2 cm × 3 = **6 cm**

26. **35 × 21 or (35 × 20) + (35 × 1); Many students find the second method easier.**

27. **A, D**

28. $\mathbf{\frac{3}{7}}$

29. **Frequency Table**

TVs	Tally	Frequency
0	|	1
1	||||	4
2	𝍸 |	6
3	𝍸 ||	7
4	||	2

Lesson 60, Warm-Up

a. $\mathbf{\frac{1}{2}}$

b. $\mathbf{\frac{2}{3}}$

c. $\mathbf{\frac{3}{4}}$

d. $\mathbf{\frac{7}{8}}$

e. **90; 30; 10**

f. **225**

g. **3**

h. **13**

Problem Solving

Lesson 60, Lesson Practice

a.

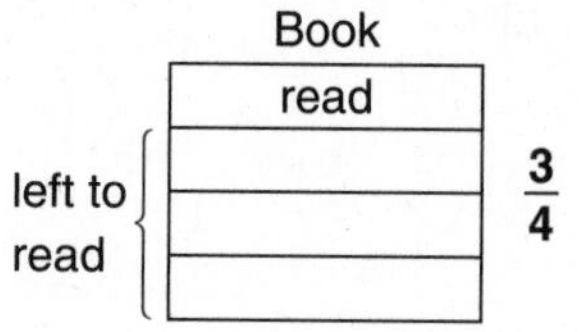

b.

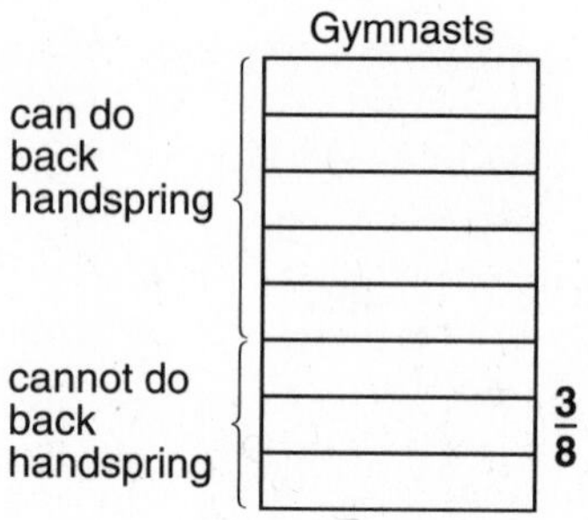

c.

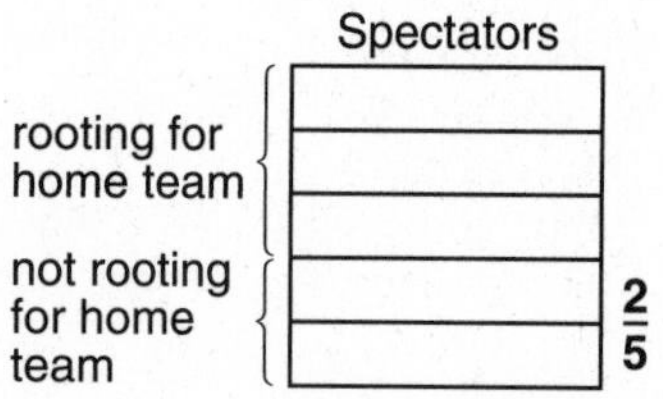

Lesson 60, Mixed Practice

1.
$$\begin{array}{r} \overset{1}{1}4 \text{ boys} \\ +\ 17 \text{ girls} \\ \hline \mathbf{31 \text{ children}} \end{array}$$

2.
$$\begin{array}{r} \$\overset{1}{2}.\overset{3}{3}9 \\ \$2.39 \\ \$4.49 \\ +\ \$0.56 \\ \hline \$9.83 \end{array} \qquad \begin{array}{r} \$\overset{9}{\cancel{10}}.\overset{9}{\cancel{0}}{}^{1}0 \\ -\ \$9.\ 8\ 3 \\ \hline \mathbf{\$0.\ 1\ 7} \end{array} \text{ or } \mathbf{17¢}$$

3. 1900 − 1800 = 100 years = **10 decades**

4. 24 ÷ 2 = **12 inches**

5.
$$\begin{array}{r} 500 \\ +\ 300 \\ \hline \mathbf{800} \end{array}$$

6. (a) **7**

(b) **4**

7. 7 pounds × 16 ounces in each pound = 112 ounces

112 ounces + 12 ounces = **124 ounces**

8. 1 mile × 4 = **4 miles**

9. $\frac{1}{6} + \frac{2}{6} + \frac{3}{6} = \frac{6}{6} = \mathbf{1}$

10. $3\frac{3}{5} + 1\frac{2}{5} = 4\frac{5}{5} = \mathbf{5}$

11. $1 - \frac{1}{8}$

$\frac{8}{8} - \frac{1}{8} = \mathbf{\frac{7}{8}}$

12. $4\frac{5}{5} - 1\frac{2}{5} = \mathbf{3\frac{3}{5}}$

13.
$$\begin{array}{r} \$3\ \overset{4}{\not{5}}.{}^{1}2\ 4 \\ -\ \$1\ 4.\ 6\ 2 \\ \hline \mathbf{\$2\ 0.\ 6\ 2} \end{array}$$

14.
$$\begin{array}{r} \$5.78 \\ \times\ \ 467 \\ \hline 40\ 46 \\ 346\ 80 \\ 2312\ 00 \\ \hline \mathbf{\$2699.26} \end{array}$$

15.
$$\begin{array}{r} \mathbf{\$4.08} \\ 9\overline{)\$36.72} \\ 36 \\ \hline 0\ 7 \\ 0 \\ \hline 72 \\ 72 \\ \hline 0 \end{array}$$

16.
$$\begin{array}{r} \mathbf{2\frac{3}{10}} \\ 10\overline{)23} \\ 20 \\ \hline 3 \end{array}$$

17. $1 - \frac{1}{8} = \mathbf{\frac{7}{8}}$

$$\begin{array}{r} \mathbf{1\ 2\frac{4}{8}\%} \text{ or } \mathbf{12\frac{1}{2}\%} \\ 8\overline{)10^{2}0} \end{array}$$

18.
$$\begin{array}{r} 374 \\ \times\ \ 360 \\ \hline 22\ 440 \\ 112\ 200 \\ \hline \mathbf{134{,}640} \end{array}$$

19.
$$\begin{array}{r} \mathbf{16\ R\ 3} \\ 40\overline{)643} \\ 40 \\ \hline 243 \\ 240 \\ \hline 3 \end{array}$$

20.
$$\begin{array}{r} 20 \\ 40\overline{)800} \end{array} \qquad \begin{array}{r} 60 \\ \times\ \ 20 \\ \hline \mathbf{1200} \end{array}$$

21.
$$\begin{array}{r} \mathbf{67} \\ 20\overline{)1340} \\ 120 \\ \hline 140 \\ 140 \\ \hline 0 \end{array}$$

22. $\frac{4}{4}$ ⊜ $\frac{5}{5}$

23. $\mathbf{\frac{8}{8}}$

24. $\mathbf{\frac{3}{10}}$

25. **25 minutes**

26. (a) $\mathbf{\frac{5}{13}}$

(b) $\frac{13}{13} - \frac{5}{13} = \mathbf{\frac{8}{13}}$

27. **D. 621 ÷ 9**

28. **12 songs**

29.
$$\begin{array}{r} {}_{1}7 \text{ CDs} \\ 13 \text{ CDs} \\ 3 \text{ CDs} \\ +\ 2 \text{ CDs} \\ \hline \mathbf{25\ CDs} \end{array}$$

INVESTIGATION 6

1. $\mathbf{\frac{2}{6}\left(\text{or } \frac{1}{3}\right)}$

2. $\mathbf{\frac{4}{6}\left(\text{or } \frac{2}{3}\right)}$

3. **10 times; Since the outcome 1 can be expected to occur $\frac{1}{6}$ of the time, I divided 60 rolls by 6.**

4. **20 times; Since each outcome can be expected to occur 10 times in 60 rolls, 2 outcomes can be expected to occur 10 + 10, or 20 times.**

5. **Answers may vary.**

6. **Answers may vary.**

7. **Answers may vary.**

8. **Yes, because region 2 is larger than region 1, and region 1 is larger than region 3.**

9. **Because region 2 is slightly smaller than $\frac{1}{2} = \frac{25}{50}$ of the spinner, $\frac{28}{50}$ overestimates the true probability of outcome 2.**

10. **Answers may vary.**

11. **Answers may vary.**

12. 30 ÷ 5 = **6 times**

13. 6 + 6 = **12 times**

14. 6 + 6 + 6 = **18 times**

15. **Answers may vary.**

Extensions

a. $\mathbf{\frac{1}{26}; \frac{5}{26}; \frac{21}{26}}$

b. **A: 7 times; B: 7 times; C: 21 times; actual results may vary.**

LESSON 61, WARM-UP

a. **10**

b. **5**

c. $\mathbf{2\frac{1}{2}; 4\frac{1}{2}; 7\frac{1}{2}}$

d. $\mathbf{\frac{2}{3}}$

e. $\mathbf{\frac{3}{4}}$

f. **50**

Problem Solving

Each rectangle's length is $\mathbf{8\frac{1}{2}}$ **in.**

Each rectangle's width is half the length of a full sheet, or $\mathbf{5\frac{1}{2}}$ **in.**

LESSON 61, LESSON PRACTICE

a. $\mathbf{\overline{ML}}$ **(or** $\mathbf{\overline{LM}}$**)**

b. 10 cm ÷ 2 = 5 cm

10 cm + 10 cm + 5 cm + 5 cm = **30 cm**

c. **Segment *BC*;**

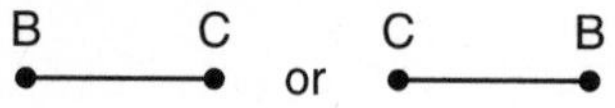

d. **Ray *CD*;**

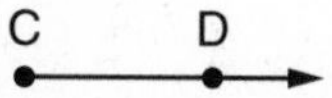

e. Line *PQ;*

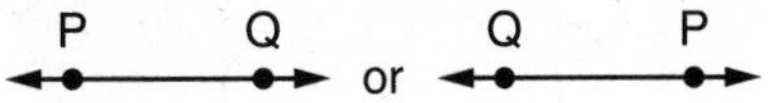

f. ∠*DMA*

g. ∠*BMD* (or ∠*DMB*)

h. \overrightarrow{MB}

i. Student should name one of the following:
∠*CMD* (or ∠*DMC*)
∠*AMB* (or ∠*BMA*)
∠*AMC* (or ∠*CMA*)
∠*BMC* (or ∠*CMB*)

LESSON 61, MIXED PRACTICE

1. 6 feet × 12 inches = 72 inches
72 inches + 3 inches = **75 inches**

2. $\dfrac{1}{6} \times \dfrac{100}{1} = \dfrac{100}{6}$

$6)\overline{100} \quad 16\dfrac{4}{6} = 16\dfrac{2}{3}\%$
$\underline{6}$
40
$\underline{36}$
4

$\dfrac{6}{6} - \dfrac{1}{6} = \dfrac{5}{6}$

3.
$$\begin{array}{r} 1\,2\,\overset{1}{\cancel{1}}\,\overset{0}{}\!{}^{1}5 \\ -\ 1\,0\,6\,6 \\ \hline 1\ 4\ 9 \text{ years} \end{array}$$

4. **7040**

5.
$$\begin{array}{r} 60 \\ \times\ \ 20 \\ \hline 1200 \end{array}$$

6. D. $\dfrac{48}{98}$

7. Factors of 12: 1, 2, 3, 4, 6, 12
Factors of 16: 1, 2, 4, 8, 16
Common factors: **1, 2, 4**

8.
$$\begin{array}{r} \overset{1}{1}2 \text{ inches} \\ \times\ \ 8 \\ \hline 96 \text{ inches} \end{array}$$

9. $\dfrac{5}{5} - \dfrac{1}{5} = \dfrac{4}{5}$

10. $\dfrac{4}{4} - \dfrac{3}{4} = \dfrac{1}{4}$

11. $3\dfrac{3}{3} - 1\dfrac{2}{3} = 2\dfrac{1}{3}$

12. $\dfrac{1}{10} + \dfrac{2}{10} + \dfrac{3}{10} + \dfrac{4}{10} = \dfrac{10}{10} = 1$

13. $5\dfrac{3}{4} + 4\dfrac{1}{4} = 9\dfrac{4}{4} = 10$

14.
$$\begin{array}{r} \overset{3}{4}\,\overset{1}{2}\,\overset{1}{6}\,{}^{1}3 \\ -\ 1\,7\,8\,4 \\ \hline 2\ 4\ 7\ 9 \end{array}$$
$Q = \mathbf{2479}$

15.
$$\begin{array}{r} \$5\overset{4}{0}.\,\overset{9}{0}\,{}^{1}0 \\ -\ \$19.\,3\,4 \\ \hline \$30.\,6\,6 \end{array}$$
$M = \mathbf{\$30.66}$

16.
$$\begin{array}{r} \overset{3}{5}8 \\ 39 \\ 24 \\ 16 \\ 52 \\ +\ 11 \\ \hline 200 \end{array}$$

17.
$$\begin{array}{r} 389 \\ \times\ 470 \\ \hline 27\,230 \\ 155\,600 \\ \hline \mathbf{182,830} \end{array}$$

18.
$$\begin{array}{r} \mathbf{605} \\ 9)\overline{5445} \\ \underline{54} \\ 04 \\ \underline{0} \\ 45 \\ \underline{45} \\ 0 \end{array}$$

19. $\begin{array}{r} \mathbf{4\frac{1}{6}} \\ 6\overline{)25} \\ \underline{24} \\ 1 \end{array}$

20. $\begin{array}{r} \mathbf{22\ R\ 14} \\ 40\overline{)894} \\ \underline{80} \\ 94 \\ \underline{80} \\ 14 \end{array}$

21. $\begin{array}{r} \mathbf{31\ R\ 13} \\ 30\overline{)943} \\ \underline{90} \\ 43 \\ \underline{30} \\ 13 \end{array}$

22. $(800 - 300) \times 20$

$500 \times 20 = \mathbf{10{,}000}$

23. $\mathbf{3\frac{7}{10}}$

24. $\mathbf{\frac{10}{20}; \frac{20}{40}}$

25. **February**

26. $\begin{array}{r} 60 \text{ mm} \\ -\ 20 \text{ mm} \\ \hline \mathbf{40 \text{ mm}} \end{array}$

27. **$\angle PWR$ (or $\angle RWP$)**

28. **3 months**

29. **8 months**

30. **22 days**

LESSON 62, WARM-UP

a. **500**

b. **250**

c. $\mathbf{2\frac{2}{3}; 3\frac{1}{3}}$

d. $\mathbf{\frac{4}{5}}$

e. $\mathbf{\frac{1}{5}}$

f. **50**

g. **1**

Problem Solving

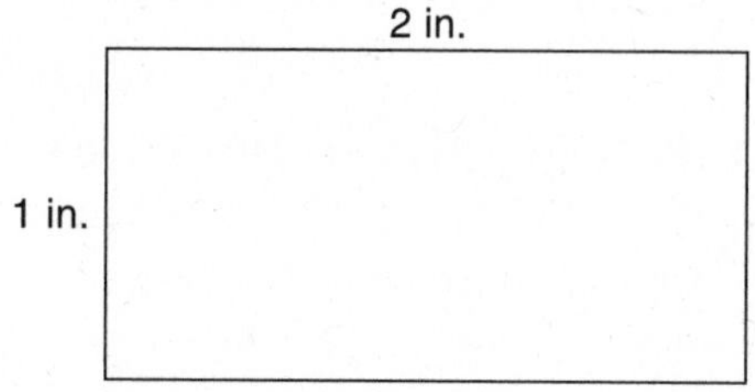

LESSON 62, LESSON PRACTICE

a. **70 + 40 = 110**

b. **60 × 20 = 1200**

c. **600 + 300 = 900**

d. **40 × 20 = 800**

e. **90 − 30 = 60**

f. **30 × 300 = 9000**

g. **700 − 400 = 300**

h. **60 ÷ 30 = 2**

i. **700 − 500 = 200**

j. **90 ÷ 30 = 3**

k.

$$\begin{array}{r} \overset{1}{\$13.00} \\ \$7.00 \\ +\ \$8.00 \\ \hline \mathbf{\$28.00} \end{array}$$

l.

$$\begin{array}{r} 60 \text{ mm} \\ 60 \text{ mm} \\ 40 \text{ mm} \\ +\ 40 \text{ mm} \\ \hline \mathbf{200 \text{ mm}} \end{array}$$

Lesson 62, Mixed Practice

1.

$$\begin{array}{r} \overset{1}{12} \\ \times\ \ 6 \text{ cookies} \\ \hline 72 \end{array} \qquad \begin{array}{r} 72 \text{ cookies} \\ -\ 20 \text{ cookies} \\ \hline \mathbf{52 \text{ cookies}} \end{array}$$

2.

$$\begin{array}{r} \mathbf{10 \text{ centuries}} \\ 100\overline{)1000} \\ \underline{100} \\ 00 \\ \underline{0} \\ 0 \end{array}$$

3.

$$\begin{array}{r} 4 \text{ ounces} \\ 7 \text{ ounces} \\ 7 \text{ ounces} \\ +\ 2 \text{ ounces} \\ \hline 20 \text{ ounces} \end{array} \qquad \begin{array}{r} \mathbf{5 \text{ ounces}} \\ 4\overline{)20} \\ \underline{20} \\ 0 \end{array}$$

4. 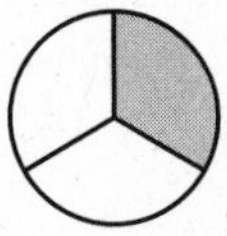 ; $\mathbf{\frac{2}{3}}$;

$$\frac{2}{3} \times \frac{100}{1} = \frac{200}{3}$$

$$\begin{array}{r} \mathbf{66\frac{2}{3}\%} \\ 3\overline{)200} \\ \underline{18} \\ 20 \\ \underline{18} \\ 2 \end{array}$$

5. $40 \times 40 = \mathbf{1600}$

6. $\frac{10}{10} - \frac{1}{10} = \mathbf{\frac{9}{10}}$

7. $\frac{8}{8} - \frac{3}{8} = \mathbf{\frac{5}{8}}$

8. $4\frac{4}{4} - 2\frac{3}{4} = \mathbf{2\frac{1}{4}}$

9. $3\frac{1}{3} + 1\frac{2}{3} = 4\frac{3}{3} = \mathbf{5}$

10. $6\frac{10}{10} - \frac{1}{10} = \mathbf{6\frac{9}{10}}$

11. $8 = 7\mathbf{\frac{6}{6}}$

12. $600 + 300 = \mathbf{900}$

13.

$$\begin{array}{r} \overset{2\,2\ \,2\,2}{89{,}786} \\ 26{,}428 \\ 57{,}814 \\ +\ 91{,}875 \\ \hline \mathbf{265{,}903} \end{array}$$

14.

$$\begin{array}{r} \$\overset{2}{\cancel{3}}\,\overset{14}{\cancel{5}},\overset{9}{\cancel{0}}{}^{1}4\,2 \\ -\ \$1\,7,\,6\,5\,1 \\ \hline \mathbf{\$1\,7,\,3\,9\,1} \end{array}$$

15.

$$\begin{array}{r} 428 \\ \times\ 396 \\ \hline 2\,568 \\ 38\,520 \\ \underline{128\,400} \\ \mathbf{169{,}488} \end{array}$$

16.

$$\begin{array}{r} \mathbf{947} \\ 5\overline{)4735} \\ \underline{45} \\ 23 \\ \underline{20} \\ 35 \\ \underline{35} \\ 0 \end{array}$$

17.

$$\begin{array}{r} \overset{2}{43} \\ \times\ \ 8 \\ \hline 344 \end{array} \qquad \begin{array}{r} 602 \\ \times\ 344 \\ \hline 2\,408 \\ 24\,080 \\ \underline{180\,600} \\ \mathbf{207{,}088} \end{array}$$

18. $\begin{array}{r} 1\frac{7}{8} \\ 8\overline{)15} \\ \underline{8} \\ 7 \end{array}$

19. $\begin{array}{r} \textbf{16 R 7} \\ 60\overline{)967} \\ \underline{60} \\ 367 \\ \underline{360} \\ 7 \end{array}$

20. $\begin{array}{r} \textbf{21 R 35} \\ 40\overline{)875} \\ \underline{80} \\ 75 \\ \underline{40} \\ 35 \end{array}$

21. (a) **C.** $\mathbf{\frac{15}{30}}$

(b) **B.** $\mathbf{\frac{7}{15}}$

(c) **A.** $\mathbf{\frac{4}{7}}$

22. $\begin{array}{r} \$24.00 \\ \$43.89 \\ \$8.67 \\ + \$0.98 \\ \hline \$77.54 \end{array}$ $\quad \begin{array}{r} \$100.00 \\ - \$77.54 \\ \hline \mathbf{\$22.46} \end{array}$

23. $\begin{array}{r} 15 \text{ mm} \\ \times \ 4 \\ \hline \mathbf{60 \text{ mm}} \end{array}$

24. $2 \times 5 = 10$
0

25. **12:50 p.m.**

26. $\mathbf{\frac{1}{2}}$

27. $\mathbf{\frac{1}{4}}$

28. **∠*WMZ* (or ∠*ZMW*)**

29. $\begin{array}{r} 12 \text{ in.} \\ 12 \text{ in.} \\ 6 \text{ in.} \\ + \ 6 \text{ in.} \\ \hline \mathbf{36 \text{ in.}} \end{array}$

30. $\begin{array}{r} \mathbf{12\frac{1}{2}} \\ 2\overline{)25} \\ \underline{2} \\ 05 \\ \underline{4} \\ 1 \end{array}$

Lesson 63, Warm-Up

a. $\mathbf{1\frac{2}{5}; 1\frac{3}{5}}$

b. $\mathbf{\frac{5}{6}}$

c. $\mathbf{\frac{9}{10}}$

d. **20**

e. **3**

Problem Solving

For the product to be greater than 400, the bottom factor must be 9. Since the ones digit of the product is 4, the ones digit of the top factor must be 6. To find the tens digit of the product, we multiply.

$\begin{array}{r} \mathbf{4\underline{6}} \\ \times \ \mathbf{\underline{9}} \\ \hline \mathbf{4\underline{1}4} \end{array}$

Lesson 63, Lesson Practice

a. $3\frac{4}{4} - \frac{1}{4} = \mathbf{3\frac{3}{4}}$

b. $2\frac{4}{4} - \frac{3}{4} = \mathbf{2\frac{1}{4}}$

c. $3\frac{4}{4} - 2\frac{1}{4} = \mathbf{1\frac{3}{4}}$

d. $1\frac{4}{4} - \frac{1}{4} = \mathbf{1\frac{3}{4}}$

e. $3\frac{2}{2} - 1\frac{1}{2} = \mathbf{2\frac{1}{2}}$

f. $5\frac{3}{3} - 1\frac{2}{3} = \mathbf{4\frac{1}{3}}$

Lesson 63, Mixed Practice

1. $\begin{array}{r} \overset{1}{3}4 \text{ cm} \\ +\ \ 7 \text{ cm} \\ \hline 41 \text{ cm} \end{array}$ $\begin{array}{r} \overset{9}{\not{1}\not{0}}{}^{1}0 \text{ cm} \\ -\ \ 4\ 1 \text{ cm} \\ \hline \mathbf{5\ 9 \text{ cm}} \end{array}$

2. 6 inches − $1\frac{1}{2}$ inches

   $5\frac{2}{2} - 1\frac{1}{2} = \mathbf{4\frac{1}{2}}$ **inches**

3. 4 × 5 = **20 quarter-pound hamburgers**

4. $\begin{array}{r} \overset{2}{1}8 \text{ books} \\ 19 \text{ books} \\ 24 \text{ books} \\ +\ \ 23 \text{ books} \\ \hline 84 \text{ books} \end{array}$ $\begin{array}{r} \mathbf{21 \text{ books}} \\ 4\overline{)84} \\ \underline{8} \\ 04 \\ \underline{4} \\ 0 \end{array}$

5. 400 + 500 = **900**

6. Factors of 14: 1, 2, 7, 14
   Factors of 21: 1, 3, 7, 21
   Common factors: **1, 7**

7. **C. Circumference**

8. $\begin{array}{r} \overset{1\ 11}{5,284,000} \\ +\ \ 6,918,500 \\ \hline \mathbf{12,202,500} \end{array}$

9. $6\frac{3}{3} - \frac{1}{3} = \mathbf{6\frac{2}{3}}$

10. $5\frac{2}{2} - 2\frac{1}{2} = \mathbf{3\frac{1}{2}}$

11. $7\frac{4}{4} - 3\frac{3}{4} = \mathbf{4\frac{1}{4}}$

12. $\frac{8}{9} + \left(\frac{2}{9} - \frac{1}{9}\right)$

    $= \frac{8}{9} + \frac{1}{9} = \frac{9}{9} = \mathbf{1}$

13. $5\frac{3}{4} - \left(3\frac{2}{4} + 1\frac{1}{4}\right)$

    $= 5\frac{3}{4} - 4\frac{3}{4} = \mathbf{1}$

14. $\begin{array}{r} \overset{3}{\not{4}}{}^{1}3,\ \overset{6}{\not{7}}\ \overset{10}{\not{1}}{}^{1}6 \\ -\ \ 1\ 9,\ 5\ 3\ 7 \\ \hline \mathbf{2\ 4,\ 1\ 7\ 9} \end{array}$

15. $\begin{array}{r} \$6.87 \\ \times\ \ 794 \\ \hline 27\ 48 \\ 618\ 30 \\ 4809\ 00 \\ \hline \mathbf{\$5454.78} \end{array}$

16. $\begin{array}{r} \mathbf{\$1.84} \\ 8\overline{)\$14.72} \\ \underline{8}\ \ \ \ \\ 6\ 7\ \ \\ \underline{6\ 4}\ \ \\ 32 \\ \underline{32} \\ 0 \end{array}$

17. $\begin{array}{r} \mathbf{2\frac{2}{9}} \\ 9\overline{)20} \\ \underline{18} \\ 2 \end{array}$

18. $\begin{array}{r} \mathbf{47 \text{ R } 11} \\ 20\overline{)951} \\ \underline{80}\ \ \\ 151 \\ \underline{140} \\ 11 \end{array}$

19. $\begin{array}{r} \mathbf{51 \text{ R } 10} \\ 50\overline{)2560} \\ \underline{250}\ \ \\ 60 \\ \underline{50} \\ 10 \end{array}$

20. $50 \times (400 + 400)$
$= 50 \times 800 = \mathbf{40{,}000}$

21. $(400 + 400) \div 40$
$= 800 \div 40$
$= \mathbf{20}$

22.
$$\begin{array}{r} {}^{2\,2\,2}\\ 4736 \\ 2849 \\ 351 \\ +\quad 78 \\ \hline \mathbf{8014} \end{array}$$

23. $\frac{8}{8} - \frac{3}{8} = \mathbf{\frac{5}{8}}$

$\frac{5}{8} \times \frac{100}{1} = \frac{500}{8}$

$$\begin{array}{r} 62\frac{4}{8} \\ 8\overline{)500} \\ \underline{48} \\ 20 \\ \underline{16} \\ 4 \end{array} = \mathbf{62\tfrac{1}{2}\%}$$

24. $\mathbf{\frac{5}{12}, \frac{5}{10}, \frac{5}{8}}$

25.
$$\begin{array}{r} 20\text{ mm} \\ \times\quad 3\phantom{\text{ mm}} \\ \hline \mathbf{60\text{ mm}} \end{array}$$

26. **Frequency Table**

Score	Tally	Frequency
51–60	|||	3
61–70	||||	4
71–80	||||	4
81–90	~~||||~~ ||	7
91–100	||	2

27. **45°**

28. **140°**

29. **Obtuse angle**

30. **∠*EOA*** **(or ∠*AOE*)**

Lesson 64, Warm-Up

a. $\mathbf{\frac{1}{3}}$

b. $\mathbf{\frac{1}{4}}$

c. $\mathbf{\frac{1}{5}}$

d. **50 hours**

e. **25**

f. **100**

Problem Solving

$7 \times 7 = 49$
$\sqrt{49} = \mathbf{7}$

Lesson 64, Lesson Practice

a. **Ones**

b. **Tens**

c. **Hundredths**

d. **Tenths**

e. **3 dollars, 8 dimes, 4 pennies**

f. **\$12.60**

g. **\$6.10**

Lesson 64, Mixed Practice

1.
$$\begin{array}{r} {}^{1\,1\;\;1\,1}\\ 116{,}521 \\ +\ 253{,}479 \\ \hline \mathbf{370{,}000} \end{array}$$

2.
$$\begin{array}{r} \$\overset{29}{\cancel{30}}.\ \overset{15}{\cancel{6}}\,{}^{1}3 \\ -\ \$17.\ 8\ 5 \\ \hline \mathbf{\$12.\ 7\ 8} \end{array}$$

3. $\begin{array}{r} 16 \\ \times\ 30 \\ \hline 480 \text{ seats} \end{array}$ $\quad$ $\begin{array}{r} 4\ \overset{7}{\cancel{8}}\,{}^{1}0 \text{ seats} \\ -\ \ 2\ 1 \text{ seats} \\ \hline \mathbf{4\ 5\ 9 \text{ seats}} \end{array}$

4. $\begin{array}{r} \mathbf{54 \text{ pages}} \\ 6\overline{)324} \\ \underline{30} \\ 24 \\ \underline{24} \\ 0 \end{array}$

5. $70 \times 50 = \mathbf{3500}$

6. $\frac{10}{10} - \frac{3}{10} = \mathbf{\frac{7}{10}}$

$\frac{7}{10} \times \frac{100}{1} = \frac{700}{10}$

$\begin{array}{r} \mathbf{70\%} \\ 10\overline{)700} \\ \underline{70} \\ 00 \\ \underline{0} \\ 0 \end{array}$

7. **2 decimal places**

8. **3 dollars, 2 dimes, 5 pennies**

9. **$4.80**

10. $\begin{array}{r} \mathbf{3\frac{1}{8}} \\ 8\overline{)25} \\ \underline{24} \\ 1 \end{array}$

11. Factors of 20: 1, 2, 4, 5, 10, 20

Factors of 30: 1, 2, 3, 5, 6, 10, 15, 30

Common factors: **1, 2, 5, 10**

12. **10:30 a.m.**

13. $\begin{array}{r} 3\ \overset{5}{\cancel{6}}\,{}^{1}0 \\ -\ 1\ 5\ 3 \\ \hline 2\ 0\ 7 \end{array}$

$a = \mathbf{207}$

14. $\begin{array}{r} 175 \\ 5\overline{)875} \\ \underline{5} \\ 37 \\ \underline{35} \\ 25 \\ \underline{25} \\ 0 \end{array}$

$m = \mathbf{175}$

15. $\frac{5}{5} - \frac{3}{5} = \frac{2}{5}$

$f = \mathbf{\frac{2}{5}}$

16. $1 - z = 1$

$1 - 1 = 0$

$z = \mathbf{0}$

17. $\begin{array}{r} \mathbf{\$5.08} \\ 6\overline{)\$30.48} \\ \underline{30} \\ 0\ 4 \\ \underline{0} \\ 48 \\ \underline{48} \\ 0 \end{array}$

18. $\begin{array}{r} \mathbf{26\ R\ 26} \\ 60\overline{)1586} \\ \underline{120} \\ 386 \\ \underline{360} \\ 26 \end{array}$

19. $\begin{array}{r} {}^{2\ 3} \\ \$4.34 \\ \$0.26 \\ \$5.58 \\ \$9.47 \\ \$6.23 \\ +\ \$0.65 \\ \hline \mathbf{\$26.53} \end{array}$

20. $5 \times 4 \times 3 \times 2 \times 1 \times 0 = \mathbf{0}$

21. $\begin{array}{r} 6\frac{3}{3} \\ -\ 3\frac{2}{3} \\ \hline \mathbf{3\frac{1}{3}} \end{array}$

22. $$\begin{array}{r} 1\frac{1}{3} \\ +\ 2\frac{2}{3} \\ \hline 3\frac{3}{3} \end{array} = \mathbf{4}$$

23. $$\begin{array}{r} 3\frac{4}{4} \\ -\ 3\frac{3}{4} \\ \hline \mathbf{\frac{1}{4}} \end{array}$$

24. $$\begin{array}{r} \overset{1}{1}2\text{ mm} \\ \times\ \ 5\quad\ \ \\ \hline \mathbf{60\text{ mm}} \end{array}$$

25. (a) $3\overline{)10}$ = 3 **R 1**
$$\begin{array}{r} 10 \\ -9 \\ \hline 1 \end{array}$$

(b) $3\overline{)100}$ = 33 **R 1**
$$\begin{array}{r} 100 \\ -9\ \ \\ \hline 10 \\ -9 \\ \hline 1 \end{array}$$

26. $1\frac{1}{4}$ in. + $1\frac{1}{4}$ in. + $1\frac{1}{4}$ in. = $\mathbf{3\frac{3}{4}}$ **in.**

27. Two of the seven letters follow Q in the alphabet, so the probability is $\mathbf{\frac{2}{7}}$.

28. $$\begin{array}{r} 3 \\ 7 \\ 9 \\ 6 \\ +\ 2 \\ \hline \mathbf{27\text{ bags}} \end{array}$$

29. $$\begin{array}{r} 9 \\ 6 \\ +\ 2 \\ \hline \mathbf{17\text{ bags}} \end{array}$$

30. **B. There will be fewer than 16 green pieces.**

LESSON 65, WARM-UP

a. **1**

b. **10**

c. **100**

d. $\mathbf{4\frac{1}{3}; 4\frac{2}{3}}$

e. **6**

f. $\mathbf{\frac{3}{5}}$

g. **8**

Problem Solving

Any right angle in an orientation other than the orientations shown in the textbook is acceptable. Possibilities include:

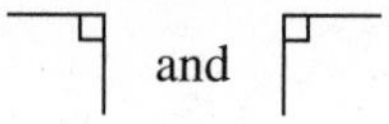

LESSON 65, ACTIVITY 1

1. **0.075 m, 0.1 m, 0.105 m, 0.123 m, 0.15 m**

2. **a.** 0.15 m > 0.123 m **b.** 0.075 m < 0.1 m
c. 0.15 m > 0.075 m **d.** 0.105 m < 0.15 m

3. a. **15 cm**
b. **123 mm**

4. a. **0.1 m**
b. **0.1 m**

5. a. **0.273 m**
b. **0.25 m**

LESSON 65, ACTIVITY 2

See student work.

Lesson 65, Lesson Practice

a. **A. 4.5 meters**

b. **1.43 m**

c. 30 ÷ 10 = **3 decimeters**

Lesson 65, Mixed Practice

1. **One possibility:**

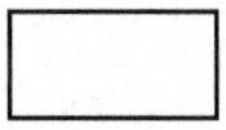

2.

$$\begin{array}{r} 12 \\ \times\ 10 \\ \hline 120 \text{ players} \end{array} \qquad \begin{array}{r} \mathbf{15 \text{ players}} \\ 8\overline{)120} \\ \underline{8} \\ 40 \\ \underline{40} \\ 0 \end{array}$$

3.

$$\begin{array}{r} 100 \text{ yards} \\ 100 \text{ yards} \\ 40 \text{ yards} \\ +\ 40 \text{ yards} \\ \hline \mathbf{280 \text{ yards}} \end{array}$$

4.

$$\begin{array}{r} \mathbf{9 \text{ inches}} \\ 4\overline{)36} \\ \underline{36} \\ 0 \end{array}$$

$$\frac{1}{4} \times \frac{100}{1} = \frac{100}{4}$$

$$\begin{array}{r} \mathbf{25\%} \\ 4\overline{)100} \\ \underline{8} \\ 20 \\ \underline{20} \\ 0 \end{array}$$

5. **45 minutes**

6.

$$\begin{array}{r} 700 \\ +\ 800 \\ \hline \mathbf{1500} \end{array}$$

7. (a) $\mathbf{\frac{1}{10}}$

(b) $\mathbf{\frac{9}{10}}$

8. **1.32 m**

9. 0.5 m = **5 decimeters**

10. $\mathbf{\frac{3}{8}, \frac{5}{10}, \frac{2}{3}, \frac{4}{4}}$

11. 1, 2, 5, 10
4 different factors

12.

$$\begin{array}{r} \mathbf{3\frac{3}{4}} \\ 4\overline{)15} \\ \underline{12} \\ 3 \end{array}$$

13. **543**

14.

$$\begin{array}{r} \overset{4}{\cancel{5}}\,\overset{9}{\cancel{0}}\,{}^{1}0 \\ -\ 3\ 9\ 5 \\ \hline \mathbf{1\ 0\ 5} \end{array}$$

15.

$$\begin{array}{r} {}^{1\,1\ \ 2\,1} \\ 36{,}195 \\ 17{,}436 \\ +\ 42{,}374 \\ \hline \mathbf{96{,}005} \end{array}$$

16.

$$\begin{array}{r} \overset{3}{\cancel{4}}\,\overset{10}{\cancel{1}},\overset{9}{\cancel{0}}\,{}^{1}2\ 6 \\ -\ 3\ 9,\ 5\ 4\ 3 \\ \hline \mathbf{1\ \ 4\ 8\ 3} \end{array}$$

17.

$$\begin{array}{r} 608 \\ \times\ 479 \\ \hline 5\ 472 \\ 42\ 560 \\ 243\ 200 \\ \hline \mathbf{291{,}232} \end{array}$$

18.

$$\begin{array}{r} \mathbf{659\ R\ 1} \\ 4\overline{)2637} \\ \underline{24} \\ 23 \\ \underline{20} \\ 37 \\ \underline{36} \\ 1 \end{array}$$

19.
$$\begin{array}{r} \mathbf{\$0.84} \\ 40\overline{)\$33.60} \\ \underline{32\ 0} \\ 1\ 60 \\ \underline{1\ 60} \\ 0 \end{array}$$

20.
$$\begin{array}{r} \mathbf{168} \\ 20\overline{)3360} \\ \underline{20} \\ 136 \\ \underline{120} \\ 160 \\ \underline{160} \\ 0 \end{array}$$

21. $3\frac{3}{8} + 5\frac{5}{8} = 8\frac{8}{8} = \mathbf{9}$

22. $4\frac{8}{8} - 3\frac{3}{8} = \mathbf{1\frac{5}{8}}$

23. $3\frac{3}{4} - 3 = \mathbf{\frac{3}{4}}$

24.
$$\begin{array}{r} 42 \\ \times\ \ 6 \\ \hline 252 \end{array} \qquad \begin{array}{r} 252 \\ \times\ \ 20 \\ \hline \mathbf{5040} \end{array}$$

25.
$$\begin{array}{r} \$5.63 \\ +\ \$12.00 \\ \hline \$17.63 \end{array} \qquad \begin{array}{r} \overset{19}{\$2\not{0}}.\overset{9}{\not{0}}{}^{1}0 \\ -\ \$17.\ 6\ 3 \\ \hline \mathbf{\$2.\ 3\ 7} \end{array}$$

26.
$$\begin{array}{r} 12 \\ \times\ \ 2 \\ \hline 24 \text{ eggs} \end{array} \qquad \begin{array}{r} 6 \text{ eggs} \\ 2\overline{)12} \\ \underline{12} \\ 0 \end{array} \qquad \begin{array}{r} \overset{1}{2}4 \text{ eggs} \\ +\ \ 6 \text{ eggs} \\ \hline \mathbf{30 \text{ eggs}} \end{array}$$

27. **1080 is divisible by all six listed numbers.**

28. $\mathbf{\frac{2}{4}\left(\text{or } \frac{1}{2}\right)}$

29. $\mathbf{\frac{3}{4}}$

30. $\frac{4}{4} = \mathbf{1}$

Lesson 66, Warm-Up

a. **Observe student actions.**

b. **Observe student actions.**

c. **Observe student actions.**

d. $\mathbf{3\frac{1}{5}; 3\frac{2}{5}}$

e. **3**

f. $\mathbf{\frac{7}{10}}$

g. **138**

h. **32**

Problem Solving

Because the product is greater than 200, both the ones digit of the top factor and the bottom factor must be large. Since the ones digit of the product is 2, the ones digits of the factors must be 8 and 9.

$$\begin{array}{r} \mathbf{2\underline{9}} \\ \times\ \ \mathbf{\underline{8}} \\ \hline \mathbf{2\underline{3}2} \end{array} \qquad \begin{array}{r} \mathbf{2\underline{8}} \\ \times\ \ \mathbf{\underline{9}} \\ \hline \mathbf{2\underline{5}2} \end{array}$$

Lesson 66, Lesson Practice

a. **276 mm, 27.6 cm; Accept answers within the range of 271–281 mm (27.1–28.1 cm).**

b. **Answers may vary. Standard copy paper is 216 mm (21.6 cm) wide.**

c. **25 mm, 2.5 cm**

d. **33 mm, 3.3 cm**

e. **18 mm, 1.8 cm**

f. **0.1**

g. **0.9**

h. **1.5**

i. **2.2**

j. **2.8**

k. **3.4**

LESSON 66, MIXED PRACTICE

1. $\begin{array}{r} \textbf{10 tens} \\ 10\overline{)100} \\ \underline{10} \\ 00 \\ \underline{00} \\ 0 \end{array}$

2. **2750**

3. $800 - 300 = \textbf{500}$

4. $\begin{array}{r} \textbf{5 pounds each} \\ 4\overline{)20} \\ \underline{20} \\ 0 \end{array}$

 $5 \times 6 = \textbf{30 pounds}$

5. $\begin{array}{r} \overset{1}{2}1 \\ + 9 \\ \hline 30 \end{array}$ $\begin{array}{r} \textbf{15 inches} \\ 2\overline{)30} \\ \underline{2} \\ 10 \\ \underline{10} \\ 0 \end{array}$

6. $\dfrac{3}{5}$ ⊘ $\dfrac{4}{9}$

7. **1**

8. (a) $\dfrac{\textbf{3}}{\textbf{10}}$

 (b) $\dfrac{\textbf{7}}{\textbf{10}}$

9. **1.4 cm**

10. 1 decimeter ⊘ 1 centimeter

11. $\begin{array}{r} \textbf{5}\tfrac{\textbf{3}}{\textbf{10}} \\ 10\overline{)53} \\ \underline{50} \\ 3 \end{array}$

12. $\begin{array}{r} \overset{1}{2}00 \text{ yards} \\ 200 \text{ yards} \\ 60 \text{ yards} \\ + 60 \text{ yards} \\ \hline \textbf{520 yards} \end{array}$

13. $\begin{array}{r} 40 \text{ mm} \\ + 35 \text{ mm} \\ \hline \textbf{75 mm} \end{array}$

14. $\begin{array}{r} \overset{2\;1}{8}\overset{1\,1}{7,864} \\ 46,325 \\ + 39,784 \\ \hline \textbf{173,973} \end{array}$

15. $\begin{array}{r} \overset{2}{3}\overset{1}{4},\overset{0}{1}\,\overset{1}{2}\,5 \\ - 1\,6,0\,8\,6 \\ \hline \textbf{1\,8,0\,3\,9} \end{array}$

16. $\begin{array}{r} \$\overset{3}{4}\,\overset{9}{\emptyset}\,\overset{9}{\emptyset}.\,\overset{9}{\emptyset}\overset{1}{0} \\ - \$3\,9\,8.\,5\,7 \\ \hline \textbf{\$1.\,4\,3} \end{array}$

17. $\begin{array}{r} \textbf{938} \\ 6\overline{)5628} \\ \underline{54} \\ 22 \\ \underline{18} \\ 48 \\ \underline{48} \\ 0 \end{array}$

18. $\begin{array}{r} 807 \\ \times 479 \\ \hline 7\,263 \\ 56\,490 \\ 322\,800 \\ \hline \textbf{386,553} \end{array}$

19. $\begin{array}{r} \$7.00 \\ \times 800 \\ \hline \textbf{\$5600.00} \end{array}$

20. $3\dfrac{2}{3} - \left(2\dfrac{1}{3} + 1 \right)$

 $= 3\dfrac{2}{3} - 3\dfrac{1}{3} = \dfrac{\textbf{1}}{\textbf{3}}$

21. $4 - \left(2 + 1\frac{1}{4}\right)$

$= 4 - 3\frac{1}{4}$

$= 3\frac{4}{4} - 3\frac{1}{4} = \mathbf{\frac{3}{4}}$

22.
$$\begin{array}{r} 36 \\ \times\ 60 \\ \hline 2160 \end{array} \qquad \begin{array}{r} 2160 \\ \times\ \ \ 7 \\ \hline \mathbf{15{,}120} \end{array}$$

23.
$$\begin{array}{r} \$8.00 \\ +\ \$2.07 \\ \hline \$10.07 \end{array} \qquad \begin{array}{r} {}^{1}\ {}^{9}\ \ {}^{9}\ \ \ \\ \$\not{2}\,\not{0}.\,\not{0}{}^{1}0 \\ -\ \$1\,0.\,0\,7 \\ \hline \mathbf{\$9.\,9\,3} \end{array}$$

24.
$$\begin{array}{r} 16 \\ -\ 10 \\ \hline \end{array}$$
6 substitutes

25. 7 + 2 = 9 games

C. 9

26. **Frequency Table**

Number of letters	Tally	Frequency
3	|	1
4	||	2
5	||	2
6	|	1
7	||	2
8	|||	3
9	|	1

27. **10.1**

28. **Check for three sides and a right angle; one possibility:**

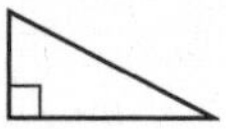

29. **90°**

30. **$\angle EOD$ (or $\angle DOE$)**

LESSON 67, WARM-UP

a. **centimeter**

b. **millimeter**

c. **8**

d. **32**

e. **7**

Problem Solving

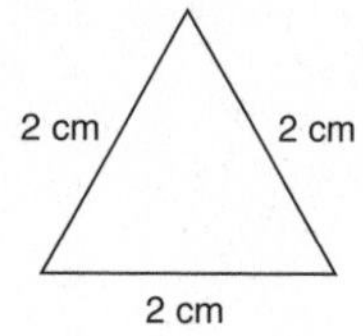

LESSON 67, ACTIVITY

1.

0.1 > 0.01
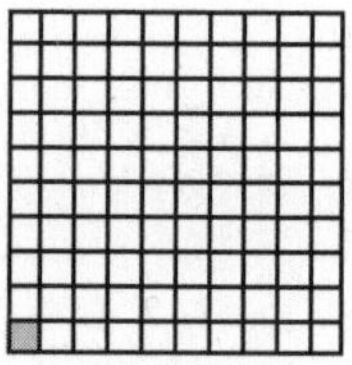

2.
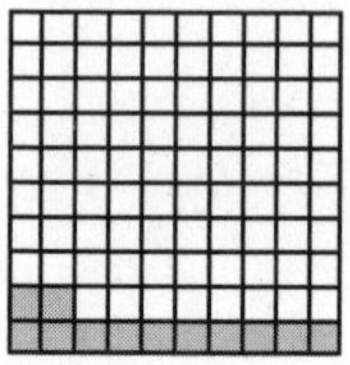
0.12 < 0.21
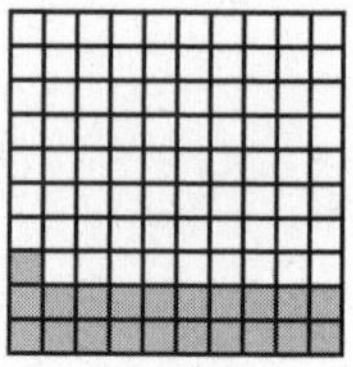

3.
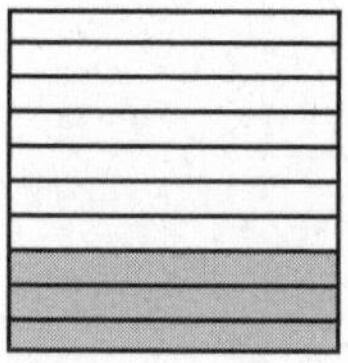
0.3 = 0.30
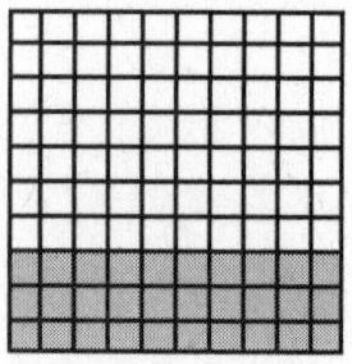

4.
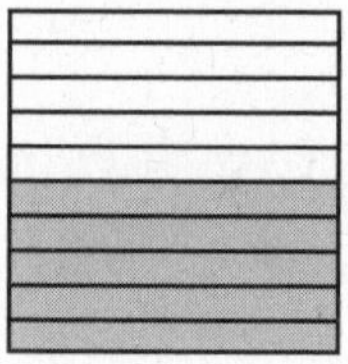
0.5 > 0.05
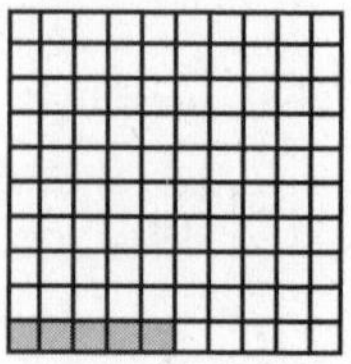

5.
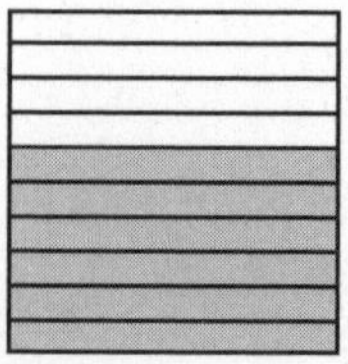
0.6 < 0.67

Lesson 67, Lesson Practice

a. $\frac{7}{10}$; **0.7**

b. $\frac{3}{10}$; **0.3**

c. $2\frac{3}{10}$; **2.3**

d. $\frac{21}{100}$; **0.21**

e. $\frac{79}{100}$; **0.79**

f. **0.9**

g. **0.39**

h. **1.7**

i. **2.99**

j. $\frac{1}{10}$

k. $\frac{3}{100}$

l. $4\frac{9}{10}$

m. $2\frac{54}{100}\left(\text{or } 2\frac{27}{50}\right)$

Lesson 67, Mixed Practice

1.
$$\begin{array}{r} \overset{2}{1}5 \\ \times\ \ 4 \\ \hline 60 \text{ books} \end{array} \qquad \begin{array}{r} \mathbf{12 \text{ books}} \\ 5\overline{)60} \\ \underline{5\ \ } \\ 10 \\ \underline{10} \\ 0 \end{array}$$

2.
$$\begin{array}{r} \mathbf{5 \text{ inches}} \\ 4\overline{)20} \\ \underline{20} \\ 0 \end{array}$$

3.
$$\begin{array}{r} \$2.13 \\ \times\ \ \ \ 2 \\ \hline \$4.26 \end{array} \qquad \begin{array}{r} \$\cancel{10}.\,\cancel{0}{}^{1}0 \\ -\ \$4.\,2\,6 \\ \hline \mathbf{\$5.\,7\,4} \end{array}$$

4. **Two and three tenths; 2.3**

5. $\frac{21}{100}$; **0.21**

6. **0.99**

7. $\frac{7}{10}$; **0.7**

8. **2.9 cm; 29 mm**

9. $\frac{41}{100}$; **0.41**

10.
$$\begin{array}{r} \mathbf{4\tfrac{3}{8}} \\ 8\overline{)35} \\ \underline{32} \\ 3 \end{array}$$

11. $\frac{3}{10}$; **0.3**

12. Factors of 12: 1, 2, 3, 4, 6, 12
Factors of 20: 1, 2, 4, 5, 10, 20
Common factors: **1, 2, 4**

13. $\frac{12}{25} + \frac{12}{25} = \mathbf{\frac{24}{25}}$

14. $3\frac{5}{8} - 1 = \mathbf{2\frac{5}{8}}$

15. $4\frac{8}{8} - 3\frac{5}{8} = \mathbf{1\frac{3}{8}}$

16.
$$\begin{array}{r} \$90.00 \\ \$9.00 \\ +\ \ \$0.01 \\ \hline \$99.01 \end{array} \qquad \begin{array}{r} \$\cancel{10}\,\cancel{0}.\,\cancel{0}{}^{1}0 \\ -\ \ \$9\,9.\,0\,1 \\ \hline \mathbf{\$0.\,9\,9} \end{array}$$

17.

$$\begin{array}{r} 872 \\ 9\overline{)7848} \\ \underline{72} \\ 64 \\ \underline{63} \\ 18 \\ \underline{18} \\ 0 \end{array}$$

18.

$$\begin{array}{r} 52 \\ 70\overline{)3640} \\ \underline{350} \\ 140 \\ \underline{140} \\ 0 \end{array}$$

19.

$$\begin{array}{r} {}^{19}\ {}^{1}0 \\ 20,\ \cancel{1}\,0\ 1 \\ -\ 19,\ 1\ 9\ 1 \\ \hline \mathbf{9\ 1\ 0} \end{array}$$

20. $10 - \left(3 + 1\dfrac{1}{3}\right)$

$= 10 - 4\dfrac{1}{3}$

$= 9\dfrac{3}{3} - 4\dfrac{1}{3} = \mathbf{5\dfrac{2}{3}}$

21. $3\dfrac{1}{4} + \left(2 - 1\dfrac{1}{4}\right)$

$= 3\dfrac{1}{4} + \left(1\dfrac{4}{4} - 1\dfrac{1}{4}\right)$

$= 3\dfrac{1}{4} + \dfrac{3}{4} = 3\dfrac{4}{4} = \mathbf{4}$

22.

$$\begin{array}{r} 24 \\ \times\ \ 8 \\ \hline 192 \end{array} \qquad \begin{array}{r} 192 \\ \times\ \ 50 \\ \hline \mathbf{9600} \end{array}$$

23. $\dfrac{\mathbf{15}}{\mathbf{30}}; \dfrac{\mathbf{25}}{\mathbf{50}}$

24.

$$\begin{array}{r} {}^{2} \\ \$1.20 \\ \$0.90 \\ \$0.90 \\ +\ \$0.50 \\ \hline \mathbf{\$3.50} \end{array}$$

25.

$$\begin{array}{r} {}^{1} \\ \$1.05 \\ +\ \$1.05 \\ \hline \$2.10 \end{array} \qquad \begin{array}{r} {}^{4} \\ \$\cancel{5}.{}^{1}0\ 0 \\ -\ \$2.\ 1\ 0 \\ \hline \mathbf{\$2.\ 9\ 0} \end{array}$$

26. (a) **Hexagon**

(b) 12 inches \div 6 = **2 inches**

27. (a) $\dfrac{\mathbf{1}}{\mathbf{11}}$

(b) $\dfrac{\mathbf{6}}{\mathbf{11}}$

28.

$$\begin{array}{cc} \begin{array}{r} 1 \\ 2 \\ 3 \\ 4 \\ 5 \\ 6 \\ 7 \\ 8 \\ {}_{4}9 \\ 10 \\ +\ 11 \\ \hline 66 \end{array} & \begin{array}{r} 6 \\ 11\overline{)66} \\ \underline{66} \\ 0 \end{array} \end{array}$$

29. 1 decimeter = 10 centimeters

$$\begin{array}{r} 10\ \text{cm} \\ \times\ \ \ 4 \\ \hline \mathbf{40\ cm} \end{array}$$

30. A. **Between 1 and 2 meters (most likely answer)**

LESSON 68, WARM-UP

a. **9**

b. $\dfrac{\mathbf{7}}{\mathbf{12}}$

c. **100**

d. **1000**

e. $4\dfrac{\mathbf{3}}{\mathbf{10}}; 5\dfrac{\mathbf{1}}{\mathbf{10}}$

f. **8**

Problem Solving

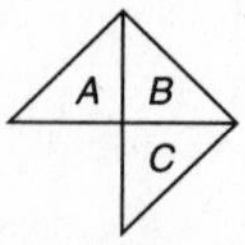

Lesson 68, Lesson Practice

a. **Eight and nine tenths**

b. **Twenty-four and forty-two hundredths**

c. **One hundred twenty-five thousandths**

d. **Ten and seventy-five thousandths**

e. **25.52**

f. **30.1**

g. **7.89**

h. **0.234**

Lesson 68, Mixed Practice

1. **7:50 a.m.**

2.
$$\begin{array}{r} 25 \\ \times\ 40 \\ \hline \textbf{1000 meters} \end{array}$$

3.
$$\begin{array}{r} \textbf{80 pages} \\ 3\overline{)240} \\ 24 \\ \hline 00 \\ 0 \\ \hline 0 \end{array}$$

$$\frac{1}{3} \times \frac{100}{1} = \frac{100}{3}$$

$$\begin{array}{r} \mathbf{33\frac{1}{3}\%} \\ 3\overline{)100} \\ 9 \\ \hline 10 \\ 9 \\ \hline 1 \end{array}$$

4.
$$\begin{array}{r} \$4 \\ 3\overline{)\$12} \\ 12 \\ \hline 0 \end{array} \qquad \begin{array}{r} \textbf{5 tickets} \\ \$4\overline{)\$20} \\ 20 \\ \hline 0 \end{array}$$

5. $\mathbf{\frac{2}{6}, \frac{1}{2}, \frac{3}{4}, \frac{5}{5}}$

6. **12**

7. $\mathbf{\frac{3}{100}}$**; 0.03**

8. **4**

9. **Two hundred seventy-nine thousandths**

10. $\mathbf{1\frac{7}{10}}$**; 1.7**

11. $\mathbf{\frac{3}{100}}$

12.
$$\begin{array}{r} \mathbf{8\frac{1}{10}} \\ 10\overline{)81} \\ 80 \\ \hline 1 \end{array}$$

13.
$$\begin{array}{r} 100\text{ mm} \\ -\ 30\text{ mm} \\ \hline \textbf{70 mm} \end{array}$$

14.
$$\begin{array}{r} {\scriptstyle 11\ \ 1} \\ 87{,}906 \\ 71{,}425 \\ +\ 57{,}342 \\ \hline \mathbf{216{,}673} \end{array}$$

15.
$$\begin{array}{r} 407 \\ \times\ 819 \\ \hline 3\ 663 \\ 4\ 070 \\ 325\ 600 \\ \hline \mathbf{333{,}333} \end{array}$$

16.
$$\begin{array}{r} \mathbf{\$1.46} \\ 6\overline{)\$8.76} \\ 6 \\ \hline 2\ 7 \\ 2\ 4 \\ \hline 36 \\ 36 \\ \hline 0 \end{array}$$

17. $\begin{array}{r} 10 \\ 6)\overline{60} \\ \underline{6} \\ 00 \\ \underline{0} \\ 0 \end{array}$ $\begin{array}{r} \mathbf{60} \\ 10)\overline{600} \\ \underline{60} \\ 00 \\ \underline{0} \\ 0 \end{array}$

18. $\begin{array}{r} \mathbf{146\ R\ 20} \\ 40)\overline{5860} \\ \underline{40} \\ 186 \\ \underline{160} \\ 260 \\ \underline{240} \\ 20 \end{array}$

19. 6×4 inches $=$ **24 inches**

20. $\begin{array}{r} {}^{3\,2} \\ {}_{2}341 \\ 5716 \\ 98 \\ 492 \\ +\ 1375 \\ \hline \mathbf{8022} \end{array}$

21. $\dfrac{7 \times 6}{42} \times \dfrac{5 \times 4}{20}$

$\begin{array}{r} 42 \\ \times\ 20 \\ \hline \mathbf{840} \end{array}$

22. $5\dfrac{1}{4} + \left(2\dfrac{4}{4} - 1\dfrac{1}{4}\right)$

$= 5\dfrac{1}{4} + 1\dfrac{3}{4} = 6\dfrac{4}{4} = \mathbf{7}$

23. $3\dfrac{1}{6} + 2\dfrac{2}{6} + 1\dfrac{3}{6} = 6\dfrac{6}{6} = \mathbf{7}$

24. $\begin{array}{r} 15 \\ 20)\overline{300} \\ \underline{20} \\ 100 \\ \underline{100} \\ 0 \end{array}$

$w = \mathbf{15}$

25. $365 \times 1 \ \boxed{=}\ 365 \div 1$

$\qquad 365 \qquad\qquad 365$

26. $\begin{array}{r} \mathbf{\$3,000} \\ 10)\overline{\$30,000} \\ \underline{30} \\ 0\ 0 \\ \underline{0} \\ 00 \\ \underline{0} \\ 00 \\ \underline{0} \\ 0 \end{array}$

27. $\triangle XYZ$

28. (a) **Heads, tails**

(b) $\dfrac{1}{2}$

29. **Six hundred twenty-five thousandths**

30. **12.75**

LESSON 69, WARM-UP

a. **$10**

b. **$1**

c. **10**

d. **7**

e. **11**

Problem Solving

8 small cubes (4 in the top layer and 4 in the bottom layer)

LESSON 69, LESSON PRACTICE

a. **Juanita**

b. $3.21\ \boxed{<}\ 32.1$

c. **2.04, 2.21, 2.4**

Lesson 69, Mixed Practice

1. $\begin{array}{r} 30 \\ \times\ 30 \\ \hline \textbf{900 tiles} \end{array}$

2. $\begin{array}{r} \overset{1}{\$6.95} \\ +\ \$0.42 \\ \hline \$7.37 \end{array}$ $\quad$ $\begin{array}{r} \$\overset{9}{\not{1}\not{0}}.\ \overset{9}{\not{0}}{}^{1}0 \\ -\ \$7.\ 3\ 7 \\ \hline \mathbf{\$2.\ 6\ 3} \end{array}$

3. $\begin{array}{r} \textbf{20 rolls} \\ 50\overline{)1000} \\ \underline{100} \\ 00 \\ \underline{0} \\ 0 \end{array}$

4. **4 times**

5. **24**

6. Factors of 10: 1, 2, 5, 10

   Factors of 15: 1, 3, 5, 15

   Common factors: **1, 5**

7. 44.4 (>) 4.44

8. **5**

9. $\mathbf{\frac{7}{10}}$**; 0.7**

10. **2.8 cm**

11. **9**

12. 4 inches $-$ $1\frac{1}{2}$ inches

    $3\frac{2}{2} - 1\frac{1}{2} = \mathbf{2\frac{1}{2}}$ **inches**

13. **Ten and five tenths**

14. **15.12**

15. $\begin{array}{r} \mathbf{468} \\ 8\overline{)3744} \\ \underline{32} \\ 54 \\ \underline{48} \\ 64 \\ \underline{64} \\ 0 \end{array}$

16. $\begin{array}{r} \overset{29}{\not{3}\not{0}},\ \overset{9}{\not{0}}\ \overset{9}{\not{0}}{}^{1}0 \\ -\ 29,\ 9\ 2\ 5 \\ \hline \mathbf{7\ 5} \end{array}$

17. $\begin{array}{r} 973 \\ \times\ 536 \\ \hline 5\,838 \\ 29\,190 \\ 486\,500 \\ \hline \mathbf{521{,}528} \end{array}$

18. $\begin{array}{r} \mathbf{37\frac{1}{2}} \\ 2\overline{)75} \\ \underline{6} \\ 15 \\ \underline{14} \\ 1 \end{array}$

19. $\begin{array}{r} \$0.65 \\ \times\quad 10 \\ \hline \mathbf{\$6.50} \end{array}$

20. $\begin{array}{r} \mathbf{\$1.92} \\ 5\overline{)\$9.60} \\ \underline{5} \\ 4\,6 \\ \underline{4\,5} \\ 10 \\ \underline{10} \\ 0 \end{array}$

21. $\begin{array}{r} \mathbf{\$1.81} \\ 30\overline{)\$54.30} \\ \underline{30} \\ 24\,3 \\ \underline{24\,0} \\ 30 \\ \underline{30} \\ 0 \end{array}$

22. $7 - \left(3 + 1\frac{1}{3}\right)$

$= 7 - 4\frac{1}{3}$

$= 6\frac{3}{3} - 4\frac{1}{3} = \mathbf{2\frac{2}{3}}$

23. $5\frac{2}{3} + \left(3\frac{1}{3} - 2\right)$

$= 5\frac{2}{3} + 1\frac{1}{3} = 6\frac{3}{3} = \mathbf{7}$

24.
```
  22
  239
  168
  197
+  95
  699 votes
```

25.
```
   1
  2̸¹3 9
− 1 9 7
    4 2 votes
```

26. (a) **1**

(b) **0**

(c) $\mathbf{\frac{3}{6}}\left(\textbf{or } \mathbf{\frac{1}{2}}\right)$

27. **Tenths**

28. $\mathbf{\frac{33}{100}}$**; 0.33; thirty-three hundredths**

29.
```
   33 1/3
3)100
   9
   10
    9
    1
```

30. **Frequency Table**

Age	Tally	Frequency
4	卌	5
5	卌	5
6	||	2
7	|	1
8	|	1

Lesson 70, Warm-Up

a. **no**

b. **yes**

c. **no**

d. **yes**

e. **25**

f. **50**

g. **75**

Problem Solving

Because the bottom factor is 4, the ones digit of the product must be 2 (no multiples of 4 end in 1 or 7). So the hundreds digit of the product is either 1 or 7. We deduce that the product is 712 (172 is too small to be the product of a three-digit number and 4). We divide 712 by 4 to find the top factor.

```
  178
×   4
  712
```

Lesson 70, Lesson Practice

a. **1.200**

b. **4.080**

c. **0.500**

d. 50 **<** 500

e. 0.4 **>** 0.04

f. 0.50 **=** 0.500

g. 0.2 **=** 0.20000

h. **2¢; $0.02**

i. **50¢; $0.50**

j. **25¢; $0.25**

k. **9¢; $0.09**

l.
$$\begin{array}{r} \overset{1}{} \\ \$0.36 \\ +\ \$0.24 \\ \hline \mathbf{\$0.60} \end{array}$$

m.
$$\begin{array}{r} \overset{0}{\not{1}}\,{}^{1}3\ 8¢ \\ -\quad 7\ 0¢ \\ \hline \mathbf{6\ 8¢} \end{array}$$

n.
$$\begin{array}{r} \$0.25 \\ -\ \$0.05 \\ \hline \mathbf{\$0.20} \end{array}$$

o.
$$\begin{array}{r} \overset{9}{\not{10}}\,{}^{1}0¢ \\ -\quad 8¢ \\ \hline \mathbf{9\ 2¢} \end{array}$$

p.
$$\begin{array}{r} \$0.65 \\ \times\quad 7 \\ \hline \mathbf{\$4.55} \end{array}$$

q.
$$\begin{array}{r} \$0.18 \\ \times\quad 20 \\ \hline \mathbf{\$3.60} \end{array}$$

LESSON 70, MIXED PRACTICE

1. 1 foot × 4 = **4 feet**

2.
$$\begin{array}{r} 1\ \overset{5}{\not{6}}\ \overset{1}{\not{2}}\,{}^{1}0 \\ -\ 1\ 4\ 9\ 2 \\ \hline \mathbf{1\ 2\ 8} \end{array}$$ **years**

3.
$$\begin{array}{r} 300 \\ \times\quad 600 \\ \hline \mathbf{180{,}000} \end{array}$$

4. $3 \times 5 = 15$

$3n = \mathbf{15}$

5. $\frac{3}{3} - \frac{1}{3} = \mathbf{\frac{2}{3}}$

$\frac{2}{3} \times \frac{100}{1} = \frac{200}{3}$

$$\begin{array}{r} \mathbf{66\frac{2}{3}\%} \\ 3\overline{)200} \\ \underline{18} \\ 20 \\ \underline{18} \\ 2 \end{array}$$

6.

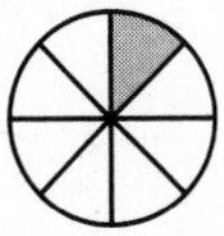

$\frac{1}{8} \times \frac{100}{1} = \frac{100}{8}$

$$\begin{array}{r} 12\frac{4}{8} \text{ or } \mathbf{12\frac{1}{2}\%} \\ 8\overline{)100} \\ \underline{8} \\ 20 \\ \underline{16} \\ 4 \end{array}$$

7.
$$\begin{array}{r} \mathbf{14\frac{2}{7}} \\ 7\overline{)100} \\ \underline{7} \\ 30 \\ \underline{28} \\ 2 \end{array}$$

8. **3**

9. $\mathbf{\frac{1}{100}}$**; 0.01**

10. **5**

11. 3 cm × 2 = 6 cm

6 cm + 3 cm = **9 centimeters**

12. **Sixteen and twenty-one hundredths**

13. **1.50**

14. 3.6 ⊜ 3.60

15.
$$\begin{array}{r} 307 \\ \times\ 593 \\ \hline 921 \\ 27\,630 \\ 153\,500 \\ \hline \mathbf{182{,}051} \end{array}$$

16.
$$\begin{array}{r} \mathbf{153} \\ 5\overline{)765} \\ 5 \\ \hline 26 \\ 25 \\ \hline 15 \\ 15 \\ \hline 0 \end{array}$$

17.
$$\begin{array}{r} \mathbf{\$1.45} \\ 60\overline{)\$87.00} \\ 60 \\ \hline 27\,0 \\ 24\,0 \\ \hline 3\,00 \\ 3\,00 \\ \hline 0 \end{array}$$

18.
$$\begin{array}{r} {}^{3\ 23} \\ 3\,517 \\ 9\,636 \\ 48 \\ 921 \\ {}_2 8\,576 \\ +\ 50{,}906 \\ \hline \mathbf{73{,}604} \end{array}$$

19. $2\frac{3}{10} + 1\frac{3}{10} + \frac{3}{10} = \mathbf{3\frac{9}{10}}$

20. $9\frac{4}{8} + \left(3\frac{8}{8} - 1\frac{7}{8}\right)$

$= 9\frac{4}{8} + 2\frac{1}{8} = \mathbf{11\frac{5}{8}}$

21.
$$\begin{array}{r} 40 \\ \times\ 50 \\ \hline 2000 \end{array} \qquad \begin{array}{r} 2000 \\ \times\ \ 60 \\ \hline \mathbf{120{,}000} \end{array}$$

22.
$$\begin{array}{r} \$84.37 \\ +\ \$12.00 \\ \hline \$96.37 \end{array} \qquad \begin{array}{r} {}^{99\ \ 9} \\ \$1\not{0}\not{0}.\not{0}{}^{1}0 \\ -\ \$96.\,3\,7 \\ \hline \mathbf{\$3.\,6\,3} \end{array}$$

23. (a) **$0.25**

(b) **25¢**

24.
$$\begin{array}{r} 35 \\ +\ 20 \\ \hline \end{array}$$
55 children

25. **Baseball**

26. (a) **Certain**

(b) **Unlikely**

(c) **Impossible**

27. **14**

28. **7 in., 9 in.**

29. **16 in.**

30. $\frac{1}{4} \times \frac{25}{25} = \frac{25}{100}$

0.25; twenty-five hundredths

INVESTIGATION 7

1. Since $43 + 6 = 49$ and $49 + 6 = 55$ and $55 + 6 = 61$, we add 6 to a term to find the next term: **67, 73, 79; arithmetic sequence.**

2. Since $2 \times 2 = 4$ and $4 \times 2 = 8$ and $8 \times 2 = 16$, we multiply a term by 2 to find the next term: **32, 64, 128; geometric sequence.**

3. Since $50 - 2 = 48$ and $48 - 2 = 46$ and $46 - 2 = 44$, we subtract 2 from a term to find the next term: **42, 40, 38; arithmetic sequence.**

4. Since $2 \times 3 = 6$ and $6 \times 3 = 18$ and $18 \times 3 = 54$, we multiply a term by 3 to find the next term: **162, 486, 1458; geometric sequence.**

5. $55 + $8 = **$63;** $63 + $8 = **$71;** $71 + $8 = **$79; arithmetic**

6. $50 \times 2 =$ **100;** $100 \times 2 =$ **200;** $200 \times 2 =$ **400; geometric**

7. **5, 9, 4**

8. **3, 6, 5**

9. **B, U, L**

10. **U, L, B**

11. **an increasing sequence of the counting numbers, with each number recorded twice in a row; 4, 4, 5, 5, …**

12. **The terms alternate between 0's and the positive even numbers. The positive even numbers increase by 2. 8, 0, 10, 0, …**

13. **the letters of the alphabet, skipping every third letter; J, K, M, N, …**

14. **the uppercase letter T, starting upright and rotating 90° clockwise for each term of the sequence; ⊣, ⊥, ⊢, …**

15. **These are the counting numbers 1, 2; followed by 1, 2, 3; followed by 1, 2, 3, 4; and so on. 1, 2, 3, 4, 5, …**

16. $1 + 3 = 4$ and $4 + 5 = 9$ and $9 + 7 = 16$ and $16 + 9 = 25$, so the increasing difference between terms (3, 5, 7, 9, …) forms a sequence of positive odd numbers. To find the next three terms we continue adding the next positive odd number to the previous term: $25 + 11 =$ **36,** $36 + 13 =$ **49,** $49 + 15 =$ **64.** (This is also a sequence of perfect squares.)

17. $2 + 1 = 3$ and $3 + 2 = 5$ and $5 + 3 = 8$ and $8 + 4 = 12$, so the increasing difference between terms (1, 2, 3, 4, …) forms a sequence of counting numbers. To find the next three terms we continue adding the next counting number to the previous term: $12 + 5 =$ **17,** $17 + 6 =$ **23,** $23 + 7 =$ **30.**

18. $5 + 8 =$ **13;** $8 + 13 =$ **21;** $13 + 21 =$ **34**

19. **C. It subtracts 4.**

20. $15 - 4 =$ **11**

21. **D. It divides by 2.**

22. $20 \div 2 =$ **10**

Lesson 71, Warm-Up

a. **yes**

b. **no**

c. $\mathbf{3\frac{1}{3}}$

d. $\mathbf{33\frac{1}{3}}$

e. **11**

Problem Solving

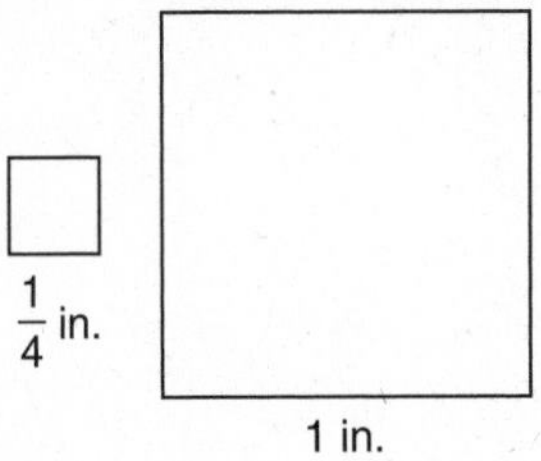

$\frac{1}{2} + \frac{1}{2} + \frac{1}{2} + \frac{1}{2} + \frac{1}{4} + \frac{1}{4} + \frac{1}{4} + \frac{1}{4}$

$+ 1 + 1 + 1 + 1 =$ **7 in.**

Lesson 71, Activity

1. 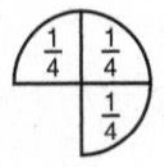

(a) $\mathbf{\frac{3}{4}}$

(b) $25\% + 25\% + 25\% =$ **75%**

(c) $0.25 + 0.25 + 0.25 =$ **0.75**

2. 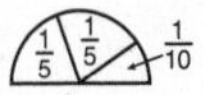

(a) $\mathbf{\frac{1}{2}}$

(b) $20\% + 20\% + 10\% =$ **50%**

(c) $0.2 + 0.2 + 0.1 =$ **0.5**

3. (a) $3\overline{)100}$ = $\mathbf{33\frac{1}{3}}$

```
   33 1/3
3)100
   9
   10
    9
    1
```

(b) $\mathbf{33\frac{1}{3}\%}$

4. (a) $3\overline{)1,000,000}$ = **333,333**   **3** is repeated.

$$\begin{array}{r} 333{,}333 \\ 3\overline{)1{,}000{,}000} \\ \underline{9} \\ 10 \\ \underline{9} \\ 10 \\ \underline{9} \\ 10 \\ \underline{9} \\ 10 \\ \underline{9} \\ 10 \\ \underline{9} \\ 1 \end{array}$$

(b) **$0.\overline{3}$**

(c) **There is a bar over the 3.**

5. (a)

$$\begin{array}{r} 125 \\ 8\overline{)1000} \\ \underline{8} \\ 20 \\ \underline{16} \\ 40 \\ \underline{40} \\ 0 \end{array}$$

(b) **0.125**

(c) **$12\frac{1}{2}\%$**

6.

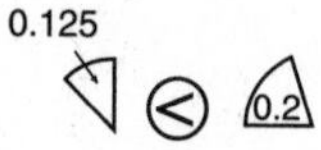

0.125 ⓧ 0.2 → 0.125 $<$ 0.2

7.

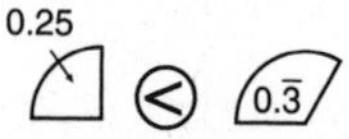

0.25 $<$ $0.\overline{3}$

8.

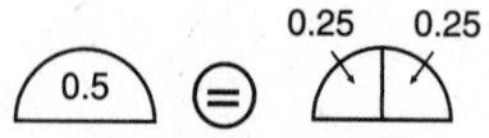

0.5 $=$ 0.25 + 0.25

9.

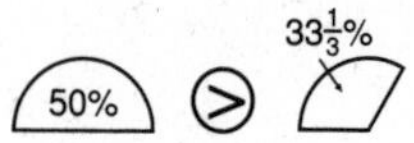

50% $>$ $33\frac{1}{3}\%$

10.

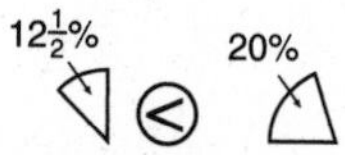

$12\frac{1}{2}\%$ $<$ 20%

Lesson 71, Lesson Practice

a.

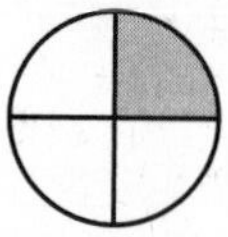

$\frac{25}{100}$ = **0.25**

b. **$\frac{20}{100}$; $\frac{2}{10}$**

c. **$\frac{33}{100}$; 33%; 0.33**

d. $\frac{1}{3}$ $>$ 0.33

Sample answer: 0.33 is not quite $\frac{1}{3}$ because 3 times 33 parts is 99 parts, which is less than the whole square.

Lesson 71, Mixed Practice

1.

$$\begin{array}{r} \overset{1}{}\ \overset{1}{} \\ \$7.98 \\ +\ \$0.49 \\ \hline \mathbf{\$8.47} \end{array}$$

2. 6 × 5 = 30 flowers

$$\begin{array}{r} 30 \\ \times\ \ 4 \\ \hline \mathbf{120\ petals} \end{array}$$

3.

$$\begin{array}{r} 6 \\ 2\overline{)12} \\ \underline{12} \\ 0 \end{array}$$

6 + 1 = **7 years old**

4. 2 half-dollars = 1 dollar
2 × 5 = **10 half-dollars**

5.

$$\begin{array}{r} 15 \\ \times\ 40 \\ \hline 600 \end{array}$$

600 + 7 = **607 nickels**

6. The factors are 1 and 7.
2 factors

7. **B. $\frac{7}{15}$**

8. 30 seconds > 28.27 seconds

Amy

9. **$2\frac{3}{10}$; 2.3**

10. **4**

11. **10.1**

12. $\frac{4}{5} \times \frac{100}{1} = \frac{400}{5}$ cents

```
   80 cents
5)400
  40
   00
   00
    0
```

13.
```
  25 mm
2)50          50
  4         + 25
  10        75 millimeters
  10
   0
```

14. 12.3 ⊜ 12.30

15.
```
   2 2
  $5.37
  $8.95
  $0.71
+ $0.39
 $15.42
```

16.
```
  59  10
  $60. 1'0
- $48. 3 7
  $11. 7 3
```

17.
```
   $9.84
 ×   150
  492 00
  984 00
$1476.00
```

18.
```
   1 1
  $1.75
+ $0.36
  $2.11
```

19.
```
  0
  1'1 5¢
−   8 0¢
    3 5¢
```

20.
```
  $0.76
×    40
 $30.40
```

21.
```
    $0.78
50)$39.00
    35 0
     4 00
     4 00
        0
```

22. $\frac{13}{100} + \frac{14}{100} = \mathbf{\frac{27}{100}}$

23. $7 - \left(6\frac{3}{5} - 1\frac{1}{5}\right)$

$= 7 - 5\frac{2}{5}$

$= 6\frac{5}{5} - 5\frac{2}{5} = \mathbf{1\frac{3}{5}}$

24. **B. It subtracts 2.**

25. 10 − 2 = **8**

26. **20¢; $0.20**

27. **Geometric**
81, 243

28. (a) $\mathbf{\frac{4}{10}}$ $\left(\textbf{or } \mathbf{\frac{2}{5}}\right)$

(b) $\mathbf{\frac{6}{10}}$ $\left(\textbf{or } \mathbf{\frac{3}{5}}\right)$

29. $\mathbf{\frac{4}{10}; \frac{40}{100}}$

30. **The figure drawn should be a rectangle or square.**

LESSON 72, WARM-UP

a. 150 minutes

b. 4200

c. yes

d. 25

e. 6; 4; 3

f. 4

Problem Solving
Adam, Barbara, Conrad
Adam, Barbara, Debby
Adam, Conrad, Debby
Barbara, Conrad, Debby

LESSON 72, LESSON PRACTICE

a.
3 cm

4 cm

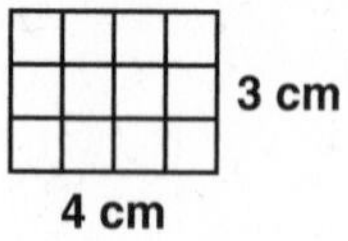

3 cm × 4 cm = **12 sq. cm**

b.
3 ft

3 ft

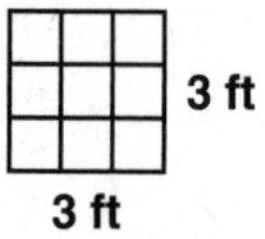

3 ft × 3 ft = **9 sq. ft**

c.
2 in.

5 in.

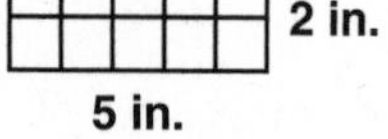

2 in. × 5 in. = **10 sq. in.**

d.
2 m

1 m

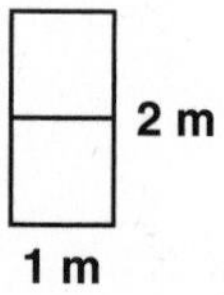

2 m × 1 m = **2 sq. m**

e. 12 ft × 10 ft = **120 sq. ft**

f. Answers may vary.

LESSON 72, MIXED PRACTICE

1. $12 \times 40¢ = t$

$$
\begin{array}{r}
12 \\
\times\ \ 40¢ \\
\hline
480¢
\end{array}
$$ or **$4.80**

2. $4C = \$10.00$

$$
\begin{array}{r}
\$2.50 \\
4)\overline{\$10.00} \\
\underline{8} \\
2\,0 \\
\underline{2\,0} \\
00 \\
\underline{0} \\
0
\end{array}
$$

3. The rule is "Count up by one, count up by three, count up by one, count up by three . . .".
16, 17, 20

4.
$$
\begin{array}{r}
80 \text{ read} \\
3)\overline{240} \\
\underline{24} \\
00 \\
\underline{0} \\
0
\end{array}
\qquad
\begin{array}{r}
\overset{1}{2}^{1}4\,0 \\
-\ \ \ 8\,0 \\
\hline
1\,6\,0 \text{ pages}
\end{array}
$$

$$\frac{3}{3} - \frac{1}{3} = \frac{2}{3}$$

$$\frac{2}{3} \times \frac{100}{1} = \frac{200}{3}$$

$$
\begin{array}{r}
66\tfrac{2}{3}\% \\
3)\overline{200} \\
\underline{18} \\
20 \\
\underline{18} \\
2
\end{array}
$$

5. 100 cm × 5 = **500 centimeters**

6. **Twelve and twenty-five hundredths**

7. $\dfrac{6}{12}$

8. **1, 2, 4, 8, 16**

9. **10.12 seconds**

10. **6**

11.
$$\begin{array}{r} \mathbf{33\tfrac{1}{3}} \\ 3\overline{)100} \\ \underline{9} \\ 10 \\ \underline{9} \\ 1 \end{array}$$

12.
$$\begin{array}{r} \overset{8}{\not{9}}\,{}^{1}0 \text{ mm} \\ -\ 3\ 5 \text{ mm} \\ \hline \mathbf{5\ 5\ mm} \end{array}$$

13.
$$\begin{array}{r} \overset{1\,2}{\$10.35} \\ \$5.18 \\ \$0.08 \\ \$11.00 \\ +\ \ \$0.97 \\ \hline \mathbf{\$27.58} \end{array}$$

14.
$$\begin{array}{r} \$\overset{7}{\not{8}}\,\overset{9}{\not{0}}.\,\overset{9}{\not{0}}\,{}^{1}0 \\ -\ \$7\,2.\,4\,7 \\ \hline \mathbf{\$7.\ 5\ 3} \end{array}$$

15.
$$\begin{array}{r} \$4.97 \\ \times\ \ \ \ \ 6 \\ \hline \mathbf{\$29.82} \end{array}$$

16.
$$\begin{array}{r} 375 \\ \times\ 548 \\ \hline 3\,000 \\ 15\,000 \\ 187\,500 \\ \hline \mathbf{205{,}500} \end{array}$$

17.
$$\begin{array}{r} \mathbf{\$5.79} \\ 7\overline{)\$40.53} \\ \underline{35} \\ 5\,5 \\ \underline{4\,9} \\ 63 \\ \underline{63} \\ 0 \end{array}$$

18.
$$\begin{array}{r} \mathbf{89} \\ 60\overline{)5340} \\ \underline{480} \\ 540 \\ \underline{540} \\ 0 \end{array}$$

19.
$$\begin{array}{r} 200 \\ 30\overline{)6000} \\ \underline{60} \\ 00 \\ \underline{0} \\ 00 \\ \underline{0} \\ 0 \end{array}$$

$m = \mathbf{200}$

20. $3\frac{3}{8} + 1\frac{1}{8} + 4\frac{4}{8} = 8\frac{8}{8} = \mathbf{9}$

21. $7\frac{3}{4} - \left(4\frac{4}{4} - 1\frac{1}{4}\right)$

$= 7\frac{3}{4} - 3\frac{3}{4} = \mathbf{4}$

22. 55.5 (>) 5.55

23. $4\frac{1}{10} + 5\frac{1}{10} + 10\frac{1}{10} = \mathbf{19\frac{3}{10}}$

24. $10 - \left(4 + 1\frac{1}{8}\right)$

$= 9\frac{8}{8} - 5\frac{1}{8} = \mathbf{4\frac{7}{8}}$

25. 4 cm + 4 cm + 2 cm + 2 cm = **12 cm**

26. 4 cm × 2 cm = **8 sq. cm**

27. **140°**

28. **One possible arrangement of sectors:**

29.
$$\begin{array}{r} \$3 \\ \$2 \\ \$5 \\ \$_{1}2 \\ \$12 \\ \$2 \\ +\ \ \$2 \\ \hline \mathbf{\$28} \end{array}$$

30. (a) **Estimate is acceptable.**

(b) **The clerk's calculation needs to be exact.**

Lesson 73, Warm-Up

a. **30 in.**

b. **60**

c. **31**

d. **103**

e. $\mathbf{\frac{3}{10}}$

f. $\mathbf{7\frac{2}{3}; 8\frac{1}{3}}$

g. **8**

Problem Solving

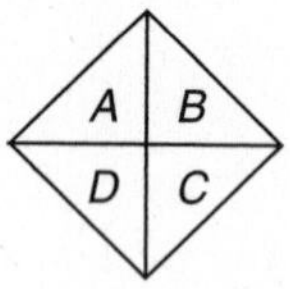

Lesson 73, Lesson Practice

a.
$$\begin{array}{r} \overset{1}{3}.4 \\ 6.7 \\ +\ \overset{1}{1}1.3 \\ \hline \mathbf{21.4} \end{array}$$

b.
$$\begin{array}{r} \overset{1}{4}.63 \\ 2.5 \\ +\ 0.46 \\ \hline \mathbf{7.59} \end{array}$$

c.
$$\begin{array}{r} \overset{1}{9}.62 \\ \overset{1}{1}2.5 \\ +\ \ 3.7 \\ \hline \mathbf{25.82} \end{array}$$

d.
$$\begin{array}{r} 3.\,\overset{5}{\cancel{6}}{}^{1}4 \\ -\ 1.\,4\,6 \\ \hline \mathbf{2.\,1\,8} \end{array}$$

e.
$$\begin{array}{r} \overset{4}{\cancel{5}}.{}^{1}3\,7 \\ -\ 1.\,6\,0 \\ \hline \mathbf{3.\,7\,7} \end{array}$$

f.
$$\begin{array}{r} 0.436 \\ -\ 0.200 \\ \hline \mathbf{0.236} \end{array}$$

g.
$$\begin{array}{r} \mathbf{4.20} \\ \mathbf{+\ 2.65} \\ \hline \mathbf{6.85} \end{array}$$

h.
$$\begin{array}{r} \mathbf{6.75} \\ \mathbf{-\ 4.50} \\ \hline \mathbf{2.25} \end{array}$$

i.
$$\begin{array}{r} \overset{1}{2}.4\text{ cm} \\ 2.4\text{ cm} \\ 2.4\text{ cm} \\ +\ 2.4\text{ cm} \\ \hline \mathbf{9.6\ cm} \end{array}$$

j.
$$\begin{array}{r} \overset{1}{0}.8\text{ mile} \\ +\ 0.8\text{ mile} \\ \hline \mathbf{1.6\ miles} \end{array}$$

Lesson 73, Mixed Practice

1. $8 \times 5 = 40$ stamps

$$\begin{array}{r} \$0.37 \\ \times\ \ \ \ 40 \\ \hline \mathbf{\$14.80} \end{array}$$

2. Brother: 18 years old

Ling: $18 \div 2 = 9$ years old

Sister: $9 - 2 =$ **7 years old**

3.
$$\begin{array}{r} \overset{1}{2}3.4\text{ seconds} \\ +\ 50.9\text{ seconds} \\ \hline \mathbf{74.3\ seconds} \end{array}$$

4.
$$\begin{array}{r} 40\text{ ft} \\ \times\ 30\text{ ft} \\ \hline \mathbf{1200\ sq.\ ft} \end{array}$$

5.$$\frac{1}{4} \times \frac{100}{1} = \frac{100}{4}$$

$$\begin{array}{r} 25\% \\ 4\overline{)100} \\ \underline{8} \\ 20 \\ \underline{20} \\ 0 \end{array}$$

6. $\frac{1}{8}$

7. Two and six tenths centimeters

8. Factors of 16: 1, 2, 4, 8, 16

Factors of 20: 1, 2, 4, 5, 10, 20

Common factors: **1, 2, 4**

9. $12 \div 3 = 4$

$2 \times 4 = 8$

$2y = \mathbf{8}$

10. $\overset{7}{\cancel{8}}.^{1}5$ centimeters
$-\ \ 3.\ 7$ centimeters
4. 8 centimeters

11. $12.0 \ \text{\textgreater} \ 1.20$

12. $\overset{4}{\cancel{5}}\overset{12}{\cancel{3}}.^{1}46$
$-\ \ \ \ 5.\ 7\ 0$
4 7. 7 6

13. $\begin{array}{r} \$6.48 \\ \times\ \ \ \ \ 9 \\ \hline \mathbf{\$58.32} \end{array}$

14. $\begin{array}{r} \overset{1}{\ }4.50 \\ +\ 6.75 \\ \hline \mathbf{11.25} \end{array}$

15. $\overset{49}{\cancel{\$5}}.\overset{}{0}{}^{1}0$
$-\ \$0.0\ 5$
$4.9 5

16. $\begin{array}{r} \mathbf{\$1.72} \\ 5\overline{)\$8.60} \\ \underline{5} \\ 3\ 6 \\ \underline{3\ 5} \\ 10 \\ \underline{10} \\ 0 \end{array}$

17. $\begin{array}{r} \mathbf{\$0.43} \\ 20\overline{)\$8.60} \\ \underline{8\ 0} \\ 60 \\ \underline{60} \\ 0 \end{array}$

18. $\begin{array}{r} 378 \\ \times\ 296 \\ \hline 2\ 268 \\ 34\ 020 \\ 75\ 600 \\ \hline \mathbf{111,888} \end{array}$

19. $\begin{array}{r} 800 \\ \times\ \ \ 500 \\ \hline \mathbf{400,000} \end{array}$

20. $\begin{array}{r} 329 \\ 30\overline{)9870} \\ \underline{90} \\ 87 \\ \underline{60} \\ 270 \\ \underline{270} \\ 0 \end{array}$

$w = \mathbf{329}$

21. $12 + 1\frac{1}{2} = \mathbf{13\frac{1}{2}}$

22. $11\frac{2}{2} - 1\frac{1}{2} = \mathbf{10\frac{1}{2}}$

23. $\frac{49}{99} + \frac{49}{99} = \mathbf{\frac{98}{99}}$

24. $2\frac{1}{2}$ hours $+\ 1\frac{1}{2}$ hours $= 3\frac{2}{2} = \mathbf{4\ hours}$

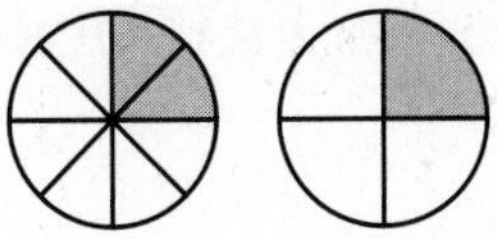

25.
$$\begin{array}{r} \$3.50 \\ \times \quad 4 \\ \hline \mathbf{\$14.00} \end{array}$$

26. 5 + 7 + 6 + 10 + 5 + 2 + 4 = 39

10
39

27. 6 + 10 + 5 + 2 + 4

27
39

28. 10 ft × 10 ft = **100 sq. ft**

29.
$$\begin{array}{r} \mathbf{\$3.00} \\ 5\overline{)\$15.00} \\ \underline{15} \\ 0\,0 \\ \underline{0} \\ 00 \\ \underline{0} \\ 0 \end{array}$$

30. **The figure drawn should be a square.**

Lesson 74, Warm-Up

a. **250 cents**

b. **32**

c. **203**

d. **36**

e. **5**

f. $\mathbf{\frac{3}{8}}$

g. **120**

h. **1**

Problem Solving

27 small cubes (9 cubes in each of the three layers)

Lesson 74, Lesson Practice

a. 1760 yd = 1 mi

$$\begin{array}{r} \mathbf{440\ yards} \\ 4\overline{)1760} \\ \underline{16} \\ 16 \\ \underline{16} \\ 00 \\ \underline{0} \\ 0 \end{array}$$

b. 10 mm = 1 cm
50 mm = **5 cm**

c. 5 feet = 12 in.

$$\begin{array}{r} 12\text{ in.} \\ \times \quad 5\phantom{\text{ in.}} \\ \hline 60\text{ in.} \end{array}$$
60 in. + 1 in. = **61 inches**

d. 1000 m = 1 km

$$\begin{array}{r} 1000 \\ \times \quad 10 \\ \hline \mathbf{10{,}000\ meters} \end{array}$$

Lesson 74, Mixed Practice

1. 6 × 10 = 60 small boxes

$$\begin{array}{r} 12 \\ \times \quad 60 \\ \hline \mathbf{720\ gizmos} \end{array}$$

2.
$$\begin{array}{r} 2.3 \\ +\ 3.5 \\ \hline \mathbf{5.8} \end{array}$$

3. $\mathbf{7g = \$3.43}$

$$\begin{array}{r} \mathbf{\$0.49} \\ 7\overline{)\$3.43} \\ \underline{2\,8} \\ 63 \\ \underline{63} \\ 0 \end{array}$$

4. $\frac{3}{6} \ \text{(=)}\ \frac{6}{12}$

$\frac{1}{2} \quad \frac{1}{2}$

5. 100 cm = 1 m
100 cm × 2 = **200 cm**

6. $\mathbf{\frac{9}{10}}$**; 0.9**

7. **Eleven and two hundredths**

8. 1 yd = 3 ft
1 ft = 12 in.

$$\begin{array}{r} 12 \text{ in.} \\ \times\ 3 \\ \hline 36 \text{ in. in a yard} \end{array}$$

$$\begin{array}{r} 36 \text{ in.} \\ \times\ 3 \\ \hline \mathbf{108\ inches} \end{array}$$

9. $4 \text{ in.} - 2\frac{1}{4} \text{ in.} = 3\frac{4}{4} - 2\frac{1}{4} = \mathbf{1\frac{3}{4}}$ **inches**

10. $$\begin{array}{r} 7 \\ +\ 1\frac{3}{4} \\ \hline \mathbf{8\frac{3}{4}} \end{array}$$

11. $$\begin{array}{r} 3\frac{5}{12} \\ -\ 3\frac{5}{12} \\ \hline \mathbf{0} \end{array}$$

12. $$\begin{array}{r} 3\frac{4}{4} \\ -\ 2\frac{1}{4} \\ \hline \mathbf{1\frac{3}{4}} \end{array}$$

13. $$\begin{array}{r} 16.20 \\ +\ \ 1.25 \\ \hline \mathbf{17.45} \end{array}$$

14. $$\begin{array}{r} \overset{29}{\cancel{30}}.{}^{1}1 \\ -\ 14.\ 2 \\ \hline \mathbf{15.\ 9} \end{array}$$

15. $$\begin{array}{r} \$12.98 \\ \times\ \ \ \ 40 \\ \hline \mathbf{\$519.20} \end{array}$$

16. $$\begin{array}{r} \mathbf{\$7.59} \\ 6\overline{)\$45.54} \\ 42 \\ \hline 3\,5 \\ 3\,0 \\ \hline 54 \\ 54 \\ \hline 0 \end{array}$$

17. $$\begin{array}{r} \mathbf{548} \\ 8\overline{)4384} \\ 40 \\ \hline 38 \\ 32 \\ \hline 64 \\ 64 \\ \hline 0 \end{array}$$

18. $$\begin{array}{r} 12 \\ \times\ 12 \\ \hline 24 \\ 120 \\ \hline \mathbf{144} \end{array}$$

19. $$\begin{array}{r} {}^{3\,2\ 2} \\ \$12.00 \\ \$0.84 \\ \$6.85 \\ \$0.09 \\ \$8.00 \\ \$98.42 \\ +\ \ \$55.26 \\ \hline \mathbf{\$181.46} \end{array}$$

20. $$\begin{array}{r} \mathbf{3\frac{3}{5}} \\ 5\overline{)18} \\ 15 \\ \hline 3 \end{array}$$

21. **2.500**

22. $$\begin{array}{r} \mathbf{6\ inches} \\ 4\overline{)24} \\ 24 \\ \hline 0 \end{array}$$

23. 6 in. × 6 in. = **36 sq. in.**

24. **60¢; $0.60**

25. **Tyler**

26. **Garvey**

27. **Valley**

28. The rule is "The letters of the alphabet backwards, skipping every other letter."
R, P

29. (a) $\mathbf{\frac{1}{3}}$

(b) $\frac{1}{3} \times \frac{100}{1} = \frac{100}{3}$

$$\begin{array}{r} \mathbf{33\frac{1}{3}\%} \\ 3\overline{)100} \\ \underline{9} \\ 10 \\ \underline{9} \\ 1 \end{array}$$

30. 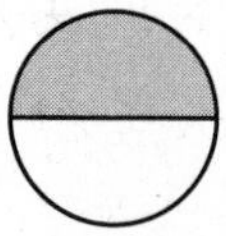

$\frac{1}{2} \times \frac{100}{1} = \frac{100}{2}$

$$\begin{array}{r} \mathbf{50\%} \\ 2\overline{)100} \\ \underline{10} \\ 00 \\ \underline{0} \\ 0 \end{array}$$

Lesson 75, Warm-Up

a. **12 in.; 3 ft**

b. **9**

c. **90**

d. **12**

e. **12**

f. **0**

Problem Solving

We know that 6 steps is about 15 ft. We see that 600 steps is 100 times as long as 6 steps, so we multiply 15 ft by 100: 10 × 15 ft = **1500 ft**

Lesson 75, Lesson Practice

a. $\frac{2}{2} = \mathbf{1}$

b. $\frac{5}{2} = \frac{2}{2} + \frac{2}{2} + \frac{1}{2}$
$= 1 + 1 + \frac{1}{2}$
$= \mathbf{2\frac{1}{2}}$

c. $\frac{5}{3} = \frac{3}{3} + \frac{2}{3}$
$= \mathbf{1\frac{2}{3}}$

d. $\frac{9}{4} = \frac{4}{4} + \frac{4}{4} + \frac{1}{4}$
$= 1 + 1 + \frac{1}{4}$
$= \mathbf{2\frac{1}{4}}$

e. $\frac{3}{2} = \frac{2}{2} + \frac{1}{2}$
$= \mathbf{1\frac{1}{2}}$

f. $\frac{3}{3} = \mathbf{1}$

g. $\frac{6}{3} = \frac{3}{3} + \frac{3}{3}$
$= 1 + 1$
$= \mathbf{2}$

h. $\frac{10}{3} = \frac{3}{3} + \frac{3}{3} + \frac{3}{3} + \frac{1}{3}$
$= 1 + 1 + 1 + \frac{1}{3}$
$= \mathbf{3\frac{1}{3}}$

i. $\frac{4}{2} = \frac{2}{2} + \frac{2}{2}$
$= 1 + 1$
$= \mathbf{2}$

j. $\frac{4}{3} = \frac{3}{3} + \frac{1}{3}$
$= 1 + \frac{1}{3}$
$= \mathbf{1\frac{1}{3}}$

k. $\dfrac{7}{3} = \dfrac{3}{3} + \dfrac{3}{3} + \dfrac{1}{3}$

$\quad = 1 + 1 + \dfrac{1}{3}$

$\quad = 2\dfrac{1}{3}$

l. $\dfrac{15}{4} = \dfrac{4}{4} + \dfrac{4}{4} + \dfrac{4}{4} + \dfrac{3}{4}$

$\quad = 1 + 1 + 1 + \dfrac{3}{4}$

$\quad = 3\dfrac{3}{4}$

m. $\dfrac{4}{5} + \dfrac{4}{5} = \dfrac{8}{5}$

$\quad = \dfrac{5}{5} + \dfrac{3}{5}$

$\quad = 1 + \dfrac{3}{5}$

$\quad = 1\dfrac{3}{5}$

n. $8\dfrac{1}{3} + 8\dfrac{1}{3} + 8\dfrac{1}{3} = 24\dfrac{3}{3} = \mathbf{25}$

o. $\dfrac{5}{8} + \dfrac{3}{8} = \dfrac{8}{8} = \mathbf{1}$

p. $7\dfrac{4}{8} + 8\dfrac{7}{8} = 15\dfrac{11}{8}$

$\quad = 15 + \dfrac{8}{8} + \dfrac{3}{8}$

$\quad = 15 + 1 + \dfrac{3}{8}$

$\quad = 16\dfrac{3}{8}$

q. $2\dfrac{1}{2}$ in. $+ 2\dfrac{1}{2}$ in. $+ 2\dfrac{1}{2}$ in. $+ 2\dfrac{1}{2}$ in. $= 8\dfrac{4}{2}$ in.

$\quad 8\dfrac{4}{2} = 8 + \dfrac{2}{2} + \dfrac{2}{2}$

$\quad = 8 + 1 + 1$

$\quad = \mathbf{10\ in.}$

LESSON 75, MIXED PRACTICE

1.
$$
\begin{array}{r} \$0.49 \\ \times \quad 10 \\ \hline \$4.90 \end{array}
\qquad
\begin{array}{r} \overset{1}{\$4.90} \\ + \ \$2.39 \\ \hline \mathbf{\$7.29} \end{array}
$$

2.
$$
\begin{array}{r} \overset{1}{12}\text{ books} \\ 13\text{ books} \\ + \ 17\text{ books} \\ \hline 42\text{ books} \end{array}
\qquad
\begin{array}{r} \mathbf{14\ books} \\ 3\overline{)42} \\ \underline{3} \\ 12 \\ \underline{12} \\ 0 \end{array}
$$

3. **31.26, 31.62, 32.16, 32.61**

$$
\begin{array}{r} 3\,2.\overset{5}{\cancel{6}}{}^{1}1 \\ - \ 3\,1.\,2\,6 \\ \hline \mathbf{1.\,3\,5} \end{array}
$$

4. **4312**

5. $1\dfrac{3}{10}$; **1.3**

6. $\dfrac{4}{3}$ $\bigcirc\!\!>$ $\dfrac{3}{4}$

7. **4.50**

8. $1\dfrac{1}{10}$; **1.1**

9. $\quad 1\text{ km} = 1000\text{ m}$

$1000\text{ m} \times 5 = \mathbf{5000\ meters}$

10. $1\dfrac{1}{4}$ in. $+ 1\dfrac{3}{4}$ in. $= 2\dfrac{4}{4} = \mathbf{3\ inches}$

11. 12 months $-$ 7 months $=$ 5 months

$\dfrac{\mathbf{5}}{\mathbf{12}}$

12.
$$
\begin{array}{r} \overset{59}{6}\!0.\overset{1}{}4\,5 \\ - \quad 6.\,7\,0 \\ \hline \mathbf{53.\,7\,5} \end{array}
$$

13.
$$
\begin{array}{r} \overset{1}{4}.80 \\ + \ 2.65 \\ \hline \mathbf{7.45} \end{array}
$$

14.
$$
\begin{array}{r} \$6.67 \\ 3\overline{)\$20.01} \\ \underline{18} \\ 2\,0 \\ \underline{1\,8} \\ 21 \\ \underline{21} \\ 0 \end{array}
$$

$d = \mathbf{\$6.67}$

15. $\begin{array}{r} 36 \\ \times\ 80 \\ \hline 2880 \end{array}$ $\begin{array}{r} 2880 \\ \times\ 9 \\ \hline \mathbf{25{,}920} \end{array}$

16. $\begin{array}{r} 506 \\ \times\ 478 \\ \hline 4\,048 \\ 35\,420 \\ 202\,400 \\ \hline \mathbf{241{,}868} \end{array}$

17. $\begin{array}{r} \mathbf{67} \\ 70\overline{)4690} \\ 420 \\ \hline 490 \\ 490 \\ \hline 0 \end{array}$

18. $\begin{array}{r} \$30.75 \\ \times\ 8 \\ \hline \mathbf{\$246.00} \end{array}$

19. $\begin{array}{r} \$\overset{1}{1}0.\overset{2}{0}0 \\ \$8.16 \\ \$0.49 \\ \$2.00 \\ +\ \$0.05 \\ \hline \mathbf{\$20.70} \end{array}$

20. $\frac{4}{5} + \frac{4}{5} = \frac{8}{5}$
$= \frac{5}{5} + \frac{3}{5}$
$= 1 + \frac{3}{5}$
$= \mathbf{1\frac{3}{5}}$

21. $\frac{5}{9} + \frac{5}{9} = \frac{10}{9}$
$= \frac{9}{9} + \frac{1}{9}$
$= 1 + \frac{1}{9}$
$= \mathbf{1\frac{1}{9}}$

22. $16\frac{2}{3} + 16\frac{2}{3} = 32\frac{4}{3}$
$= 32 + \frac{3}{3} + \frac{1}{3}$
$= 32 + 1 + \frac{1}{3}$
$= \mathbf{33\frac{1}{3}}$

23. 1 ft = 12 in.

$\begin{array}{r} 12\text{ in.} \\ \times\ 4 \\ \hline \mathbf{48\text{ in.}} \end{array}$

$\frac{1}{4} \times \frac{100}{1} = \frac{100}{4}$

$\begin{array}{r} \mathbf{25\%} \\ 4\overline{)100} \\ 8 \\ \hline 20 \\ 20 \\ \hline 0 \end{array}$

24. (a) 1 ft × 1 ft = **1 sq. ft**

(b) 12 in. × 12 in. = **144 sq. in.**

25. **19 answers**

26. $\begin{array}{r} \overset{1}{\not 2}{}^{1}0 \\ -\ 1\ 6 \\ \hline \end{array}$ **4 questions**

27. **Quarter**

28. 1 cm + 1 cm + 2 cm + 2 cm = **6 cm (or 60 mm)**

29. 1 cm × 2 cm = **2 sq. cm (or 200 sq. mm)**

30. **40°**

Lesson 76, Warm-Up

a. **10 mm; 100 cm**

b. **yes**

c. **no**

d. **$2.50**

e. $\mathbf{\frac{5}{12}}$

f. **9**

Problem Solving

The bottom factor must be 8 because $7 \times 8 = 56$. The 5 of 56 is regrouped. Since the product is between 500 and 600, we know that the product of 8 and the tens digit of the top factor must be between 45 and 54 (because the product plus 5 must be in the 50's). The only multiple of 8 in that range is 48, so the tens digit of the top factor is 6, and the tens digit of the product is 3.

$$\begin{array}{r} \mathbf{6}7 \\ \times\ 8 \\ \hline 5\mathbf{3}6 \end{array}$$

Lesson 76, Lesson Practice

a. 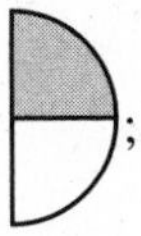 ; $\mathbf{\frac{1}{4}}$

b. $\mathbf{\frac{1}{10}; \frac{1}{10}; \frac{1}{100};}$

$\frac{1}{10} \times \frac{1}{10} = \mathbf{\frac{1}{100}}$

c. $\frac{3}{4} \times \frac{1}{2} = \mathbf{\frac{3}{8}}$

d. $\frac{1}{2} \times \frac{1}{3} = \mathbf{\frac{1}{6}}$

e. $\frac{2}{5} \times \frac{2}{3} = \mathbf{\frac{4}{15}}$

f. $\frac{1}{3} \times \frac{2}{3} = \mathbf{\frac{2}{9}}$

g. $\frac{3}{5} \times \frac{1}{2} = \mathbf{\frac{3}{10}}$

h. $\frac{2}{3} \times \frac{2}{3} = \mathbf{\frac{4}{9}}$

i. $\frac{1}{2} \times \frac{2}{2} = \mathbf{\frac{2}{4}} \left(\textbf{or } \mathbf{\frac{1}{2}}\right)$

j. $\frac{1}{2} \times \frac{1}{3} = \mathbf{\frac{1}{6}}$

k. $\frac{1}{2}$ in. $\times$ $\frac{1}{2}$ in. $=$ $\mathbf{\frac{1}{4}}$ **sq. in.**

Lesson 76, Mixed Practice

1. **17 + *m* = 36**

$$\begin{array}{r} \overset{2}{\not 3}{}^{1}6 \\ -\ 1\ 7 \\ \hline \mathbf{1\ 9} \end{array}$$ **miles**

2. **3*m* = 57**

$$\begin{array}{r} \mathbf{19} \\ 3\overline{)57} \\ \underline{3} \\ 27 \\ \underline{27} \\ 0 \end{array}$$ **miles**

3.
$$\begin{array}{r} \overset{8}{\not 9}.{}^{1}26 \\ -\ 6.\ 34 \\ \hline \mathbf{2.\ 92} \end{array}$$

4. Factors of 6: 1, 2, 3, 6
Factors of 12: 1, 2, 3, 4, 6, 12
Common factors: **1, 2, 3, 6**

5.
$$\begin{array}{r} 6 \\ 3\overline{)18} \\ \underline{18} \\ 0 \end{array}$$

$6 \times 2 = 12$
$2n = \mathbf{12}$

6. 10 cm $\times$ 10 cm $=$ **100 sq. cm**

7. 4.5 ⊜ 4.500

8. $\mathbf{\frac{3}{8}, \frac{1}{2}, \frac{2}{3}, \frac{5}{5}, \frac{4}{3}}$

9. (a)
$$\begin{array}{r} \mathbf{32} \\ 2\overline{)64} \\ \underline{6} \\ 04 \\ \underline{4} \\ 0 \end{array}$$ **squares**

(b)
$$\begin{array}{r} \mathbf{16} \\ 2\overline{)32} \\ \underline{2} \\ 12 \\ \underline{12} \\ 0 \end{array}$$ **squares**

(c) $\frac{1}{2} \times \frac{1}{2} = \mathbf{\frac{1}{4}}$

(d) $\frac{1}{4} \times \frac{100}{1} = \frac{100}{4}$

$$\begin{array}{r} \mathbf{25\%} \\ 4\overline{)100} \\ \underline{8} \\ 20 \\ \underline{20} \\ 0 \end{array}$$

10. $$\begin{array}{r} \overset{6}{\not{7}}\,{}^{1}8 \text{ mm} \\ -\ 2\ 9 \text{ mm} \\ \hline \mathbf{4\ 9\ mm} \end{array}$$

11. $$\begin{array}{r} \overset{1}{\not{2}}\,{}^{1}4.\ 8\ 6 \\ -\ \ 9.\ 7\ 0 \\ \hline \mathbf{1\ 5.\ 1\ 6} \end{array}$$

12. $$\begin{array}{r} \overset{8}{\not{9}}.{}^{1}0\ 6 \\ -\ 3.\ 9\ 0 \\ \hline \mathbf{5.\ 1\ 6} \end{array}$$

13. $$\begin{array}{r} \$4.50 \\ 8\overline{)\$36.00} \\ \underline{32} \\ 4\,0 \\ \underline{4\,0} \\ 00 \\ \underline{0} \\ 0 \end{array}$$

$m = \mathbf{\$4.50}$

14. $$\begin{array}{r} 152 \\ 50\overline{)7600} \\ \underline{50} \\ 260 \\ \underline{250} \\ 100 \\ \underline{100} \\ 0 \end{array}$$

$w = \mathbf{152}$

15. $$\begin{array}{r} \$16.08 \\ \times \quad 9 \\ \hline \mathbf{\$144.72} \end{array}$$

16. $$\begin{array}{r} 638 \\ \times \ 570 \\ \hline 44\ 660 \\ 319\ 000 \\ \hline \mathbf{363{,}660} \end{array}$$

17. $$\begin{array}{r} 3\frac{1}{3} \\ +\ 1\frac{2}{3} \\ \hline 4\frac{3}{3} \end{array} = \mathbf{5}$$

18. $$\begin{array}{r} 1\frac{2}{3} \\ +\ 1\frac{2}{3} \\ \hline 2\frac{4}{3} \end{array} = 2 + \frac{3}{3} + \frac{1}{3}$$
$$= 2 + 1 + \frac{1}{3}$$
$$= \mathbf{3\frac{1}{3}}$$

19. $$\begin{array}{r} 3\frac{5}{5} \\ -\ 1\frac{2}{5} \\ \hline \mathbf{2\frac{3}{5}} \end{array}$$

20. $\frac{1}{2} \times \frac{3}{5} = \mathbf{\frac{3}{10}}$

21. $\frac{1}{3} \times \frac{2}{3} = \mathbf{\frac{2}{9}}$

22. $\frac{1}{2} \times \frac{6}{6} = \mathbf{\frac{6}{12}}\ \left(\mathbf{or\ \frac{1}{2}}\right)$

23. **B. It doubles the number.**

24. $$\begin{array}{r} 12 \\ \times \ 2 \\ \hline \mathbf{24} \end{array}$$

25. $$\begin{array}{r} \mathbf{10} \\ 2\overline{)20} \\ \underline{2} \\ 00 \\ \underline{0} \\ 0 \end{array}$$

26. $\frac{3}{8}$ in. $\times$ $\frac{3}{4}$ in. $=$ $\mathbf{\frac{9}{32}}$ **sq. in.**

27. Length: $\frac{3}{4}$ in. + $\frac{3}{4}$ in. = $\frac{6}{4}$ in.

= $1\frac{2}{4}$ = $1\frac{1}{2}$ in.

Width: $\frac{3}{8}$ in. + $\frac{3}{8}$ in. = $\frac{6}{8}$ in.

= $\frac{3}{4}$ in.

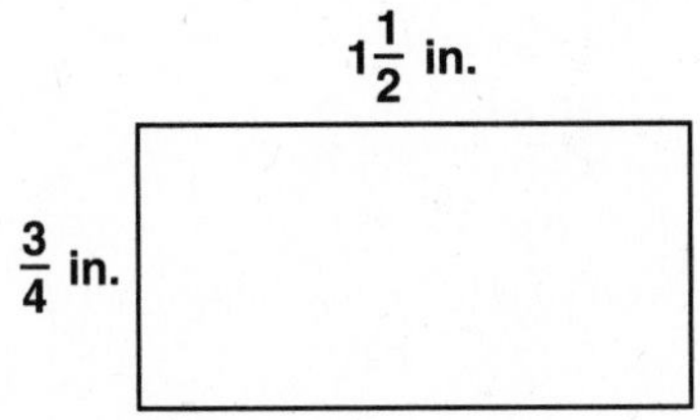

28. (a) **1**

(b) **3** and **4**

29. (a) $\mathbf{\frac{1}{5}}$

(b) $\mathbf{\frac{1}{4}}$

(c) $\mathbf{\frac{1}{20}}$

(d) $\frac{1}{5} \times \frac{1}{4} = \mathbf{\frac{1}{20}}$

30. **1, 2, 4, 5, 10, 20, 25, 50, 100**

Lesson 77, Warm-Up

a. **100 cm; 1000 mm**

b. **25**

c. **250**

Problem Solving

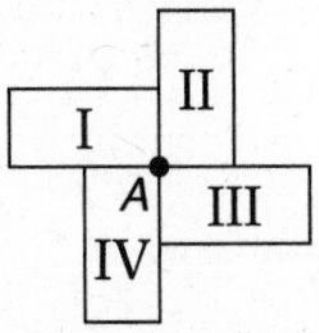

Lesson 77, Lesson Practice

a. 1 lb = 16 oz
16 oz ÷ 2 = **8 oz**

b. 1 kg = 1000 g
1000 g ÷ 2 = **500 g**

c. 1 lb = 16 oz
16 oz × 10 = **160 oz**

d. 2000 lb = 1 tn
2000 lb × 16 = **32,000 lb**

e. 1 g = 1000 mg
1000 ÷ 500 = **2 tablets**

Lesson 77, Mixed Practice

1.
$$\begin{array}{r} 1\,\overset{8}{\cancel{9}}\,{}^{1}0\,9 \\ -\quad 7\,4 \\ \hline \mathbf{1\,8\,3\,5} \end{array}$$

2.
$$\begin{array}{r} \overset{1\,1}{16.9} \\ +\ 23.7 \\ \hline \mathbf{40.6} \end{array}$$

3. **1.23, 1.32, 2.13, 13.2**

4. (a) $4\overline{)36}$ = **9 students**
36
0

(b) $3\overline{)9}$ = **3 students**
9
0

(c) $\mathbf{\frac{3}{36}}$

5. **2000 lb**

6. $\mathbf{\frac{11}{100}}$**; 0.11; 11%**

7. 1 lb = 16 oz
16 oz × 2 = **32 oz**

8. $1 \text{ kg} = 1000 \text{ g}$
$1000 \text{ g} \times 3 = \textbf{3000 grams}$

9.
$$\overset{1}{}3.5 \text{ cm}$$
$$\underline{+\ 4.6 \text{ cm}}$$
$$\textbf{8.1 cm}$$

10. $\dfrac{3}{4} + \dfrac{3}{4} + \dfrac{3}{4} = \dfrac{9}{4}$

$\dfrac{9}{4} = \dfrac{4}{4} + \dfrac{4}{4} + \dfrac{1}{4}$

$\phantom{\dfrac{9}{4}} = 1 + 1 + \dfrac{1}{4}$

$\phantom{\dfrac{9}{4}} = \mathbf{2\dfrac{1}{4}}$

11. $\dfrac{3}{3} + \dfrac{2}{2} = 1 + 1$

$\phantom{\dfrac{3}{3} + \dfrac{2}{2}} = \mathbf{2}$

12. $3\dfrac{5}{8} + 4\dfrac{6}{8} = 7\dfrac{11}{8}$

$7\dfrac{11}{8} = 7 + \dfrac{8}{8} + \dfrac{3}{8}$

$\phantom{7\dfrac{11}{8}} = 7 + 1 + \dfrac{3}{8}$

$\phantom{7\dfrac{11}{8}} = \mathbf{8\dfrac{3}{8}}$

13.
$$\overset{3\ 2}{}{}_{3}463$$
$$2{,}875$$
$$2{,}489$$
$$8{,}897$$
$$\underline{+\ 7{,}963}$$
$$\textbf{22,687}$$

14. $\dfrac{1}{2} \times \dfrac{5}{6} = \mathbf{\dfrac{5}{12}}$

15. $\dfrac{2}{3} \times \dfrac{3}{4} = \mathbf{\dfrac{6}{12}} \left(\text{or } \mathbf{\dfrac{1}{2}}\right)$

16. $\dfrac{1}{2} \times \dfrac{2}{2} = \mathbf{\dfrac{2}{4}} \left(\text{or } \mathbf{\dfrac{1}{2}}\right)$

17.
$$\overset{39\ \ ^{1}0}{}{4\!\!\!/0\ 1\!\!.^{1}3}$$
$$\underline{-\ 26\ 4.\ 7}$$
$$\textbf{13 6. 6}$$

18.
$$\$5.67$$
$$\underline{\times\ \ \ \ \ \ 80}$$
$$\textbf{\$453.60}$$

19.
$$347$$
$$\underline{\times\ 249}$$
$$3\ 123$$
$$13\ 880$$
$$\underline{69\ 400}$$
$$\textbf{86,403}$$

20.
$$50$$
$$\underline{\times\ \ 50}$$
$$\textbf{2500}$$

21.
$$\begin{array}{l} \$5.00 \\ \underline{+\ \$0.04} \\ \$5.04 \end{array} \qquad \overset{\textbf{\$0.84}}{6\overline{)\$5.04}}$$
$$\begin{array}{r} \underline{4\ 8} \\ 24 \\ \underline{24} \\ 0 \end{array}$$

22.
$$\overset{\textbf{8,034 R 3}}{8\overline{)64{,}275}}$$
$$\begin{array}{r} \underline{64} \\ 0\ 2 \\ \underline{0} \\ 27 \\ \underline{24} \\ 35 \\ \underline{32} \\ 3 \end{array}$$

23.
$$\overset{63}{60\overline{)3780}}$$
$$\begin{array}{r} \underline{360} \\ 180 \\ \underline{180} \\ 0 \end{array}$$
$w = \mathbf{63}$

24. **40 mm**

25. $40 \text{ mm} + 40 \text{ mm} + 20 \text{ mm}$
$+\ 20 \text{ mm} = \textbf{120 mm}$

26. $40 \text{ mm} \times 20 \text{ mm} = \textbf{800 sq. mm}$

27. **3, 5, 7, 3**

28. (a) $\mathbf{\dfrac{1}{12}}$

(b) $\mathbf{\dfrac{1}{3}}$

(c) $\mathbf{\dfrac{1}{36}}$

(d) $\dfrac{1}{12} \times \dfrac{1}{3} = \mathbf{\dfrac{1}{36}}$

29. **B. 1 gram**

30. (a) $\mathbf{\frac{3}{16}}$

(b) $\mathbf{\frac{3}{16}}$ **sq. in.**

Lesson 78, Warm-Up

a. **16 oz; 2000 lb; 4000 lb; 6000 lb; 8000 lb**

b. **no**

c. **yes**

d. $\mathbf{2\frac{1}{2}}$

e. **0**

Problem Solving

32 1-inch cubes

Lesson 78, Lesson Practice

a.

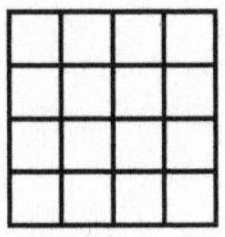

b. $2 \times 2 \times 2 = \mathbf{8}$

c. $\mathbf{3 \times 3 \times 3 \times 3 = 81}$

d. $\mathbf{2 \times 2 \times 2 \times 2 \times 2 = 32}$

e. $\mathbf{11 \times 11 = 121}$

f. $2\overline{)10}$ = 5; 10; 0

$5 \times 5 = 25$

$m^2 = \mathbf{25}$

g. $\mathbf{(2 \times 10^5) + (5 \times 10^4)}$

h. $\mathbf{(3 \times 10^6) + (6 \times 10^5)}$

i. $\mathbf{(6 \times 10^4) + (5 \times 10^2)}$

Lesson 78, Mixed Practice

1. $\frac{1}{2} \times \frac{1}{3} = \mathbf{\frac{1}{6}}$

$\frac{1}{6} \times \frac{100}{1} = \frac{100}{6}$

$6\overline{)100}$ = $16\frac{4}{6}$ = $\mathbf{16\frac{2}{3}\,\%}$

6
40
36
4

2. $\$1\,3\,0\,0$ (regrouped: $\$0\,12\,10\,0$)
$-\ \$8\,6\,0$
$4 4 0

3. $4\overline{)316}$ = **79 pages**

28
36
36
0

4. 2000 lb = 1 ton

2000 lb ÷ 2 = **1000 lb**

5. 1 lb = 16 oz

16 oz ÷ 2 = **8 oz**

6. **B.**

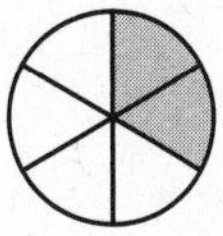

7. **C.** $\mathbf{\frac{16}{30}}$

8. **22 mm; 2.2 cm**

9. Factors of 6: 1, 2, 3, 6

Factors of 8: 1, 2, 4, 8

Common factors: **1, 2**

10. 6.4 cm (regrouped: 5.14 cm)
$-$ 3.9 cm
2.5 cm

11. $\frac{2}{3} + \frac{2}{3} + \frac{2}{3} = \frac{6}{3}$

$\frac{6}{3} = \frac{3}{3} + \frac{3}{3}$
$= 1 + 1$
$= \mathbf{2}$

12. $\frac{3}{3} - \frac{2}{2} = 1 - 1 = \mathbf{0}$

13. $9\frac{4}{10} + 4\frac{9}{10} = 13\frac{13}{10}$

$13\frac{13}{10} = 13 + \frac{10}{10} + \frac{3}{10}$
$= 13 + 1 + \frac{3}{10}$
$= \mathbf{14\frac{3}{10}}$

14.
```
   4.6
+ 3.27
  7.87
```
7.87

15.
```
   39   9
   $40. 0¹0
 − $13. 4 8
   $26. 5 2
```
$26.52

16.
```
  $20.50
 ×     8
 $164.00
```
$164.00

17.
```
     $6.30
 9)$56.70
    54
     2 7
     2 7
       00
        0
        0
```
$6.30

18.
```
    13
  × 13
    39
   130
   169
```
169

19.
```
      58 R 10
 80)4650
    400
     650
     640
      10
```
58 R 10

20.
```
     19 3/5
 5)98
   5
   48
   45
    3
```
$\mathbf{19\frac{3}{5}}$

21. $\frac{3}{4} \times \frac{1}{2} = \mathbf{\frac{3}{8}}$

22. $\frac{3}{2} \times \frac{3}{4} = \mathbf{\frac{9}{8}}$ $\left(\textbf{or } \mathbf{1\frac{1}{8}}\right)$

23. $\frac{1}{3} \times \frac{2}{2} = \mathbf{\frac{2}{6}}$ $\left(\textbf{or } \mathbf{\frac{1}{3}}\right)$

24.
```
   ¹
   1.5
 + 1.5
   3.0 miles
```
3.0 miles

25. **8:07 a.m.**

26. **7, 3, 5, 7**

27. (a) $\mathbf{\frac{2}{7}}$

(b) $\mathbf{\frac{3}{7}}$

(c) $\mathbf{\frac{5}{7}}$

28. 10 cm × 5 cm = **50 sq. cm**

29. **C. Rhombus**

30. $\mathbf{(2 \times 10^7) + (5 \times 10^6)}$

Lesson 79, Warm-Up

a. **100 cm; 1000 m**

b. **5**

c. **50**

d. $\mathbf{3\frac{1}{5}}$

e. **$0.50**

f. $\mathbf{\frac{1}{2}}$

Problem Solving

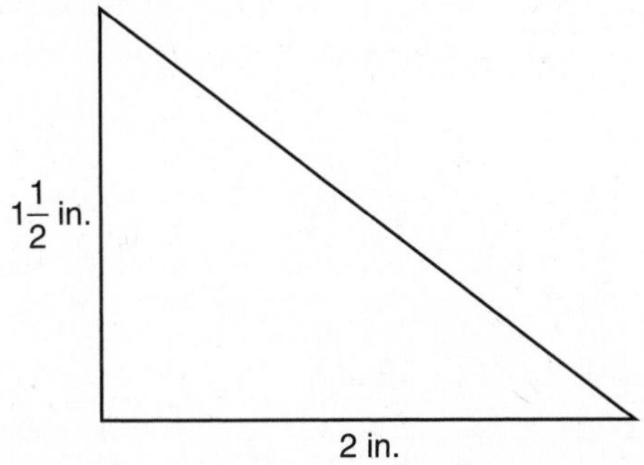

Lesson 79, Lesson Practice

a. $\frac{3}{4} \times \mathbf{\frac{3}{3}} = \frac{9}{12}$

b. $\frac{2}{3} \times \mathbf{\frac{2}{2}} = \frac{4}{6}$

c. $\frac{1}{3} \times \mathbf{\frac{4}{4}} = \frac{4}{12}$

d. $\frac{1}{4} \times \mathbf{\frac{25}{25}} = \frac{25}{100}$

e. $\frac{1}{3} \times \frac{3}{3} = \mathbf{\frac{3}{9}}$

f. $\frac{2}{3} \times \frac{5}{5} = \mathbf{\frac{10}{15}}$

g. $\frac{3}{5} \times \frac{2}{2} = \mathbf{\frac{6}{10}}$

h. $\mathbf{\frac{3}{6}; \frac{2}{6}}$

$\frac{3}{6} + \frac{2}{6} = \mathbf{\frac{5}{6}}$

i. $\frac{3}{5} \times \frac{20}{20} = \mathbf{\frac{60}{100}}$**; 60%**

Lesson 79, Mixed Practice

1.
$$\begin{array}{r} \textbf{40 days} \\ 50\overline{)2000} \\ \underline{200} \\ 00 \\ \underline{0} \\ 0 \end{array}$$

2. 12 in. + 6 in. = **18 inches**

3.
$$\begin{array}{r} \overset{0}{\cancel{\$1}}{}^{1}0.95 \\ -\ \$6.30 \\ \hline \$4.65 \end{array} \qquad \begin{array}{r} \$4.65 \\ \times\ \ 3 \\ \hline \mathbf{\$13.95} \end{array}$$

4.
$$\begin{array}{r} \overset{1\,1\ \,1}{10.15} \\ +\ 29.89 \\ \hline 40.04 \end{array}$$

Forty and four hundredths

5. $\frac{2}{3} \times \mathbf{\frac{3}{3}} = \frac{6}{9}$

6.

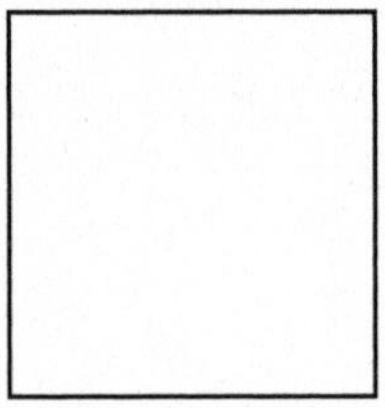

1 in.; 1 square inch

7. Factors of 9: 1, 3, 9
Factors of 12: 1, 2, 3, 4, 6, 12
Common factors: **1, 3**

8. $\frac{3}{4} = \mathbf{\frac{9}{12}}$

$\frac{2}{3} = \mathbf{\frac{8}{12}}$

$\frac{9}{12} + \frac{8}{12} = \frac{17}{12}$

$\frac{17}{12} = \frac{12}{12} + \frac{5}{12}$

$= 1 + \frac{5}{12}$

$= \mathbf{1\frac{5}{12}}$

9.
$$\begin{array}{r} \overset{8}{\cancel{9}}.{}^{1}1\text{ cm} \\ -\ 4.2\text{ cm} \\ \hline \mathbf{4.9\text{ cm}} \end{array}$$

10. $1\frac{1}{5} + 2\frac{2}{5} + 3\frac{3}{5} = 6\frac{6}{5}$

$$6\frac{6}{5} = 6 + \frac{5}{5} + \frac{1}{5}$$
$$= 6 + 1 + \frac{1}{5}$$
$$= \mathbf{7\frac{1}{5}}$$

11. $5 - \left(3\frac{5}{8} - 3\right)$

$$4\frac{8}{8} - \frac{5}{8} = \mathbf{4\frac{3}{8}}$$

12.
$$\begin{array}{r} \overset{9}{\cancel{\$10}}.{}^{1}0\,0 \\ -\ \$0.\,1\,0 \\ \hline \mathbf{\$9.\,9\,0} \end{array}$$

13.
$$\begin{array}{r} \mathbf{\$2.50} \\ 4\overline{)\$10.00} \\ \underline{8} \\ 2\,0 \\ \underline{2\,0} \\ 00 \\ \underline{0} \\ 0 \end{array}$$

14.
$$\begin{array}{r} \$0.64 \\ \times\quad 9 \\ \hline \mathbf{\$5.76} \end{array}$$

15.
$$\begin{array}{r} \overset{29}{\cancel{30}}.{}^{1}4 \\ -\ 24.\,6 \\ \hline 5.\,8 \end{array}$$

$M = \mathbf{5.8}$

16.
$$\begin{array}{r} 2.4 \\ +\ 6.35 \\ \hline 8.75 \end{array}$$

$W = \mathbf{8.75}$

17.
$$\begin{array}{r} 728 \\ 9\overline{)6552} \\ \underline{63} \\ 25 \\ \underline{18} \\ 72 \\ \underline{72} \\ 0 \end{array}$$

$n = \mathbf{728}$

18.
$$\begin{array}{r} \mathbf{6{,}265\ R\ 4} \\ 7\overline{)43{,}859} \\ \underline{42} \\ 1\,8 \\ \underline{1\,4} \\ 45 \\ \underline{42} \\ 39 \\ \underline{35} \\ 4 \end{array}$$

19.
$$\begin{array}{r} 15 \\ \times\ 15 \\ \hline 75 \\ \underline{150} \\ \mathbf{225} \end{array}$$

20.
$$\begin{array}{r} \mathbf{51\ R\ 57} \\ 80\overline{)4137} \\ \underline{400} \\ 137 \\ \underline{80} \\ 57 \end{array}$$

21. $\frac{1}{2} \times \frac{1}{5} = \mathbf{\frac{1}{10}}$

22. $\frac{3}{4} \times \frac{2}{2} = \mathbf{\frac{6}{8}}\ \left(\textbf{or } \mathbf{\frac{3}{4}}\right)$

23. $\frac{3}{5} \times \frac{5}{4} = \mathbf{\frac{15}{20}}\ \left(\textbf{or } \mathbf{\frac{3}{4}}\right)$

24. **D. 350**

25.
$$\begin{array}{r} 200 \\ 350 \\ +\ 400 \\ \hline \end{array}$$
950 ice cream cones

26. $\mathbf{\frac{5}{6}}$

27. $240 + 12 = \mathbf{252}$

28. **C. 15 kilograms**

29.
$$\begin{array}{r} \overset{1}{1}.5 \text{ cm} \\ \times \quad 3 \\ \hline \mathbf{4.5 \text{ cm}} \end{array}$$

30. 500 mg ⓧ 1.0 g

$\frac{1}{2}$ g   1.0 g

Lesson 80, Warm-Up

a. **1000 g; 500 g**

b. **4**

c. **40**

d. $\mathbf{5\frac{1}{3}}$

e. **$5.00**

f. **0**

Problem Solving

Each term is the sum of the two preceding terms.

So the next three terms are

8 + 13 = **21** and 13 + 21 = **34**

and 34 + 21 = **55.**

Lesson 80, Lesson Practice

a. **11, 13, 17, 19**

b. **1, 3, 7, 21; composite; 21 has more than two factors.**

c. **1**

d. **Possibilities include:**

```
×××
×××   ×××××××××
×××
```

Lesson 80, Mixed Practice

1.
$$\begin{array}{r} 12 \\ \times \ \$0.20 \\ \hline \$2.40 \end{array} \qquad \begin{array}{r} \$\overset{1}{\not 2}.\overset{13}{\not 4}{}^{1}0 \\ - \ \$0.9\,6 \\ \hline \mathbf{\$1.4\,4} \end{array}$$

2.
$$\begin{array}{r} \mathbf{500} \text{ pounds} \\ 4\overline{)2000} \\ \underline{20} \\ 00 \\ \underline{0} \\ 00 \\ \underline{0} \\ 0 \end{array}$$

3. Factors of 8: 1, 2, 4, 8

Factors of 12: 1, 2, 3, 4, 6, 12

Common factors: **1, 2, 4**

4. **13, 17, 19**

5. $\frac{3}{4} \times \mathbf{\frac{3}{3}} = \frac{9}{12}$

6. $\mathbf{\frac{3}{6}; \frac{4}{6}}$

$\frac{3}{6} + \frac{4}{6} = \frac{7}{6}$

$\frac{7}{6} = \frac{6}{6} + \frac{1}{6}$

$= 1 + \frac{1}{6}$

$= \mathbf{1\frac{1}{6}}$

7. **2 factors**

8. $\mathbf{\frac{3}{8}, \frac{6}{12}, \frac{4}{6}, \frac{5}{6}, \frac{7}{7}}$

9.
$$\begin{array}{r} \mathbf{220} \text{ yards} \\ 8\overline{)1760} \\ \underline{16} \\ 16 \\ \underline{16} \\ 00 \\ \underline{0} \\ 0 \end{array}$$

10.
$$\begin{array}{r} \mathbf{42 \text{ mm}} \\ 2\overline{)84} \\ \underline{8} \\ 04 \\ \underline{4} \\ 0 \end{array}$$

11.
$$
\begin{array}{r}
{}^{11}\;{}^{1} \\
\$8.43 \\
\$0.68 \\
\$15.00 \\
+\;\;\$0.05 \\
\hline
\mathbf{\$24.16}
\end{array}
$$

12.
$$
\begin{array}{r}
6.505 \\
-\;1.400 \\
\hline
\mathbf{5.105}
\end{array}
$$

13.
$$
\begin{array}{r}
\$1\overset{1}{2}.\,\overset{9}{\cancel{0}}{}^{1}0 \\
-\;\$0.\,1\;2 \\
\hline
\mathbf{\$11.\,8\;8}
\end{array}
$$

14.
$$
\begin{array}{r}
\$18.07 \\
\times\;\;\;\;\;6 \\
\hline
\mathbf{\$108.42}
\end{array}
$$

15.
$$
\begin{array}{r}
\$12.72 \\
6\overline{)\$76.32} \\
\underline{6} \\
16 \\
\underline{12} \\
4\;3 \\
\underline{4\;2} \\
12 \\
\underline{12} \\
0
\end{array}
$$

$w = \mathbf{\$12.72}$

16. $2 \times 2 \times 2 \times 2 \times 2 \times 2 = \mathbf{64}$

17.
$$
\begin{array}{r}
\mathbf{68\ R\ 31} \\
70\overline{)4791} \\
\underline{420} \\
591 \\
\underline{560} \\
31
\end{array}
$$

18.
$$
\begin{array}{r}
\mathbf{52\tfrac{1}{7}} \\
7\overline{)365} \\
\underline{35} \\
15 \\
\underline{14} \\
1
\end{array}
$$

19. $\dfrac{3}{4} \times \dfrac{3}{4} = \mathbf{\dfrac{9}{16}}$

20. $\dfrac{3}{2} \times \dfrac{3}{2} = \mathbf{\dfrac{9}{4}} \left(\text{or } \mathbf{2\dfrac{1}{4}}\right)$

21. $\dfrac{3}{10} \times \dfrac{10}{10} = \mathbf{\dfrac{30}{100}}$

22. $3\dfrac{2}{3} + 1\dfrac{2}{3} = 4\dfrac{4}{3}$

$4\dfrac{4}{3} = 4 + \dfrac{3}{3} + \dfrac{1}{3}$

$= 4 + 1 + \dfrac{1}{3}$

$= \mathbf{5\dfrac{1}{3}}$

23. $4\dfrac{5}{5} - \dfrac{1}{5} = \mathbf{4\dfrac{4}{5}}$

24. $\dfrac{7}{10} - \dfrac{7}{10} = \mathbf{0}$

25. **1:10 a.m.**

26. **2**

27. $(1 \times 10^8) + (5 \times 10^7)$ **km**

28. **Geometric**
32, 64

29. $\mathbf{\dfrac{1}{2}}$

30. $\mathbf{\dfrac{8}{10}}; \mathbf{\dfrac{80}{100}}$

INVESTIGATION 8

1.

Number of Siblings of Students Surveyed

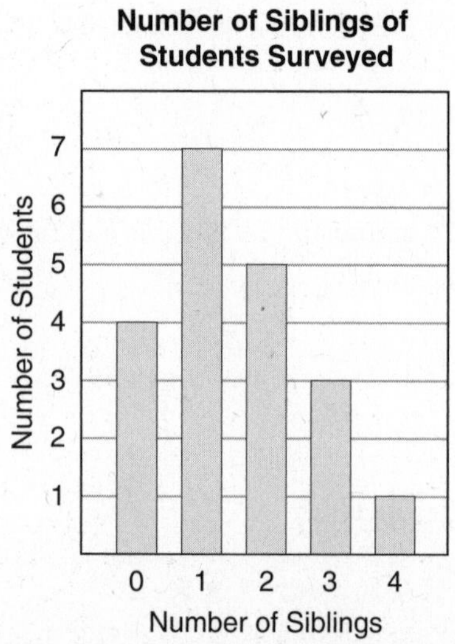

2.

Frequency Table

Weight	Freq.
11–13 lb	3
14–16 lb	4
17–19 lb	5
20–22 lb	7
23–25 lb	3

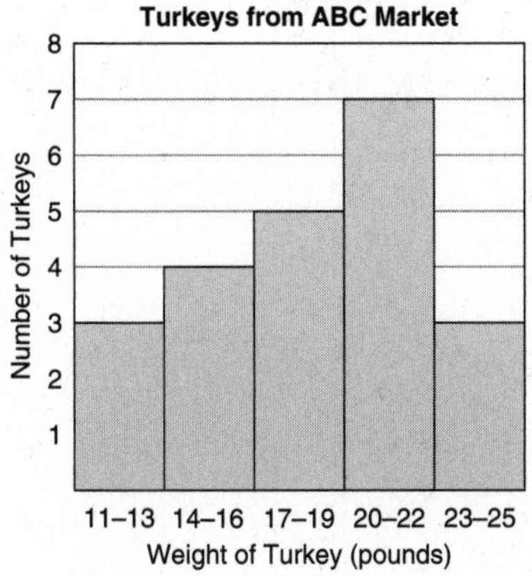

3.

Stem	Leaf
1	2 3 4 8 9 9
2	0 2 2 3 5

4.

Stem	Leaf
1	1 2 3 4 4 6 6 7 8 8 9 9
2	0 0 0 1 2 2 2 3 3 5

5.

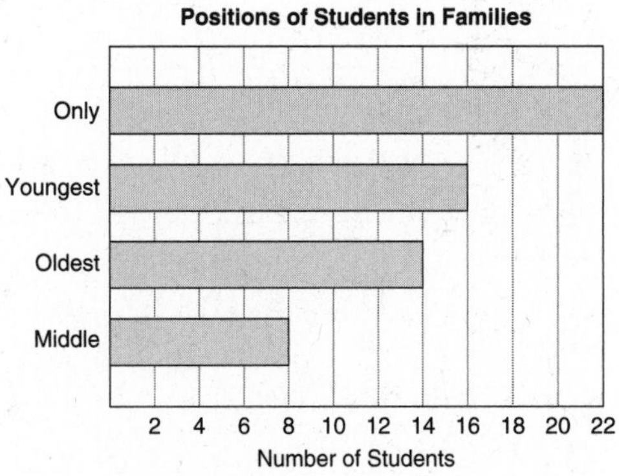

6. $10 \times 4 =$ **40 children**

7. $6 \times 4 =$ **24 children**

8. **The pictograph should have half as many icons in each category.**

9. **sleeping; eating**

10. **reading**

11. **watching television; playing; sleeping**

12. **Look for diameters in the circle to identify activities that together form about one half of the graph.**

13. $7 + 3 =$ **10 families**

14. $8 + 7 =$ **15 boys**

15. $7 + 6 =$ **13 soccer players**

16. **7 boys**

17. **6 soccer players**

18. $30 - 15 =$ **15 girls**

19. $15 - 6 =$ **9 girls**

Extensions

a. **See student work.**

b. **See student work.**

c. **See student work.**

Lesson 81, Warm-Up

a. **3 ft; 5280 ft**

b. $\mathbf{7\frac{1}{2}}$

c. **75**

d. **$30**

e. **15 minutes**

f. **100**

Problem Solving

Lesson 81, Lesson Practice

a. $\dfrac{8 \div 4}{12 \div 4} = \mathbf{\dfrac{2}{3}}$

b. **B. $\dfrac{3}{8}$**

c. $\dfrac{2}{3} \times \dfrac{1}{2} = \dfrac{2}{6}$

$\dfrac{2 \div 2}{6 \div 2} = \mathbf{\dfrac{1}{3}}$

d. $\dfrac{3}{10} + \dfrac{3}{10} = \dfrac{6}{10}$

$\dfrac{6 \div 2}{10 \div 2} = \mathbf{\dfrac{3}{5}}$

e. $\dfrac{3}{8} - \dfrac{1}{8} = \dfrac{2}{8}$

$\dfrac{2 \div 2}{8 \div 2} = \mathbf{\dfrac{1}{4}}$

f. $\dfrac{3}{9} = \dfrac{3 \div 3}{9 \div 3} = \dfrac{1}{3}$

$1\dfrac{3}{9} = \mathbf{1\dfrac{1}{3}}$

g. $\dfrac{6}{9} = \dfrac{6 \div 3}{9 \div 3} = \dfrac{2}{3}$

$2\dfrac{6}{9} = \mathbf{2\dfrac{2}{3}}$

h. $\dfrac{5}{10} = \dfrac{5 \div 5}{10 \div 5} = \dfrac{1}{2}$

$2\dfrac{5}{10} = \mathbf{2\dfrac{1}{2}}$

i. $1\dfrac{1}{4} + 2\dfrac{1}{4} = 3\dfrac{2}{4}$

$\dfrac{2}{4} = \dfrac{2 \div 2}{4 \div 2} = \dfrac{1}{2}$

$3\dfrac{2}{4} = \mathbf{3\dfrac{1}{2}}$

j. $1\dfrac{1}{8} + 5\dfrac{5}{8} = 6\dfrac{6}{8}$

$\dfrac{6}{8} = \dfrac{6 \div 2}{8 \div 2} = \dfrac{3}{4}$

$6\dfrac{6}{8} = \mathbf{6\dfrac{3}{4}}$

k. $5\dfrac{5}{12} - 1\dfrac{1}{12} = 4\dfrac{4}{12}$

$\dfrac{4}{12} = \dfrac{4 \div 4}{12 \div 4} = \dfrac{1}{3}$

$4\dfrac{4}{12} = \mathbf{4\dfrac{1}{3}}$

Lesson 81, Mixed Practice

1. $\begin{array}{r} \overset{0\,1}{\cancel{1}}\overset{2}{\cancel{3}}{}^{1}5 \\ -\ \ 9\,8 \\ \hline \mathbf{3\,7} \end{array}$

2. $\begin{array}{r} \overset{1\,2}{109} \\ 98 \\ +\ 135 \\ \hline 342 \end{array}$ $\quad \begin{array}{r} \mathbf{114} \\ 3\overline{)342} \\ \underline{3} \\ 04 \\ \underline{3} \\ 12 \\ \underline{12} \\ 0 \end{array}$

3. $\begin{array}{r} \overset{1}{1}2 \text{ in.} \\ \times\ \ 5 \phantom{\text{ in.}} \\ \hline 60 \text{ in.} \end{array}$ $\quad 60 \text{ in.} + 4 \text{ in.} = \textbf{64 inches}$

4. $\begin{array}{r} \overset{2}{\cancel{3}}{}^{1}2.6 \\ -\ 26.5 \\ \hline \mathbf{6.1} \end{array}$

5. $\mathbf{\dfrac{8}{12}; \dfrac{3}{12}}$

$\dfrac{8}{12} + \dfrac{3}{12} = \mathbf{\dfrac{11}{12}}$

6. **23, 29**

7. $\dfrac{10 \div 2}{12 \div 2} = \mathbf{\dfrac{5}{6}}$

8. (a) $\begin{array}{r} \textbf{6 runners} \\ 4\overline{)24} \\ \underline{24} \\ 0 \end{array}$

(b) $6 \div 2 =$ **3 girls**

(c) $\dfrac{3}{24} = \mathbf{\dfrac{1}{8}}$

9. 20 mm + 20 mm + 10 mm
+ 10 mm = **60 mm**

10. 20 mm × 10 mm = **200 sq. mm**

11.
$$2\overline{)48}$$ quotient 24 mm
$$\frac{4}{08}$$
$$\frac{8}{0}$$

48 mm
+ 24 mm
72 mm

12.
 3.40
+ 6.25
9.65

13.
 $\overset{5}{6}.\overset{1}{2}5$
− 3.40
2.85

14.

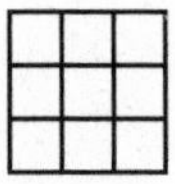

15.
 $14.50
6)$87.00
 6
 27
 24
 30
 30
 00
 0
 0

16.
 60 R 38
40)2438
 240
 38
 0
 38

17.
 586 $\frac{6}{9}$ = **586 $\frac{2}{3}$**
9)5280
 45
 78
 72
 60
 54
 6

18.
 $\overset{1}{\$}5.80$
+ $0.28
$6.08

 $\overset{9}{\$}\overset{9}{1}\overset{1}{0}.\overset{}{\cancel{0}}\overset{1}{0}$
− $6.08
$3.92

19. $5\frac{3}{5} + \left(3\frac{5}{5} - 1\frac{3}{5}\right)$

= $5\frac{3}{5} + 2\frac{2}{5}$ = $7\frac{5}{5}$ = **8**

20. $\frac{3}{6} = \frac{3 \div 3}{6 \div 3} = \frac{1}{2}$

21. $\frac{4}{3} \times \frac{1}{2} = \frac{4}{6}$

$\frac{4 \div 2}{6 \div 2} = \frac{2}{3}$

22. $\frac{10}{7} \times \frac{7}{10} = \frac{70}{70} = 1$

23. **9 months**

24.
 $6.50
× 30
$195.00

25.
 $195
− $135
$60

26. (a) $\frac{3}{6} = \frac{3 \div 3}{6 \div 3} = \frac{1}{2}$

(b) **Various answers are possible, including the probability of rolling an odd number with one number cube.**

27. (a) **4 students**

(b) **13 students**

28. ;
All squares are rectangles, but some rectangles are not squares.

29. $\frac{15}{100}$; $\frac{15 \div 5}{100 \div 5} = \frac{3}{20}$

30. $\frac{1}{2} \times \frac{1}{2}$ \bigcirc $\frac{1}{2}$
$\frac{1}{4}$

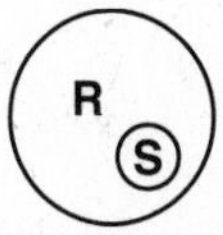

LESSON 82, WARM-UP

a. **81**

b. **144**

c. **16 ounces; 32 ounces**

d. $\mathbf{\frac{1}{2}, \frac{1}{3}, \frac{1}{4}, \frac{1}{5}}$

e. $\mathbf{33\frac{1}{3}}$

f. **15 minutes**

g. **12**

Problem Solving

$\frac{4}{10} = \mathbf{\frac{2}{5}}$

LESSON 82, LESSON PRACTICE

a. Factors of 6: ①, 2, ③, 6
Factors of 9: ①, ③, 9
GCF is **3**

b. Factors of 6: ①, ②, ③, ⑥
Factors of 12: ①, ②, ③, 4, ⑥, 12
GCF is **6**

c. Factors of 15: ①, 3,⑤, 15
Factors of 100: ①, 2, 4, ⑤, 10, 20, 25, 50, 100
GCF is **5**

d. Factors of 6: ①, ②, 3, 6
Factors of 10: ①, ②, 5, 10
GCF is **2**

e. Factors of 12: ①, 2, ③, 4, 6, 12
Factors of 15: ①, ③, 5, 15
GCF is **3**

f. Factors of 7: ①, 7
Factors of 10: ①, 2, 5, 10
GCF is **1**

g. $\frac{6 \div 3}{9 \div 3} = \mathbf{\frac{2}{3}}$

h. $\frac{6 \div 6}{12 \div 6} = \mathbf{\frac{1}{2}}$

i. $\frac{15 \div 5}{100 \div 5} = \mathbf{\frac{3}{20}}$

LESSON 82, MIXED PRACTICE

1.
```
    $5.75
 6)$34.50
   30
    4 5
    4 2
      30
      30
       0
```

2.
```
    400
 ×  500
 200,000
```
(**200,000**)

3. **3752**

4. **A. 4 to 5 feet**

5.
```
   7
   8¹0. 4 8
 − 6 5. 1 4
   1 5. 3 4
```
(**15.34**)

6.
```
   1
   12 inches
 ×  8
   96 inches
```
(**96 inches**)

7. **B. 21**

8. (a) Factors of 20: ①, ②, 4, ⑤, ⑩, 20
Factors of 30: ①, ②, 3, ⑤, 6, ⑩, 15, 30
GCF is **10**

(b) $\frac{20 \div 10}{30 \div 10} = \mathbf{\frac{2}{3}}$

9. $\frac{3}{4} \times \frac{12}{1} = \frac{36}{4}$

```
   9 in.
 4)36
   36
    0
```
(**9 in.**)

10. $4 \text{ in.} - \frac{3}{4} \text{ in.} = 3\frac{4}{4} - \frac{3}{4} = \mathbf{3\frac{1}{4}}$ **inches**

11. (a) $\dfrac{1}{3} \times \dfrac{12}{1} = \dfrac{12}{3}$

$12 \div 3 = \mathbf{4}$

(b) $\dfrac{2}{3} \times \dfrac{12}{1} = \dfrac{24}{3}$

$24 \div 3 = \mathbf{8}$

12. $\dfrac{6 \div 6}{12 \div 6} = \dfrac{\mathbf{1}}{\mathbf{2}}$

13. $2^3 \; \text{\small \textcircled{\textless}} \; 3^2$

$\underset{8}{2 \times 2 \times 2} \qquad \underset{9}{3 \times 3}$

14. $\dfrac{5}{7} + \dfrac{3}{7} = \dfrac{8}{7}$

$\dfrac{8}{7} = \dfrac{7}{7} + \dfrac{1}{7}$

$= \mathbf{1\dfrac{1}{7}}$

15. $\dfrac{4}{4} - \dfrac{2}{2}$

$1 - 1 = \mathbf{0}$

16. $\dfrac{2}{3} \times \dfrac{3}{3} = \dfrac{6}{9}$

17.
$$
\begin{array}{r}
\overset{2\,2\,2}{976.5} \\
470.4 \\
436.7 \\
+\;\;\;98.6 \\
\hline
\mathbf{1982.2}
\end{array}
$$

18.
$$
\begin{array}{r}
\$\overset{39}{\cancel{40}}.\,\overset{9}{\cancel{0}}{}^{1}0 \\
-\;\$32.\,8\,5 \\
\hline
\mathbf{\$7.\,1\,5}
\end{array}
$$

19.
$$
\begin{array}{r}
\$8.47 \\
\times\;\;\;\;\;\;70 \\
\hline
\mathbf{\$592.90}
\end{array}
$$

20.
$$
\begin{array}{r}
\mathbf{7{,}285\ R\ 5} \\
6\overline{)43{,}715} \\
\underline{42}\phantom{{,}715} \\
1\,7 \\
\underline{1\,2} \\
51 \\
\underline{48} \\
35 \\
\underline{30} \\
5
\end{array}
$$

21.
$$
\begin{array}{r}
\mathbf{88} \\
30\overline{)2640} \\
\underline{240} \\
240 \\
\underline{240} \\
0
\end{array}
$$

22.
$$
\begin{array}{r}
367 \\
\times\;\;418 \\
\hline
2\,936 \\
3\,670 \\
146\,800 \\
\hline
\mathbf{153{,}406}
\end{array}
$$

23. $3\dfrac{1}{4} + 3\dfrac{1}{4} = 6\dfrac{2}{4}$

$\dfrac{2 \div 2}{4 \div 2} = \dfrac{1}{2}$

$6\dfrac{2}{4} = \mathbf{6\dfrac{1}{2}}$

24.
$$
\begin{array}{r}
\mathbf{\$4.66} \\
4\overline{)\$18.64} \\
\underline{16} \\
2\,6 \\
\underline{2\,4} \\
24 \\
\underline{24} \\
0
\end{array}
$$

25.
$$
\begin{array}{r}
\overset{5}{\cancel{6}}{}^{1}7 \text{ inches} \\
-\;5\,9 \text{ inches} \\
\hline
\mathbf{8 \text{ inches}}
\end{array}
$$

26.
$$
\begin{array}{r}
\overset{1}{1}2 \text{ inches} \\
\times\;\;\;\;5 \\
\hline
60 \text{ inches}
\end{array}
$$

James

27.
$$
\begin{array}{r}
\overset{1}{6}7 \text{ in.} \\
60 \text{ in.} \\
+\;59 \text{ in.} \\
\hline
186 \text{ in.}
\end{array}
\qquad
\begin{array}{r}
\mathbf{62 \text{ in.}} \\
3\overline{)186} \\
\underline{18} \\
06 \\
\underline{6} \\
0
\end{array}
$$

28. $\dfrac{6}{8} = \dfrac{6 \div 2}{8 \div 2} = \dfrac{\mathbf{3}}{\mathbf{4}}$

29. **D.**

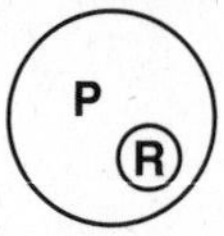

All rectangles are parallelograms, but some parallelograms are not rectangles.

30. $\mathbf{\frac{22}{100}}$; $\frac{22 \div 2}{100 \div 2} = \mathbf{\frac{11}{50}}$

Lesson 83, Warm-Up

a. **138**

b. **245**

c. $\mathbf{\frac{1}{4}, \frac{1}{2}, \frac{3}{4}}$

d. **36 in.; 5280 ft**

e. $\mathbf{66\frac{2}{3}}$

f. **36**

Problem Solving

Bao's straw: $3\frac{3}{4} - \frac{1}{4} =$ $\mathbf{3\frac{1}{2}}$ **inches long**

Sherry's straw: $3\frac{3}{4} + \frac{1}{2} =$ $\mathbf{4\frac{1}{4}}$ **inches long**

Lesson 83, Lesson Practice

a. **Rectangular solid**

b. **Cylinder**

c. **Cone**

d. **Rectangular solid**

e. **4 triangular faces**

f. **1 rectangular face**

g. **8 edges**

h. **5 vertices**

Lesson 83, Mixed Practice

1. **3:15 p.m.**

2. **2 dimes**

3. **23,287,420**

4. (a) $\frac{1}{3} \times \frac{24}{1} = \frac{24}{3}$

   $24 \div 3 = \mathbf{8}$

   (b) $\frac{2}{3} \times \frac{24}{1} = \frac{48}{3}$

$$\begin{array}{r} \mathbf{16} \\ 3\overline{)48} \\ \underline{3} \\ 18 \\ \underline{18} \\ 0 \end{array}$$

5. **11, 13, 17, 19**

6. (a) Factors of 4: ①,②,④

   Factors of 8: ①,②,④, 8

   GCF is **4**

   (b) $\frac{4 \div 4}{8 \div 4} = \mathbf{\frac{1}{2}}$

7. (a) **Cube**

   (b) **6 faces**

8. **C. Sphere**

9. **1.7**

10. **D. Diameter**

11.
$$\begin{array}{r} \scriptstyle 1 \\ 3.62 \\ +\ 4.50 \\ \hline \mathbf{8.12} \end{array}$$

12.
$$\begin{array}{r} \scriptstyle 2\ \ {}^{1}6\ \ 9 \\ \not{3}.\ \not{7}\ \not{0}\ {}^{1}4 \\ -\ 2.\ 9\ 1\ 8 \\ \hline \mathbf{0.\ 7\ 8\ 6} \end{array}$$

13.
$$\begin{array}{r} 21 \\ \times\ 21 \\ \hline 21 \\ 420 \\ \hline \mathbf{441} \end{array}$$

14.
$$\begin{array}{r} \$6.25 \\ \times \quad 4 \\ \hline \mathbf{\$25.00} \end{array}$$

15.
$$\begin{array}{r} \$2.43 \\ 6\overline{)\$14.58} \\ 12 \\ \hline 2\,5 \\ 2\,4 \\ \hline 18 \\ 18 \\ \hline 0 \end{array}$$

$w = \mathbf{\$2.43}$

16. $\mathbf{\frac{4}{12}; \frac{9}{12}}$

$$\frac{4}{12} + \frac{9}{12} = \frac{13}{12}$$

$$\frac{13}{12} = \frac{12}{12} + \frac{1}{12}$$

$$= \mathbf{1\frac{1}{12}}$$

17. $\frac{6}{8} = \frac{6 \div 2}{8 \div 2} = \mathbf{\frac{3}{4}}$

18. $\frac{3}{4} \times \frac{3}{3} = \mathbf{\frac{9}{12}}$

19.
$$\begin{array}{r} 4\frac{1}{6} \\ +\ 2\frac{1}{6} \\ \hline 6\frac{2}{6} \end{array} = \mathbf{6\frac{1}{3}}$$

20.
$$\begin{array}{r} 3\frac{3}{4} \\ +\ 1\frac{1}{4} \\ \hline 4\frac{4}{4} \end{array} = \mathbf{5}$$

21.
$$\begin{array}{r} 4\frac{4}{4} \\ -\ 1\frac{1}{4} \\ \hline \mathbf{3\frac{3}{4}} \end{array}$$

22. 0.1 (>) 0.01

23. $\frac{1}{2} + \frac{1}{2} + \frac{1}{2} = \frac{3}{2}$

$$\frac{3}{2} = \frac{2}{2} + \frac{1}{2}$$

$$= \mathbf{1\frac{1}{2}}$$

24.
$$\begin{array}{r} \overset{1}{\$9.00} \\ \$9.00 \\ \$4.50 \\ +\ \$3.75 \\ \hline \mathbf{\$26.25} \end{array}$$

25. $\$18 \div 2 = \mathbf{\$9.00}$

26. **C.** 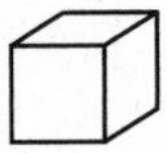

27.
$$\begin{array}{r} 14\ \text{ft} \\ \times\ \ 10\ \text{ft} \\ \hline \mathbf{140\ sq.\ ft} \end{array}$$

28. (a) **Rent**

(b) $\mathbf{\frac{1}{4}}$

29.
$$\begin{array}{r} \overset{1}{2}.4\ \text{cm} \\ \times\ \ 4\ \text{cm} \\ \hline \mathbf{9.6\ cm} \end{array}$$

30. $\mathbf{(1 \times 10^5) + (8 \times 10^4) + (6 \times 10^3)}$

Lesson 84, Warm-Up

a. **288**

b. **272**

c. $\mathbf{\frac{1}{3}, \frac{1}{2}, \frac{2}{3}}$

d. **2000 lb; 1000 lb**

e. **$50**

f. **25**

g. **9**

Problem Solving

When the stack is 4 in. tall, there are about 1440 pages in the stack. One foot (12 in.) is three times taller than 4 in., so there would be about 3 times more pages in the stack. So John will have read about 1440×3, or **4320 pages,** when the stack is one foot tall.

LESSON 84, LESSON PRACTICE

a. $3 + 7 + 9 + 9 + 4 = 32$

$$5\overline{)32} = 6\frac{2}{5}$$
$$\begin{array}{r} 32 \\ -30 \\ \hline 2 \end{array}$$

Mean: $6\frac{2}{5}$

3, 4, ⑦, 9, 9

Median: 7

Mode: 9

$9 - 3 = 6$

Range: 6

b. $16 + 2 + 5 + 7 + 11 + 13 = 54$

$$6\overline{)54} = 9$$

Mean: 9

2, 5, 7, 11, 13, 16

$11 + 7 = 18$

$18 \div 2 = 9$

Median: 9

Mode: none

$16 - 2 = 14$

Range: 14

c. $3 + 10 + 2 + 10 + 10 + 1 + 3 + 10 = 49$

$$8\overline{)49} = 6\frac{1}{8}$$
$$\begin{array}{r} 49 \\ -48 \\ \hline 1 \end{array}$$

Mean: $6\frac{1}{8}$

1, 2, 3, 3, 10, 10, 10, 10

$10 + 3 = 13$

$13 \div 2 = 6\frac{1}{2}$

Median: $6\frac{1}{2}$

Mode: 10

$10 - 1 = 9$

Range: 9

d. $13 + 10 + 10 + 11 + 11 + 10 + 11 = 76$

$$7\overline{)76} = 10\frac{6}{7}$$
$$\begin{array}{r} 76 \\ -7 \\ \hline 06 \\ -0 \\ \hline 6 \end{array}$$

Mean: $10\frac{6}{7}$

10, 10, 10, 11, 11, 11, 13

Median: 11

Mode: 10 and 11

$13 - 10 = 3$

Range: 3

LESSON 84, MIXED PRACTICE

1. $3\overline{)\$2418} = $ **\$806**
$$\begin{array}{r} 2418 \\ -24 \\ \hline 01 \\ -0 \\ \hline 18 \\ -18 \\ \hline 0 \end{array}$$

2.

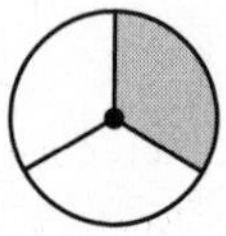

$\frac{1}{3} \times \frac{100}{1} = \frac{100}{3}$

$$3\overline{)100} = \mathbf{33\frac{1}{3}\%}$$
$$\begin{array}{r} 100 \\ -9 \\ \hline 10 \\ -9 \\ \hline 1 \end{array}$$

3. 1 kg = **1000 grams**

4.
$$\begin{array}{r} 700 \\ \times\ 500 \\ \hline \mathbf{350{,}000} \end{array}$$

5. **D. 12:10**

6. **0.01, 0.1, 1.0, 1.01**

7. (a) Factors of 8: ①, ②, ④, 8

Factors of 12: ①, ②, 3, ④, 6, 12

GCF is **4**

(b) $\frac{8 \div 4}{12 \div 4} = \mathbf{\frac{2}{3}}$

8. (a) $\frac{1}{4} \times \frac{80}{1} = \frac{80}{4}$

$$4\overline{)80} = \mathbf{20}$$

8
00
0
0

(b) $\frac{3}{4} \times \frac{80}{1} = \frac{240}{4}$

$$4\overline{)240} = \mathbf{60}$$

24
00
0
0

9. $\frac{1}{2} \times \mathbf{\frac{3}{3}} = \frac{3}{6}$

10. $\frac{4}{6} = \frac{4 \div 2}{6 \div 2} = \mathbf{\frac{2}{3}}$

11. $\mathbf{1\frac{3}{10}}$**; 1.3**

12.

$$\begin{array}{r} \overset{1}{9}.9 \\ 6.14 \\ 7.5 \\ +\ 8.31 \\ \hline \mathbf{31.85} \end{array}$$

13.

$$\begin{array}{r} \$\not{1}\overset{9}{\not{0}}.\overset{9}{\not{0}}{}^{1}0 \\ -\ \$0.5\ 9 \\ \hline \mathbf{\$9.4\ 1} \end{array}$$

14. $30\overline{)672}$ = **22 R 12**

60
72
60
12

15.

$$\begin{array}{r} \$0.68 \\ \times\ \ 5 \\ \hline \mathbf{\$3.40} \end{array}$$

16. $5\overline{)\$3.40}$ = **$0.68**

3 0
40
40
0

17. $9\frac{3}{3} - 3\frac{1}{3} = \mathbf{6\frac{2}{3}}$

18. $\frac{3}{4} \times \frac{5}{4} = \mathbf{\frac{15}{16}}$

19. Cone

20. B. $\overline{PO}$

21. C. $\angle BOD$

22.

$$\begin{array}{r} \$\overset{2}{1}.\overset{3}{5}5 \\ \$0.69 \\ \$0.69 \\ +\ \$1.99 \\ \hline \mathbf{\$4.92} \end{array}$$

23. $8\overline{)\$8.80}$ = **$1.10**

8
0 8
8
00
0
0

24. $0.97

25. $0.97

26.

$$\begin{array}{r} \mathbf{\$1.99} \\ \mathbf{-\ \$0.69} \\ \hline \mathbf{\$1.30} \end{array}$$

27.

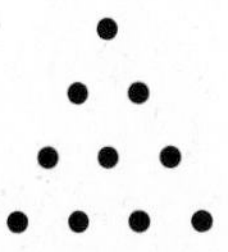

This figure is made of ten dots. The fourth triangular number is **10.**

28. $1 \text{ in.} + 1\frac{1}{4} \text{ in.} + \frac{3}{4} \text{ in.} = 2\frac{4}{4} = \textbf{3 inches}$

29. $\mathbf{\frac{90}{100}}$

$\frac{90 \div 10}{100 \div 10} = \mathbf{\frac{9}{10}}$

0.9

30.
$$\begin{array}{r} \overset{1}{\cancel{2}}\,{}^{1}1\,2°\text{F} \\ -\quad 3\,2°\text{F} \\ \hline \mathbf{1\,8\,0°F} \end{array}$$

Lesson 85, Warm-Up

a. **6 pounds 12 ounces**

b. **13 pounds 2 ounces**

c. **16 ounces; 8 ounces**

d. $\mathbf{\frac{1}{5}, \frac{2}{5}, \frac{3}{5}, \frac{4}{5}}$

e. **$50**

f. **100**

g. **100**

Problem Solving

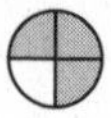

Lesson 85, Lesson Practice

a. 1 gal = 4 qt

$\frac{1}{4}$ gal = 1 qt

Quart

b. 2 pt = 1 qt

4 qt = 1 gal

1 gal = **8 pt**

c. 1000 mL = 1 L

2 L = **2000 mL**

d. 1 cup = $\frac{1}{2}$ pt

1 pt = 16 oz

1 cup = **8 ounces**

Lesson 85, Mixed Practice

1. **One possibility:**

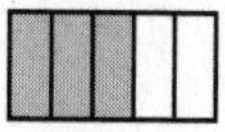

$\frac{3}{5} \times \frac{100}{1} = \frac{300}{5}$

$300 \div 5 = \mathbf{60\%}$

2. **401**

3. **2.5 cm; 25 mm**

4. $20 \times 4 = $ **80 times**

5. **A.** $\overline{BC}$

6. (a) Factors of 6: ①, 2, ③, 6
Factors of 9: ①, ③, 9
GCF: **3**

(b) $\frac{6 \div 3}{9 \div 3} = \mathbf{\frac{2}{3}}$

7. (a) $\frac{1}{5} \times \frac{60}{1} = \frac{60}{5}$

$$\begin{array}{r} \mathbf{12} \\ 5\overline{)60} \\ \underline{5} \\ 10 \\ \underline{10} \\ 0 \end{array}$$

(b) $\frac{2}{5} \times \frac{60}{1} = \frac{120}{5}$

$$\begin{array}{r} \mathbf{24} \\ 5\overline{)120} \\ \underline{10} \\ 20 \\ \underline{20} \\ 0 \end{array}$$

8. $1\frac{1}{4} \text{ in.} + 2\frac{1}{4} \text{ in.} = 3\frac{2}{4} \text{ in.}$

$\frac{2}{4} = \frac{2 \div 2}{4 \div 2} = \frac{1}{2}$

$3\frac{2}{4} \text{ in.} = \mathbf{3\frac{1}{2} \text{ in.}}$

9. **0, 0.01, 0.1, 1.0**

10. 1 qt = 2 pt

4 qt = **8 pints**

11. 1 L = 1000 mL

3 L = **3000 mL**

12. $6\overline{)100}$ = **$16\frac{4}{6}$**

```
   16 4/6
6)100
  6
  40
  36
   4
```

$\frac{4 \div 2}{6 \div 2} = \frac{2}{3}$

$16\frac{4}{6} = \mathbf{16\frac{2}{3}}$

13.
```
   $17.56
   $12.00
 +  $0.95
   $30.51
```
$30.51

14.
```
   4.324
 − 1.910
   2.414
```
2.414

15.
```
      396
   ×  405
    1 980
  158 400
  160,380
```
160,380

16.
```
   $1.25
 ×    20
  $25.00
```
$25.00

17.
```
   400 R 5
9)3605
  36
   00
    0
    05
     0
     5
```
400 R 5

18.
```
    $0.25
10)$2.50
    2 0
      50
      50
       0
```
$0.25

19. $\frac{15}{20} = \frac{15 \div 5}{20 \div 5} = \mathbf{\frac{3}{4}}$

20. $3 - \left(2\frac{2}{3} - 1\right)$

$= 2\frac{3}{3} - 1\frac{2}{3} = \mathbf{1\frac{1}{3}}$

21. $\mathbf{\frac{6}{10}, \frac{5}{10}}$

$\frac{6}{10} + \frac{5}{10} = \frac{11}{10}$

$\frac{11}{10} = \frac{10}{10} + \frac{1}{10}$

$= \mathbf{1\frac{1}{10}}$

22.
```
   6.15
 + 5.12
  11.27
```
11.27

23. **3**

24. **4 feet 11 inches**

25. **5 feet 5 inches**

26. **4, 8, 16, 32**

27. 50 ÷ 2 = 25

About 25 times

28. 100 − 80 = **20**

29. **The mode is 90 because the score 90 occurs more often than any other score.**

30. **The median is 90 because it is the middle score when the scores are arranged in order.**

LESSON 86, WARM-UP

a. **$18.75**

b. **$16.25**

c. **4 quarters; 4 quarts**

d. $\mathbf{\frac{1}{2}, \frac{1}{3}, \frac{1}{4}, \frac{1}{5}}$

e. **$15**

f. **99**

Problem Solving

5 ft

3 ft

B	G	B	G	B
G	B	G	B	G
B	G	B	G	B

8 blue squares, 7 gold squares

LESSON 86, LESSON PRACTICE

a. $\frac{1}{3} \times \frac{4}{1} = \frac{4}{3} = \mathbf{1\frac{1}{3}}$

b. $\frac{3}{5} \times \frac{2}{1} = \frac{6}{5} = \mathbf{1\frac{1}{5}}$

c. $\frac{2}{3} \times \frac{2}{1} = \frac{4}{3} = \mathbf{1\frac{1}{3}}$

d. $\frac{1}{5} \times \frac{4}{1} = \mathbf{\frac{4}{5}}$

e. $\frac{1}{6} \times \frac{5}{1} = \mathbf{\frac{5}{6}}$

f. $\frac{2}{3} \times \frac{5}{1} = \frac{10}{3} = \mathbf{3\frac{1}{3}}$

LESSON 86, MIXED PRACTICE

1. **One possibility:**

2.
$$\begin{array}{r} 7000 \\ -\ 3000 \\ \hline \mathbf{4000} \end{array}$$

3. **6 + 4 = 10**

4. 1 L = 1000 mL
   2 L = **2000 mL**

5. $\mathbf{\frac{33}{100}}$**; 0.33; 33%**

6. (a) $\frac{1}{3} \times \frac{120}{1} = \frac{120}{3}$
   $120 \div 3 = \mathbf{40}$

   (b) $\frac{2}{3} \times \frac{120}{1} = \frac{240}{3}$
   $240 \div 3 = \mathbf{80}$

7. **B.** $\overline{RT}$

8. $\mathbf{\frac{7}{16}, \frac{9}{18}, \frac{5}{8}, \frac{6}{6}, \frac{8}{7}}$

9. $\mathbf{2\frac{3}{5}}$

10. $\frac{2}{3} \times \frac{2}{1} = \frac{4}{3} = \mathbf{1\frac{1}{3}}$

11. $\frac{3}{4} \times \frac{4}{1} = \frac{12}{4} = \mathbf{3}$

12. $3 - \left(2\frac{3}{5} - 1\frac{1}{5}\right)$
    $= 2\frac{5}{5} - 1\frac{2}{5} = \mathbf{1\frac{3}{5}}$

13.
$$\begin{array}{r} \overset{1}{4}.70 \\ 3.63 \\ +\ 2.00 \\ \hline \mathbf{10.33} \end{array}$$

14.
$$\begin{array}{r} \overset{29\ \ {}^{1}0}{\cancel{30}\,\cancel{1}.{}^{1}4} \\ -\ 143.5 \\ \hline \mathbf{157.9} \end{array}$$

15.
$$\begin{array}{r} 476 \\ \times\ \ 890 \\ \hline 42\,840 \\ 380\,800 \\ \hline \mathbf{423{,}640} \end{array}$$

16. $\begin{array}{r} \mathbf{87} \\ 4\overline{)348} \\ \underline{32} \\ 28 \\ \underline{28} \\ 0 \end{array}$

17. $\begin{array}{r} \mathbf{87} \\ 40\overline{)3480} \\ \underline{320} \\ 280 \\ \underline{280} \\ 0 \end{array}$

18. $\begin{array}{r} \mathbf{\$7.06} \\ 6\overline{)\$42.36} \\ \underline{42} \\ 0\,3 \\ \underline{0} \\ 36 \\ \underline{36} \\ 0 \end{array}$

19. $\begin{array}{r} 22 \\ \times\ 22 \\ \hline 44 \\ \underline{440} \\ \mathbf{484} \end{array}$

20. (a) Factors of 60: ①, ②, 3, ④, ⑤, 6, ⑩, 12, 15, ⑳, 30, 60
Factors of 100:
①, ②, ④, ⑤, ⑩, ⑳, 25, 50, 100
GCF is **20**

(b) $\frac{60 \div 20}{100 \div 20} = \mathbf{\frac{3}{5}}$

21. $\mathbf{\frac{9}{12}; \frac{8}{12}}$

$\frac{9}{12} - \frac{8}{12} = \mathbf{\frac{1}{12}}$

22. **3**

23. (a) **Cube**

(b) **8 vertices**

24. 12 − 7 = **5 books**

25. $\begin{array}{r} 180 \\ \times\ 10 \\ \hline \end{array}$
1800 pages

26. 7, 8, ⑨, 10, 11
9

27. $\frac{4}{6} = \frac{4 \div 2}{6 \div 2} = \mathbf{\frac{2}{3}}$

28. $\frac{1}{4} = \mathbf{25\%}$

29. 1 quart (<) 1 liter

30.

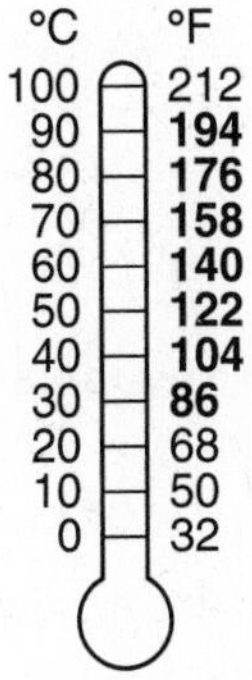

Lesson 87, Warm-Up

a. **7 ft**

b. **13 ft 4 in.**

c. **4 quarts; 8 quarts; 12 quarts**

d. $\mathbf{\frac{1}{5}, \frac{1}{6}, \frac{1}{7}, \frac{1}{8}}$

e. **6**

Problem Solving
(3, 1, 2) and **(3, 2, 1)**

Lesson 87, Lesson Practice

a. 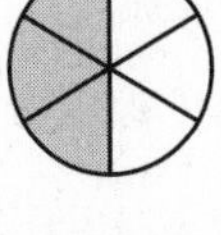 **; 3**

b.  **; 4**

c. **1**

d. **4**

e. **2**

f. **2**

LESSON 87, MIXED PRACTICE

1. 10 ft + 10 ft + 20 ft + 20 ft = **60 feet**

2. 10 ft × 20 ft = **200 sq. ft**

3. **C. 34.12; thirty-four and twelve hundredths**

4. $\mathbf{0, 0.3, \frac{1}{2}, 1}$

5.
16 oz = 1 pt
2 pt = 1 qt = 32 ounces
2 qt = **64 ounces**

6. (a) $\mathbf{\frac{1}{4}}$

(b) **4 quarters**

(c) 4 × 3 = **12 quarters**

7. $\mathbf{\frac{7}{10}; 0.7; 70\%}$

8. 2 × 3 + 5
6 + 5 = 11
B. 11

9. $\frac{3}{4} \times \frac{84}{1} = \frac{252}{4}$

$$\begin{array}{r} \mathbf{63} \text{ **millimeters**} \\ 4\overline{)252} \\ \underline{24} \\ 12 \\ \underline{12} \\ 0 \end{array}$$

10. $\mathbf{\frac{3}{6}, \frac{2}{6}}$

$\frac{3}{6} - \frac{2}{6} = \mathbf{\frac{1}{6}}$

11. (a) Factors of 20: ①, ②, 4, ⑤, ⑩, 20
Factors of 50: ①, ②, ⑤, ⑩, 25, 50
GCF is **10.**

(b) $\frac{20 \div 10}{50 \div 10} = \mathbf{\frac{2}{5}}$

12. $\frac{3}{5} \times \frac{4}{1} = \frac{12}{5} = \mathbf{2\frac{2}{5}}$

13. **6**

14. $3\frac{7}{8} - 1\frac{1}{8} = 2\frac{6}{8} = \mathbf{2\frac{3}{4}}$

15.
$$\begin{array}{r} \mathbf{45} \\ 50\overline{)2250} \\ \underline{200} \\ 250 \\ \underline{250} \\ 0 \end{array}$$

16.
$$\begin{array}{r} \mathbf{45} \\ 5\overline{)225} \\ \underline{20} \\ 25 \\ \underline{25} \\ 0 \end{array}$$

17.
$$\begin{array}{r} {}^{2\,3\,3} \\ 5365 \\ 428 \\ 3997 \\ 659 \\ 7073 \\ \underline{+\ \ 342} \\ \mathbf{17{,}864} \end{array}$$

18.
$$\begin{array}{r} \mathbf{\$2.05} \\ 4\overline{)\$8.20} \\ \underline{8} \\ 0\,2 \\ \underline{0} \\ 20 \\ \underline{20} \\ 0 \end{array}$$

19.
$$\begin{array}{r} 20 \\ \underline{\times\ 20} \\ \mathbf{400} \end{array}$$

20.
$$\begin{array}{r} \$12.75 \\ \underline{\times\qquad 80} \\ \mathbf{\$1020.00} \end{array}$$

21. $8\overline{)100}$ = $12\frac{4}{8}$ = $\mathbf{12\frac{1}{2}}$

$$\begin{array}{r} 12\frac{4}{8} \\ 8\overline{)100} \\ \underline{8} \\ 20 \\ \underline{16} \\ 4 \end{array}$$

22. **2**

23. **3**

24. $\begin{array}{r} 41 \text{ miles} \\ +\ 49 \text{ miles} \\ \hline 90 \text{ miles} \end{array}$ $\qquad \begin{array}{r} 90 \text{ miles} \\ -\ 40 \text{ miles} \\ \hline \mathbf{50 \text{ miles}} \end{array}$

25. $\begin{array}{r} 49 \text{ miles} \\ +\ 20 \text{ miles} \\ \hline \mathbf{69 \text{ miles}} \end{array}$

26. 10 + 2 = **12 children**

27. 5 + 2 = **7 children**

28. **2 children**

29. 5 + 2 + 10 = **17 children**

30. 20 − 17 = **3 children**

Lesson 88, Warm-Up

a. **170**

b. **170**

c. $\mathbf{\frac{1}{4}, \frac{5}{6}, \frac{3}{4}}$

d. **16 ounces; 16 ounces; The two have the same number of ounces.**

e. **100**

Problem Solving

12 1-inch cubes

Lesson 88, Lesson Practice

a. **Я**

b. **ꓤ**

c. **Reflection**

d. **Rotation**

e. **Translation**

f. **Reflection and translation, or rotation and reflection, or rotation and reflection and translation**

Lesson 88, Mixed Practice

1. $\frac{1}{4} \times \frac{2}{1} = \frac{2}{4} = \mathbf{\frac{1}{2}}$ **mile**

2. **April 4, 2070**

3. $\begin{array}{r} 4.25 \\ -\ 3.12 \\ \hline \mathbf{1.13} \end{array}$

4. (a) **10 dimes**

   (b) 10 dimes × 5 = **50 dimes**

5. $\frac{2}{3} \times \frac{150}{1} = \frac{300}{3}$

   300 ÷ 3 = **100**

6. 1 gal = 4 qt

   $\frac{1}{2}$ gal = **2 quarts**

7. **B. Spoke**

8. $\mathbf{\frac{2}{6}}$

   $\frac{5}{6} - \frac{2}{6} = \frac{3}{6} = \mathbf{\frac{1}{2}}$

9. (a) $\frac{6}{8} = \mathbf{\frac{3}{4}}$

(b) $\frac{3}{4} \times \frac{100}{1} = \frac{300}{4}$

$$\begin{array}{r} \mathbf{75\%} \\ 4\overline{)300} \\ \underline{28} \\ 20 \\ \underline{20} \\ 0 \end{array}$$

10. $\frac{2}{3} \times \frac{84}{1} = \frac{168}{3}$

$$\begin{array}{r} \mathbf{56\text{ mm}} \\ 3\overline{)168} \\ \underline{15} \\ 18 \\ \underline{18} \\ 0 \end{array}$$

11. $\frac{3}{5} + \frac{3}{5} + \frac{3}{5} \; \textcircled{=} \; 3 \times \frac{3}{5}$

12. **A.** $\angle ABC$

13. $\frac{1}{8} \times \frac{3}{1} = \mathbf{\frac{3}{8}}$

14. **3**

15. (a) **4**

(b) $\frac{1}{6} \times \frac{4}{1} = \frac{4}{6} = \mathbf{\frac{2}{3}}$

16. $\frac{1}{4} + \frac{1}{4} = \frac{2}{4} = \mathbf{\frac{1}{2}}$

17. $\frac{7}{8} - \frac{1}{8} = \frac{6}{8} = \mathbf{\frac{3}{4}}$

18. $4\frac{10}{10} - 1\frac{3}{10} = \mathbf{3\frac{7}{10}}$

19.
$$\begin{array}{r} \$6.\overset{1}{5}7 \\ \$_{1}0.38 \\ +\ \$16.00 \\ \hline \mathbf{\$22.95} \end{array}$$

20.
$$\begin{array}{r} \overset{3}{\not4}\,\overset{11}{\not2}\,\overset{10}{\not1}.\,{}^{1}0\,5 \\ -\ 1\,2\,5.\,7\,0 \\ \hline \mathbf{2\,9\,5.\,3\,5} \end{array}$$

21.
$$\begin{array}{r} 30 \\ \times\ 30 \\ \hline \mathbf{900} \end{array}$$

22.
$$\begin{array}{r} 340 \\ \times\ 607 \\ \hline 2\,380 \\ 204\,000 \\ \hline \mathbf{206{,}380} \end{array}$$

23.
$$\begin{array}{r} \mathbf{\$0.85} \\ 9\overline{)\$7.65} \\ \underline{7\,2} \\ 45 \\ \underline{45} \\ 0 \end{array}$$

24.
$$\begin{array}{r} 50\text{ min} \\ +\ 40\text{ min} \\ \hline \mathbf{90\text{ minutes}} \end{array}$$

25. **2:40 p.m.**

26. 30 ft × 30 ft = **900 sq. ft**

27. $\mathbf{3\frac{1}{2}, 4, 4\frac{1}{2}}$

28. $\frac{10}{15} = \mathbf{\frac{2}{3}}$

29. **B.** Ⓟ Ⓣ
No parallelogram is a trapezoid, and no trapezoid is a parallelogram.

30. $\mathbf{\frac{3}{8}}$ **sq. in.**

Lesson 89, Warm-Up

a. **136**

b. **136**

c. $\mathbf{\frac{1}{3}, \frac{1}{2}, \frac{2}{3}}$

d. **16 ounces; 2 pints; 32 ounces**

e. **7**

Problem Solving

1 pint + 1 quart = 2 cups + 2 pints
= 2 cups + (2 × 2 cups) = **6 cups**

LESSON 89, LESSON PRACTICE

a. **1**

b. **2**

c. **4**

d. **7**

e. $\sqrt{36}$ ⓐ 3^2 — $\sqrt{36} < 3^2$; $6 \quad 9$

f. $\sqrt{25} - \sqrt{16}$

$5 - 4 = \mathbf{1}$

LESSON 89, MIXED PRACTICE

1. $$\begin{array}{r} \overset{1}{1}776 \\ +\ \ 50 \\ \hline \mathbf{1826} \end{array}$$

2. **500 days**

$$\begin{array}{r} 500 \\ \$20\overline{)\$10{,}000} \\ 10\,0 \\ \hline 0\,0 \\ 0 \\ \hline 00 \\ 00 \\ \hline 0 \end{array}$$

3. **C. 20**

4. **0, 0.5, 1, 1.1, $\frac{3}{2}$**

5. (a) **2 cartons**

(b) **6 cartons**

6. **1,354,760**

7. $\mathbf{\frac{3}{6}}$

$\frac{5}{6} - \frac{3}{6} = \frac{2 \div 2}{6 \div 2} = \mathbf{\frac{1}{3}}$

8. (a) $\frac{3 \div 3}{9 \div 3} = \mathbf{\frac{1}{3}}$

(b) $\frac{1}{3} \times \frac{100}{1} = \frac{100}{3}$

$$\begin{array}{r} \mathbf{33\tfrac{1}{3}\%} \\ 3\overline{)100} \\ 9 \\ \hline 10 \\ 9 \\ \hline 1 \end{array}$$

9. (a) **Rectangular solid**

(b) **12 edges**

10. **3.1 cm; 31 mm**

11. $\frac{2}{5} \times \frac{3}{1} = \frac{6}{5} = \mathbf{1\frac{1}{5}}$

12. $\frac{2}{5} + \frac{2}{5} + \frac{2}{5} = \frac{6}{5} = \mathbf{1\frac{1}{5}}$

13. $1\frac{1}{4} + 1\frac{1}{4} = 2\frac{2}{4} = \mathbf{2\frac{1}{2}}$

14. $3\frac{5}{6} - 1\frac{1}{6} = 2\frac{4}{6} = \mathbf{2\frac{2}{3}}$

15. $$\begin{array}{r} \overset{2}{4}\overset{2}{2}.\overset{1}{6}0 \\ 49.76 \\ 28.70 \\ +\ 53.18 \\ \hline \mathbf{174.24} \end{array}$$

16. $$\begin{array}{r} \$\overset{1}{2}.\overset{1}{4}8 \\ +\ \$0.57 \\ \hline \$3.05 \end{array} \qquad \begin{array}{r} \$\overset{9}{\cancel{10}}.\overset{9}{\cancel{0}}{}^{1}0 \\ -\ \ \$3.\ 0\ 5 \\ \hline \mathbf{\$6.\ 9\ 5} \end{array}$$

17. $$\begin{array}{r} 42 \\ \times\ \ 5 \\ \hline 210 \end{array} \qquad \begin{array}{r} 210 \\ \times\ \ 36 \\ \hline 1260 \\ 6300 \\ \hline \mathbf{7560} \end{array}$$

18. $$\begin{array}{r} \$6.15 \\ \times\ \ \ \ 10 \\ \hline \mathbf{\$61.50} \end{array}$$

19. $\begin{array}{r} 69 \\ 40\overline{)2760} \\ \underline{240} \\ 360 \\ \underline{360} \\ 0 \end{array}$

20. $\begin{array}{r} 69 \\ 4\overline{)276} \\ \underline{24} \\ 36 \\ \underline{36} \\ 0 \end{array}$

$W = \mathbf{69}$

21. **5**

22. $\frac{1}{2} \times \frac{6}{8} = \frac{6}{16} = \mathbf{\frac{3}{8}}$

23. $\begin{array}{r} \mathbf{37\ R\ 1} \\ 10\overline{)371} \\ \underline{30} \\ 71 \\ \underline{70} \\ 1 \end{array}$

24. $10 - 4 = 6$
6 years old

25. $29 + 6 =$ **35 years old**

26. $\sqrt{25} - \sqrt{9}$
$5 - 3 = \mathbf{2}$

27. $3^2 + 4^2 = 9 + 16 = \mathbf{25}$

28. **1. dime: heads,**
quarter: heads
2. dime: tails,
quarter: heads
3. dime: tails,
quarter: tails

29. (a) $\mathbf{\frac{1}{2}}$

(b) $\mathbf{\frac{1}{4}}$

(c) $\mathbf{\frac{1}{8}}$

(d) $\frac{1}{2} \times \frac{1}{4} = \mathbf{\frac{1}{8}}$

30.

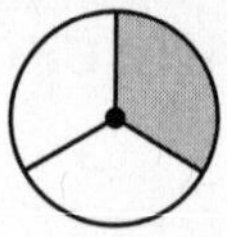

$\frac{1}{3} \times \frac{100}{1} = \frac{100}{3}$

$\begin{array}{r} \mathbf{33\frac{1}{3}\%} \\ 3\overline{)100} \\ \underline{9} \\ 10 \\ \underline{9} \\ 1 \end{array}$

Lesson 90, Warm-Up

a. **276**

b. **276**

c. $\mathbf{\frac{1}{2}, \frac{1}{3}, \frac{1}{4}}$

d. **1000 mL; quart**

e. **25**

Problem Solving

Rectangle's proportions may vary. One possibility:

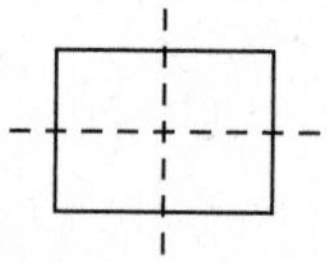

Lesson 90, Lesson Practice

a. $\frac{4 \div 4}{12 \div 4} = \mathbf{\frac{1}{3}}$

b. $\frac{6 \div 6}{18 \div 6} = \mathbf{\frac{1}{3}}$

c. $\frac{16 \div 8}{24 \div 8} = \mathbf{\frac{2}{3}}$

d. $\frac{4 \div 4}{16 \div 4} = \mathbf{\frac{1}{4}}$

e. $\frac{12 \div 4}{16 \div 4} = \mathbf{\frac{3}{4}}$

f. $\frac{60 \div 20}{100 \div 20} = \mathbf{\frac{3}{5}}$

g. $\frac{7}{16} + \frac{1}{16} = \frac{8}{16}$

$\frac{8 \div 8}{16 \div 8} = \mathbf{\frac{1}{2}}$

h. $\frac{3}{4} \times \frac{4}{5} = \frac{12}{20}$

$\frac{12 \div 4}{20 \div 4} = \mathbf{\frac{3}{5}}$

i. $\frac{19}{24} - \frac{1}{24} = \frac{18}{24}$

$\frac{18 \div 6}{24 \div 6} = \mathbf{\frac{3}{4}}$

j. $\frac{25}{100} = \mathbf{\frac{1}{4}}$

k. $\frac{60}{100} = \mathbf{\frac{3}{5}}$

l. $\frac{90}{100} = \mathbf{\frac{9}{10}}$

Lesson 90, Mixed Practice

1. **0**

2. $\mathbf{\frac{5}{10}; \frac{6}{10}}$

$\frac{5}{10} + \frac{6}{10} = \frac{11}{10} = \mathbf{1\frac{1}{10}}$

3. (a) **4 qt**

(b) 4 qt × 6 = **24 qt**

4. $\begin{array}{r} \overset{1}{1}4.7 \\ +\ \ 4.4 \\ \hline \mathbf{19.1} \end{array}$

5. **0.5; $\frac{1}{2}$; 50%**

6. **Sphere**

7. **B. $\overline{CD}$**

8. **Right triangle**

9. **One possibility: rotation (about the midpoint of $\overline{DB}$)**

10. $\frac{5}{6} + \frac{5}{6} = \frac{10}{6} = 1\frac{4}{6} = \mathbf{1\frac{2}{3}}$

11. $\frac{5}{6} \times \frac{2}{1} = \frac{10}{6} = 1\frac{4}{6} = \mathbf{1\frac{2}{3}}$

12. **4**

13. $\frac{1}{12} + \frac{7}{12} = \frac{8}{12} = \mathbf{\frac{2}{3}}$

14. $6\frac{2}{3} - \left(3\frac{3}{3} - \frac{1}{3}\right)$

$= 6\frac{2}{3} - 3\frac{2}{3} = \mathbf{3}$

15. $\frac{2}{3} \times \frac{3}{4} = \frac{6}{12} = \mathbf{\frac{1}{2}}$

16. $\begin{array}{r} \overset{1}{2}6.40 \\ +\ \ 2.64 \\ \hline \mathbf{29.04} \end{array}$

17. $\begin{array}{r} \overset{7}{\not{8}}.{}^{1}3\ 6 \\ -\ 4.\ 7\ 0 \\ \hline \mathbf{3.\ 6\ 6} \end{array}$

18. $\begin{array}{r} 40 \\ \times\ \ 40 \\ \hline \mathbf{1600} \end{array}$

19. $\sqrt{81} = \mathbf{9}$

20. $$\begin{array}{r} \mathbf{48} \\ 10\overline{)480} \\ \underline{40} \\ 80 \\ \underline{80} \\ 0 \end{array}$$

21. $$\begin{array}{r} 48 \\ 5\overline{)240} \\ \underline{20} \\ 40 \\ \underline{40} \\ 0 \end{array}$$

$n = \mathbf{48}$

22. **3**

23. $\frac{3}{4} \times \frac{3}{1} = \frac{9}{4} = \mathbf{2\frac{1}{4}}$

24. $\frac{3}{5} \times \mathbf{\frac{20}{20}} = \frac{60}{100}$

25. $$\begin{array}{r} {}^{1} \\ 75 \\ 35 \\ 35 \\ \underline{+\ 50} \\ \mathbf{195} \end{array}$$ **tokens**

26. $$\begin{array}{r} {}^{39} \\ \cancel{40}{}^{1}0 \text{ tokens} \\ \underline{-\ 19\,5 \text{ tokens}} \\ \mathbf{20\,5 \text{ tokens}} \end{array}$$

27. $\frac{6}{16} = \mathbf{\frac{3}{8}}$

28. (a) **5 in.**

(b) 5 in. × 4 in. = **20 in.**

29.

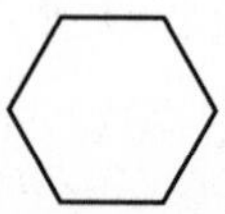

;

The pattern is a sequence of regular polygons with the number of sides in each polygon increasing by one.

30. (a) **1, 2, 4, 8, 16**

(b) **Odd**

(c) **4**

(d) **4**

INVESTIGATION 9

1.

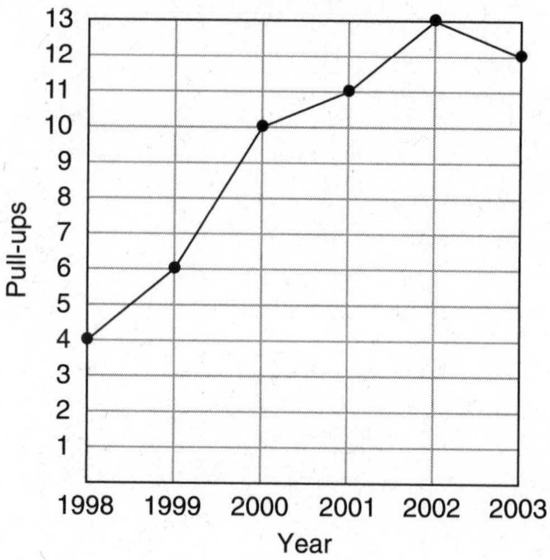

2. **2003**

3. **2000; We can tell because the line is steepest between 1999 and 2000.**

4. **$10,000; $9000**

5. **2000; $14,000**

6. **1998;** $12,000 − $7000 = **$5000**

7. **2000;** $14,000 − $10,000 = **$4000**

8. **an increase of about $4000**

9. **250,000 people**

10. **700,000 people**

11. 700,000 − 350,000 = **350,000 people**

12. 675,000 − 350,000 = **325,000 people**

13. **1984**

14. **the 1990s**

15. Naomi's scores improved, then declined, while Takeshi's scores never declined (they always improved or stayed the same).

Legend

Naomi ●——●

Takeshi ○——○

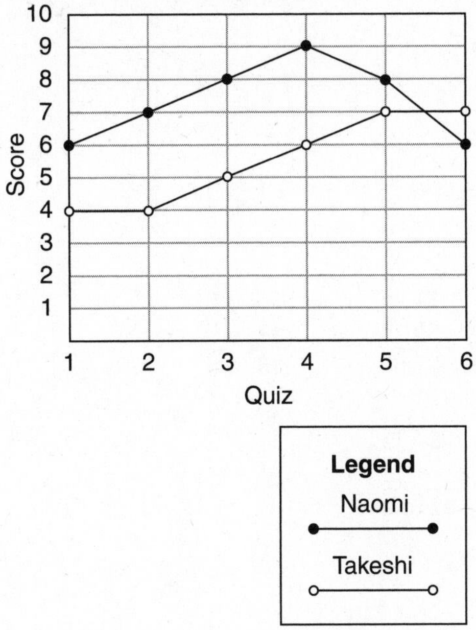

LESSON 91, WARM-UP

a. XIII

b. 8

Problem Solving

Year	First Day
2001	Monday
2002	Tuesday
2003	Wednesday
2004	Thursday
2005	Saturday
2006	Sunday
2007	Monday
2008	Tuesday
2009	Thursday
2010	Friday
2011	Saturday
2012	Sunday
2013	Tuesday
2014	Wednesday
2015	Thursday

LESSON 91, LESSON PRACTICE

a. $\dfrac{6}{4} \longrightarrow 4\overline{)6}$ $\quad 1\frac{2}{4}$

$\dfrac{4}{2}$

$1\frac{2}{4} = \mathbf{1\frac{1}{2}}$

b. $\dfrac{10}{6} \longrightarrow 6\overline{)10}$ $\quad 1\frac{4}{6}$

$\dfrac{6}{4}$

$1\frac{4}{6} = \mathbf{1\frac{2}{3}}$

c. $2\frac{8}{6} \longrightarrow 2 + 1\frac{2}{6} = 3\frac{2}{6}$

$3\frac{2}{6} = \mathbf{3\frac{1}{3}}$

d. $3\frac{10}{4} \longrightarrow 3 + 2\frac{2}{4} = 5\frac{2}{4}$

$5\frac{2}{4} = \mathbf{5\frac{1}{2}}$

e. $\dfrac{10}{4} \longrightarrow 4\overline{)10}$ $\quad 2\frac{2}{4}$

$\dfrac{8}{2}$

$2\frac{2}{4} = \mathbf{2\frac{1}{2}}$

f. $\dfrac{12}{8} \longrightarrow 8\overline{)12}$ $\quad 1\frac{4}{8}$

$\dfrac{8}{4}$

$1\frac{4}{8} = \mathbf{1\frac{1}{2}}$

g. $4\frac{14}{8} \longrightarrow 4 + 1\frac{6}{8} = 5\frac{6}{8}$

$5\frac{6}{8} = \mathbf{5\frac{3}{4}}$

h. $1\frac{10}{8} \longrightarrow 1 + 1\frac{2}{8} = 2\frac{2}{8}$

$2\frac{2}{8} = \mathbf{2\frac{1}{4}}$

i. $1\frac{5}{6} + 1\frac{5}{6} = 2\frac{10}{6} \longrightarrow 2 + 1\frac{4}{6} = 3\frac{4}{6}$

$3\frac{4}{6} = \mathbf{3\frac{2}{3}}$

j. $2\frac{3}{4} + 4\frac{3}{4} = 6\frac{6}{4} \longrightarrow 6 + 1\frac{2}{4} = 7\frac{2}{4}$

$7\frac{2}{4} = \mathbf{7\frac{1}{2}}$

k. $\frac{5}{3} \times \frac{3}{2} = \frac{15}{6} \longrightarrow 6\overline{)15}$ $= 2\frac{3}{6}$

$$\begin{array}{r} 2\frac{3}{6} \\ 6\overline{)15} \\ \underline{12} \\ 3 \end{array}$$

$2\frac{3}{6} = \mathbf{2\frac{1}{2}}$

l. $\frac{5}{8}$ **in.** + $\frac{5}{8}$ **in.** + $\frac{5}{8}$ **in.** + $\frac{5}{8}$ **in.** = $\frac{20}{8}$ **in.**

$\frac{20}{8}$ **in.** = $2\frac{4}{8}$ **in.**

$2\frac{4}{8}$ **in.** = $2\frac{1}{2}$ **inches**

LESSON 91, MIXED PRACTICE

1. 12 feet ÷ 2 = 6 feet

$$\begin{array}{r} 10 \\ \times\ 6 \text{ feet} \\ \hline \mathbf{60 \text{ feet}} \end{array}$$

2.

$$\begin{array}{r} \$6.50 \\ \times\ \ 5 \\ \hline \mathbf{\$32.50} \end{array}$$

3. **154,343,515**

4. (a) **4 laps**

(b) 4 × 5 = **20 laps**

5. $\frac{3}{4} = \mathbf{\frac{6}{8}}$

$\frac{5}{8} + \frac{6}{8} = \frac{11}{8} = \mathbf{1\frac{3}{8}}$

6. $1\frac{2}{6} = \mathbf{1\frac{1}{3}}$

7. **C.** $\overline{RT}$

8. $\frac{1}{2}$ of 2 ⊜ $2 \times \frac{1}{2}$

1   1

9. **Cylinder**

10.

$$\begin{array}{r} \overset{1}{3}.2 \text{ cm} \\ 1.8 \text{ cm} \\ +\ 1.8 \text{ cm} \\ \hline \mathbf{6.8 \text{ cm}} \end{array}$$

11. $1\frac{3}{4} + 1\frac{3}{4} = 2\frac{6}{4}$

$2 + 1\frac{2}{4} = 3\frac{2}{4} = \mathbf{3\frac{1}{2}}$

12. $5\frac{7}{8} - 1\frac{3}{8} = 4\frac{4}{8} = \mathbf{4\frac{1}{2}}$

13. $\frac{3}{1} \times \frac{3}{8} = \frac{9}{8} = \mathbf{1\frac{1}{8}}$

14.

$$\begin{array}{r} \$\overset{1}{1}.25 \\ +\ \$0.35 \\ \hline \$1.60 \end{array} \qquad \begin{array}{r} \$\overset{9}{\not{1}\not{0}}.{}^{1}0\,0 \\ -\ \ \$1.\,6\,0 \\ \hline \mathbf{\$8.\,4\,0} \end{array}$$

15.

$$\begin{array}{r} \$4.32 \\ \times\ \ 5 \\ \hline \mathbf{\$21.60} \end{array}$$

16.

$$\begin{array}{r} 416 \\ \times\ 740 \\ \hline 16\,640 \\ +\ 291\,200 \\ \hline \mathbf{307{,}840} \end{array}$$

17.

$$\begin{array}{r} 2.3 \\ +\ 0.65 \\ \hline 2.95 \end{array} \qquad \begin{array}{r} \overset{3}{\not{4}}.\,\overset{14}{\not{5}}{}^{1}1 \\ -\ 2.\,9\,5 \\ \hline \mathbf{1.\,5\,6} \end{array}$$

18.

$$\begin{array}{r} \mathbf{120} \\ 8\overline{)960} \\ \underline{8}\ \ \\ 16\ \\ \underline{16}\ \\ 00 \\ \underline{0} \\ 0 \end{array}$$

19.

$$\begin{array}{r} \mathbf{120} \\ 80\overline{)9600} \\ \underline{80}\ \ \\ 160\ \\ \underline{160}\ \\ 00 \\ \underline{0} \\ 0 \end{array}$$

20.

$$\begin{array}{r} \mathbf{\$2.40} \\ 5\overline{)\$12.00} \\ \underline{10}\ \ \ \ \\ 2\,0\ \ \\ \underline{2\,0}\ \ \\ 00 \\ \underline{0} \\ 0 \end{array}$$

21. $\frac{5}{2} \times \frac{2}{3} = \frac{10}{6} = 1\frac{4}{6} = \mathbf{1\frac{2}{3}}$

22. **2**

23. $\frac{2}{3} \times \frac{6}{1} = \frac{12}{3} = \mathbf{4}$

24. **B. It multiplies by 3.**

25. $12 \div 3 = \mathbf{4}$

26. **1, 4, 4, 1, 4**

27. **List: 6, 6, 6, 7, 8, 8, 9**
7

28. **6**

29. $9 - 6 = \mathbf{3}$

30. $6 + 6 + 6 + 7 + 8 + 8 + 9 = 50$

$$\begin{array}{r} \mathbf{7\frac{1}{7}} \\ 7\overline{)50} \\ \underline{49} \\ 1 \end{array}$$

Lesson 92, Warm-Up

a. $\mathbf{\frac{3}{4}, \frac{2}{3}, \frac{1}{2}}$

b. **$500**

c. **$1000**

d. $\mathbf{33\frac{1}{3}}$

e. **111**

f. **XVII**

g. **12**

Problem Solving
(B, A, C), (B, C, A), (C, A, B), (C, B, A)

Lesson 92, Lesson Practice

a.
$$\begin{array}{r} \mathbf{23} \\ 11\overline{)253} \\ \underline{22} \\ 33 \\ \underline{33} \\ 0 \end{array}$$

b.
$$\begin{array}{r} \mathbf{12\ R\ 1} \\ 21\overline{)253} \\ \underline{21} \\ 43 \\ \underline{42} \\ 1 \end{array}$$

c.
$$\begin{array}{r} \mathbf{13} \\ 31\overline{)403} \\ \underline{31} \\ 93 \\ \underline{93} \\ 0 \end{array}$$

d.
$$\begin{array}{r} \mathbf{21\ R\ 1} \\ 12\overline{)253} \\ \underline{24} \\ 13 \\ \underline{12} \\ 1 \end{array}$$

e.
$$\begin{array}{r} \mathbf{25} \\ 12\overline{)300} \\ \underline{24} \\ 60 \\ \underline{60} \\ 0 \end{array}$$

f.
$$\begin{array}{r} \mathbf{22\ R\ 4} \\ 23\overline{)510} \\ \underline{46} \\ 50 \\ \underline{46} \\ 4 \end{array}$$

g.
$$\begin{array}{r} \mathbf{12\ players} \\ 12\overline{)144} \\ \underline{12} \\ 24 \\ \underline{24} \\ 0 \end{array}$$

h.
$$\begin{array}{r} \mathbf{22\ R\ 22} \\ 30\overline{)682} \\ \underline{60} \\ 82 \\ \underline{60} \\ 22 \end{array}$$

i. $$\begin{array}{r} \mathbf{22\ R\ 5} \\ 32\overline{)709} \\ \underline{64} \\ 69 \\ \underline{64} \\ 5 \end{array}$$

j. $$\begin{array}{r} \mathbf{20\ R\ 20} \\ 43\overline{)880} \\ \underline{86} \\ 20 \\ \underline{0} \\ 20 \end{array}$$

k. $$\begin{array}{r} \mathbf{42} \\ 22\overline{)924} \\ \underline{88} \\ 44 \\ \underline{44} \\ 0 \end{array}$$

l. $$\begin{array}{r} \mathbf{34\ R\ 2} \\ 22\overline{)750} \\ \underline{66} \\ 90 \\ \underline{88} \\ 2 \end{array}$$

m. $$\begin{array}{r} \mathbf{6} \\ 21\overline{)126} \\ \underline{126} \\ 0 \end{array}$$

n. $$\begin{array}{r} \mathbf{31\ R\ 3} \\ 21\overline{)654} \\ \underline{63} \\ 24 \\ \underline{21} \\ 3 \end{array}$$

o. $$\begin{array}{r} \mathbf{22\ R\ 8} \\ 41\overline{)910} \\ \underline{82} \\ 90 \\ \underline{82} \\ 8 \end{array}$$

p. $$\begin{array}{r} \mathbf{61\ R\ 9} \\ 21\overline{)1290} \\ \underline{126} \\ 30 \\ \underline{21} \\ 9 \end{array}$$

LESSON 92, MIXED PRACTICE

1. **Possibilities include:**

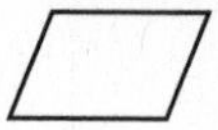 and 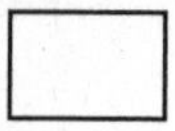

2. $2\frac{1}{2}$ hours

$$60 \text{ min.} \times 2 = 120 \text{ min}$$
$$120 \text{ min.} + 30 \text{ min.} = \mathbf{150\ minutes}$$

3. **3.9**

4. $$\begin{array}{r} 12 \\ 2\overline{)24} \\ \underline{2} \\ 04 \\ \underline{4} \\ 0 \end{array} \qquad \begin{array}{r} 12 \\ \underline{\times\ 3} \\ \mathbf{36\ eggs} \end{array}$$

5. $4 \times 4 \times 4 =$ **64 cubes**

6. (a) **3 apples**

(b) $3 \times 4 =$ **12 apples**

7. **Pyramid; 8 edges**

8. $\frac{50}{100} = \mathbf{0.50; \frac{1}{2}; 50\%}$

9. **D. 0.05**

10. $40 \text{ mm} \div 2 = 20 \text{ mm}$

$40 \text{ mm} + 20 \text{ mm} + 20 \text{ mm} =$ **80 mm**

11. $$\begin{array}{r} 8.7 \\ \underline{+\ 6.25} \\ \mathbf{14.95} \end{array}$$

12. $$\begin{array}{r} \overset{0}{\not{1}}{}^{1}2.75 \\ \underline{-\ \ 4.20} \\ \mathbf{8.55} \end{array}$$

13. $4 \times 4 \times 4 =$ **64**

14. $$\begin{array}{r} \$125 \\ \underline{\times\ \ \ 8} \\ \mathbf{\$1000} \end{array}$$

15. $10 - 8 =$ **2**

16. $13\overline{)293}$ = **22 R 7**

```
    22 R 7
13)293
   26
    33
    26
     7
```

17. $24\overline{)510}$ = **21 R 6**

```
    21 R 6
24)510
   48
    30
    24
     6
```

18. $3\frac{5}{8} + 1\frac{7}{8} = 4\frac{12}{8} = 5\frac{4}{8} = \mathbf{5\frac{1}{2}}$

19. $4\frac{5}{5} - 1\frac{2}{5} = \mathbf{3\frac{3}{5}}$

20. $\frac{1}{3} \times \frac{5}{1} = \frac{5}{3} = \mathbf{1\frac{2}{3}}$

21. $\frac{3}{4} \times \frac{4}{3} = \frac{12}{12} = \mathbf{1}$

22. $\frac{6}{10} = \frac{3}{5}$

$\frac{3}{5} \div \frac{1}{5} = \mathbf{3}$

23. $\frac{2}{5} = \mathbf{\frac{4}{10}}$

$\frac{4}{10} + \frac{1}{10} = \frac{5}{10} = \mathbf{\frac{1}{2}}$

24. 4 ft × 3 ft = **12 sq. ft**

25. (a) **Unlikely**

(b) **Likely**

(c) **Impossible**

26. **Round each price to the nearest dollar and then add.**

```
  $2.00
  $1.00
  $4.00
+ $1.00
  $8.00
```

27.

```
   1 3
  $2.19
  $1.19
  $3.87
+ $1.39
  $8.64
```

28. $3\overline{)\$8.64}$ = **$2.88**

```
  $2.88
3)$8.64
   6
   2 6
   2 4
     24
     24
      0
```

29. **4, 4, 1, 4, 4**

30.

```
       7
  2 1. 8 ¹1 seconds
- 2 1. 3 4 seconds
  0. 4 7 second
```

Lesson 93, Warm-Up

a. $\mathbf{\frac{1}{4}, \frac{1}{3}, \frac{1}{2}}$

b. **238**

c. **109**

d. **$50**

e. **$5**

f. **$0.50**

g. **4**

h. **XI**

i. **9**

Problem Solving

First Cube	Second Cube
1	4
2	3
3	2
4	1

Lesson 93, Lesson Practice

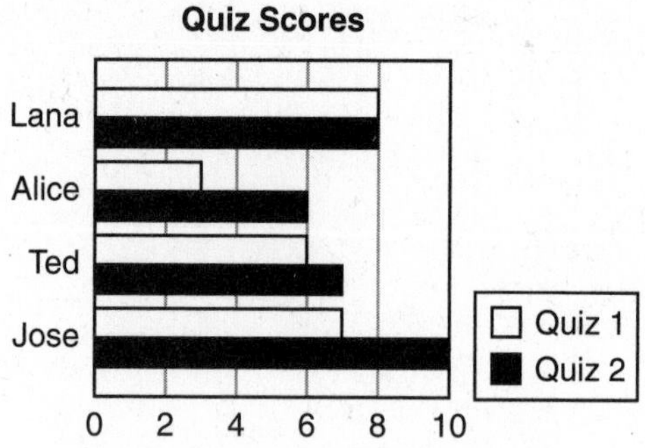

Lesson 93, Mixed Practice

1. 2 pt = 1 qt
2 × 2 = 4 pounds
About 4 pounds

2.
$$\begin{array}{r} \$0.85 \\ 3\overline{)\$2.55} \\ \underline{2\,4} \\ 15 \\ \underline{15} \\ 0 \end{array} \qquad \begin{array}{r} \$0.85 \\ \times \quad 4 \\ \hline \mathbf{\$3.40} \end{array}$$

3.
$$\begin{array}{r} \mathbf{150} \\ 2\overline{)300} \\ \underline{2} \\ 10 \\ \underline{10} \\ 00 \\ \underline{0} \\ 0 \end{array}$$
150 marbles

4. $\frac{5}{10} = \mathbf{0.5}$

$\frac{5}{10} = \mathbf{\frac{1}{2}}$

$\frac{50}{100} = \mathbf{50\%}$

5. (a) **5 plums**

(b) 5 × 3 = **15 plums**

6. **12 − 9 = 3**

7. $\frac{2}{3}$ of 3 ⊜ $3 \times \frac{2}{3}$

8. 18 ÷ 3 = 6
n = 6
2(6) + 5 = 12 + 5 = 17
B. 17

9. **Cube**

10. **D. $\angle AOC$**

11.
$$\begin{array}{r} 1\frac{3}{5} \\ +\ 2\frac{4}{5} \\ \hline 3\frac{7}{5} \end{array} = \mathbf{4\frac{2}{5}}$$

12.
$$\begin{array}{r} 4\frac{5}{8} \\ -\ \frac{1}{8} \\ \hline 4\frac{4}{8} \end{array} = \mathbf{4\frac{1}{2}}$$

13.
$$\begin{array}{r} 6\frac{5}{6} \\ -\ 1\frac{5}{6} \\ \hline \mathbf{5} \end{array}$$

14. **8**

15. $\frac{8}{10} \times \frac{5}{10} = \frac{40}{100} = \mathbf{\frac{2}{5}}$

16. $\frac{1}{5} \times \frac{10}{1} = \frac{10}{5} = \mathbf{2}$

17.
$$\begin{array}{r} 5.67 \\ -\ 0.80 \\ \hline 4.87 \end{array} \qquad \begin{array}{r} 12.34 \\ -\ 4.87 \\ \hline \mathbf{7.47} \end{array}$$

18.
$$\begin{array}{r} \$20.00 \\ -\ \$6.55 \\ \hline \$13.45 \end{array} \qquad \begin{array}{r} \mathbf{\$2.69} \\ 5\overline{)\$13.45} \\ \underline{10} \\ 3\,4 \\ \underline{3\,0} \\ 45 \\ \underline{45} \\ 0 \end{array}$$

19.
$$\begin{array}{r} \$0.56 \\ \times \quad 10 \\ \hline \mathbf{\$5.60} \end{array}$$

20.
$$\begin{array}{r} 78 \\ \times \quad 6 \\ \hline 468 \end{array} \qquad \begin{array}{r} 468 \\ \times \quad 900 \\ \hline \mathbf{421{,}200} \end{array}$$

21. **31 R 9**
31)970
93
40
31
9

22. 81 − 3 = **78**

23. $\frac{3}{4} = \mathbf{\frac{9}{12}}; \frac{1}{6} = \mathbf{\frac{2}{12}}$
$\frac{9}{12} + \frac{2}{12} = \mathbf{\frac{11}{12}}$

24. **3 cm**

25. 3 cm − 1 cm = 2 cm
3 cm + 3 cm + 2 cm + 2 cm = **10 cm**

26. 3 cm × 2 cm = **6 sq. cm**

27. (a) **ш, Ǝ, m**

(b) **4**

(c) **Rotation**

28. 15
4)60
4
20
20
0
About 15 times

29. $\frac{1}{4} \times \frac{100}{1} = \frac{100}{4}$

25%
4)100
8
20
20
0

30. $\frac{1}{2} \times \frac{5}{5} = \frac{5}{10}$
0.5

Lesson 94, Warm-Up

a. $\mathbf{\frac{1}{2}, \frac{1}{3}, \frac{1}{4}}$

b. **5**

c. **10**

d. $\mathbf{7\frac{1}{2}}$

e. **0**

f. **XX**

g. **14**

Problem Solving

There are two pints in one quart, two quarts in one half gallon, and two half gallons in one gallon. So there are 8 pints in one gallon.
8 pints − 1 quart = 8 pints − 2 pints
= **6 pints**

Lesson 94, Lesson Practice

a. **41 R 13**
19)792
76
32
19
13

b. **20**
30)600
60
00
0
0

c. **4 R 5**
29)121
116
5

d. **31 R 1**
29)900
87
30
29
1

e. **17 R 13**
48)829
48
349
336
13

f.
```
     41 R 21
29)1210
   116
    50
    29
    21
```

g.
```
     32
28)896
   84
    56
    56
     0
```

h.
```
     43 R 8
18)782
   72
    62
    54
     8
```

i.
```
     30 R 30
39)1200
   117
    30
     0
    30
```

Lesson 94, Mixed Practice

1. **11, 31, 41**

2.
```
   24
×   8
  192
```
(**192**)

3.
```
   220 yards
3)660 yards        220 yards
  6              ×       6
  06              1320 yards
   6
   00
    0
    0
```
(**1320 yards**)

4. **4.9**

5.
```
   19 trombone players
4)76
  4
  36
  36
   0
```
(**19 trombone players**)

6. (a) $\mathbf{\frac{1}{10}}$

(b) **10 dimes**

(c) 10 × 4 = **40 dimes**

7. **B. 786 ÷ 19**

8. (a) **4**

(b) **3**

9. **Cone**

10. **B. Right**

11.
```
   1 1 2
  $63.75
   $1.48
   $0.59
+  $5.00
  $70.82
```
(**$70.82**)

12.
```
  101        9 10
−  10       1̸0̸ 1̸¹0
   91     −    9 1
               9 1 9
```
(**919**)

13.
```
  $3.48
×     7
 $24.36
```
(**$24.36**)

14.
```
   25
×  25
  125
  500
  625
```
(**625**)

15.
```
     41 R 7
19)786
   76
    26
    19
     7
```

16. 6 + 8 = **14**

17.
```
     31 R 22
38)1200
   114
    60
    38
    22
```

18. $\frac{5}{6} + \frac{5}{6} + \frac{5}{6} = \frac{15}{6} = 2\frac{3}{6} = \mathbf{2\frac{1}{2}}$

19. $\frac{5}{6} \times \frac{3}{1} = \frac{15}{6} = 2\frac{3}{6} = \mathbf{2\frac{1}{2}}$

20. $\frac{8 \div 4}{12 \div 4} = \mathbf{\frac{2}{3}}$

21. $3 - \left(1\frac{4}{4} - \frac{1}{4}\right)$

$2\frac{4}{4} - 1\frac{3}{4} = \mathbf{1\frac{1}{4}}$

22. $\frac{1}{3} \times \frac{3}{4} = \frac{3}{12} = \mathbf{\frac{1}{4}}$

23. $\frac{2}{3} = \mathbf{\frac{8}{12}}$

$\frac{11}{12} - \frac{8}{12} = \frac{3}{12} = \mathbf{\frac{1}{4}}$

24. 67 in. − 60 in. = **7 inches**

25. 12 inches
× 5
60 inches
12th birthday

26. **C. Pentagon**

27. (a) 3 ft + 3 ft + 3 ft + 3 ft = **12 ft**

(b) 1 yd + 1 yd + 1 yd + 1 yd = **4 yd**

28. (a) 3 ft × 3 ft = **9 sq. ft**

(b) 1 yd × 1 yd = **1 sq. yd**

29. 1 yd ⊜ 3 ft

30. 1 sq. yd ⊜ 9 sq. ft

Lesson 95, Warm-Up

a. $\mathbf{2\frac{1}{2}}$

b. $\mathbf{1\frac{2}{3}}$

c. $\mathbf{1\frac{1}{4}}$

d. **3**

e. **6**

f. **9**

g. **XV**

h. **26**

Problem Solving

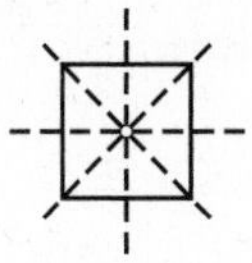

Lesson 95, Lesson Practice

a. $\mathbf{\frac{5}{4}}$

b. $\mathbf{\frac{5}{6}}$

c. $\mathbf{\frac{1}{3}}$

d. $\mathbf{\frac{8}{7}}$

e. $\mathbf{\frac{8}{3}}$

f. $\mathbf{\frac{1}{5}}$

g. $\mathbf{\frac{10}{3}}$

h. $\mathbf{\frac{12}{5}}$

i. $\mathbf{\frac{1}{2}}$

j. $\mathbf{\frac{5}{1}}$

k. $\mathbf{\frac{1}{10}}$

l. $\mathbf{\frac{1}{1}}$

m. $\mathbf{\frac{5}{3}}$ **or** $\mathbf{1\frac{2}{3}}$

n. $\mathbf{\frac{5}{4}}$ **or** $\mathbf{1\frac{1}{4}}$

o. **The product is 1. See student work.**

LESSON 95, MIXED PRACTICE

1.
$$\begin{array}{r} \overset{1}{3}5 \text{ lb} \\ 42 \text{ lb} \\ +\ 34 \text{ lb} \\ \hline 111 \text{ lb} \end{array} \qquad \begin{array}{r} \mathbf{37\ lb} \\ 3\overline{)111 \text{ lb}} \\ 9 \\ \hline 21 \\ 21 \\ \hline 0 \end{array}$$

2.
$3 \times 4 = 12$
$12 + 2 + 5 = \mathbf{19\ bones}$

3. $\frac{25}{100} = \mathbf{0.25, \frac{1}{4}; 25\%}$

4. $\frac{2}{3} \times \frac{3}{2} = \frac{6}{6} = \mathbf{1}$

5. (a) $\mathbf{\frac{1}{4}}$

(b) **4 quarters**

(c) $4 \times 5 = \mathbf{20\ quarters}$

6. $\mathbf{\frac{4}{3}}$

$\frac{3}{4} \times \frac{4}{3} = \frac{12}{12} = \mathbf{1}$

7. **B. 500 ÷ 25**

8. (a) $\mathbf{\frac{1}{6}}$

(b) $\mathbf{\frac{4}{1}}$ (or **4**)

9. **C. ∠*KNL***

10.
$$\begin{array}{r} \$2\overset{19}{\not{2}}\not{0}.\overset{9}{\not{0}}{}^{1}0 \\ -\ \$4.72 \\ \hline \$15.28 \end{array} \qquad \begin{array}{r} \mathbf{\$1.91} \\ 8\overline{)\$15.28} \\ 8 \\ \hline 72 \\ 72 \\ \hline 08 \\ 8 \\ \hline 0 \end{array}$$

11.
$$\begin{array}{r} \$1.25 \\ \times\ 160 \\ \hline 7500 \\ 12500 \\ \hline \mathbf{\$200.00} \end{array}$$

12.
$$\begin{array}{r} 1.4 \\ +\ 0.28 \\ \hline 1.68 \end{array} \qquad \begin{array}{r} 2\overset{4}{\not{5}}.\overset{13}{\not{4}}{}^{1}5 \\ -\ 1.68 \\ \hline \mathbf{23.77} \end{array}$$

13.
$$\begin{array}{r} 100 \\ \times\ 100 \\ \hline \mathbf{10{,}000} \end{array}$$

14.
$$\begin{array}{r} \mathbf{4\ R\ 16} \\ 31\overline{)140} \\ 124 \\ \hline 16 \end{array}$$

15.
$$\begin{array}{r} 21 \\ 27\overline{)567} \\ 54 \\ \hline 27 \\ 27 \\ \hline 0 \end{array}$$

$x = \mathbf{21}$

16. $\frac{15 \div 5}{25 \div 5} = \mathbf{\frac{3}{5}}$

17. $1\frac{5}{6} + 1\frac{5}{6} = 2\frac{10}{6} = 3\frac{4}{6} = \mathbf{3\frac{2}{3}}$

18. $4\frac{5}{6} - 1\frac{1}{6} = 3\frac{4}{6} = \mathbf{3\frac{2}{3}}$

19. $\frac{3}{8} \times \frac{24}{1} = \frac{72}{8} = \mathbf{9}$

20. $\frac{3}{1} \times \frac{4}{5} = \frac{12}{5} = \mathbf{2\frac{2}{5}}$

21. **9**

22. $\frac{3}{4} = \mathbf{\frac{9}{12}}; \frac{1}{6} = \mathbf{\frac{2}{12}}$

$\frac{9}{12} - \frac{2}{12} = \mathbf{\frac{7}{12}}$

23. $10\overline{)123}$ → $\mathbf{12\frac{3}{10}}$

10
23
20
3

24. **14 toothpicks**

25. **12 small squares**

26. **8°C**

27.

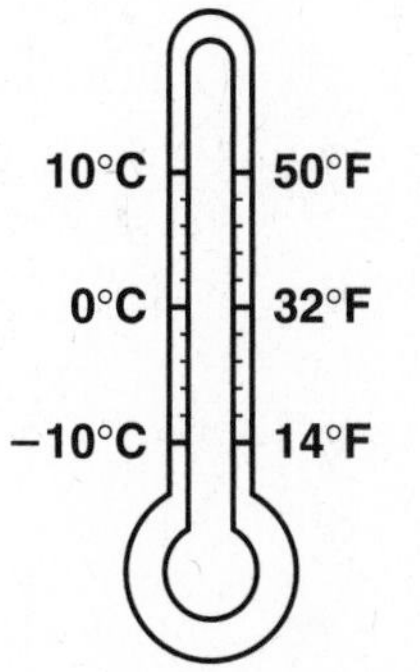

50° − 18° = 32°
32° − 18° = 14°

28. 10 − 6 = **4**

29. **C. Unlikely**

30. **D.**

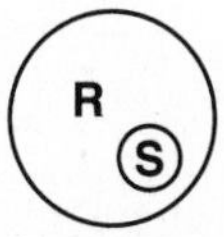

Every square is a rhombus, but some rhombuses are not squares.

Lesson 96, Warm-Up

a. $\mathbf{3\frac{1}{2}}$

b. $\mathbf{\frac{6}{5}}$

c. **3**

d. **6**

e. **9**

f. **XXXIV**

g. **23**

Problem Solving

We look for a pattern. Bead numbers that have a remainder of 0 when divided by 3 are blue. Bead numbers that have a remainder of 1 when divided by 3 are red. Bead numbers that have a remainder of 2 when divided by 3 are white. 100 ÷ 3 = 33 R 1, so the one hundredth bead will be **red.**

Lesson 96, Lesson Practice

a. $\frac{1}{3} \times \frac{2}{1} = \mathbf{\frac{2}{3}}$

b. $\frac{2}{3} \times \frac{4}{3} = \mathbf{\frac{8}{9}}$

c. $\frac{2}{3} \times \frac{4}{1} = \frac{8}{3} = \mathbf{2\frac{2}{3}}$

d. $\frac{1}{2} \times \frac{3}{1} = \frac{3}{2} = \mathbf{1\frac{1}{2}}$

e. $\frac{3}{4} \times \frac{3}{2} = \frac{9}{8} = \mathbf{1\frac{1}{8}}$

f. $\frac{3}{1} \times \frac{4}{3} = \frac{12}{3} = \mathbf{4}$

g. $\frac{2}{1} \times \frac{3}{1} = \frac{6}{1} = \mathbf{6}$

h. $\frac{3}{1} \times \frac{3}{2} = \frac{9}{2} = \mathbf{4\frac{1}{2}}$

i. $\frac{10}{1} \times \frac{6}{5} = \frac{60}{5} = \mathbf{12}$

j. $\frac{3}{4} \div \frac{1}{3}$

$\frac{3}{4} \times \frac{3}{1} = \frac{9}{4} = \mathbf{2\frac{1}{4}}$

k. $\frac{1}{3} \div \frac{3}{4}$

$\frac{1}{3} \times \frac{4}{3} = \mathbf{\frac{4}{9}}$

Lesson 96, Mixed Practice

1.

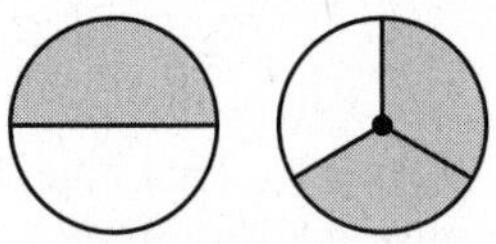

2. $\frac{1}{2} \times \frac{1}{2} = \mathbf{\frac{1}{4}}$

$\frac{1}{4} \times \frac{100}{1} = \frac{100}{4}$

$$\begin{array}{r} \mathbf{25\%} \\ 4\overline{)100} \\ \underline{8} \\ 20 \\ \underline{20} \\ 0 \end{array}$$

3. $\frac{2}{3} \times \frac{12}{1} = \frac{24}{3} = \mathbf{8}$

4.
$$\begin{array}{r} 700 \\ \times\ 500 \\ \hline \mathbf{350{,}000} \end{array}$$

5. **93,814,200**

6. **B.** $\mathbf{3 \div \frac{1}{10}}$

7. $\frac{1}{4} = \mathbf{\frac{5}{20}}$; $\frac{1}{5} = \mathbf{\frac{4}{20}}$

$\frac{5}{20} + \frac{4}{20} = \mathbf{\frac{9}{20}}$

8. (a) $\frac{1}{1} \times \frac{10}{1} = \frac{10}{1} = \mathbf{10}$

(b) $\frac{3}{1} \times \frac{10}{1} = \frac{30}{1} = \mathbf{30}$

9. **3, 6, 9, 12**

10. $\frac{3}{4} \times \frac{10}{1} = \frac{30}{4}$

$$\begin{array}{r} 7\frac{2}{4} \\ 4\overline{)30} \\ \underline{-28} \\ 2 \end{array} \quad = \mathbf{7\frac{1}{2}\ tons}$$

11. 3 cm + 4 cm = 7 cm

10 cm − 7 cm = **3 cm**

12. $\frac{75}{100} = \mathbf{0.75}$

$\frac{75}{100} = \mathbf{\frac{3}{4}}$

$\frac{75}{100} = \mathbf{75\%}$

13. $\frac{1}{3} \times \frac{4}{1} = \frac{4}{3} = \mathbf{1\frac{1}{3}}$

14. $\frac{1}{4} \times \frac{3}{1} = \mathbf{\frac{3}{4}}$

15. $\frac{3}{1} \times \frac{2}{1} = \frac{6}{1} = \mathbf{6}$

16.
$$\begin{array}{r} 3.75 \\ -\ 1.40 \\ \hline 2.35 \end{array}$$

$m = \mathbf{2.35}$

17.
$$\begin{array}{r} \overset{1}{3}.75 \\ +\ 1.40 \\ \hline 5.15 \end{array}$$

$m = \mathbf{5.15}$

18. $\frac{1}{10} \times \mathbf{\frac{10}{10}} = \frac{10}{100}$

19.
$$\begin{array}{r} \$0.47 \\ \times\ \ 20 \\ \hline \mathbf{\$9.40} \end{array}$$

$$\begin{array}{r} \mathbf{37\ R\ 13} \\ 15\overline{)568} \\ \underline{45} \\ 118 \\ \underline{105} \\ 13 \end{array}$$

21.
$$\begin{array}{r} \mathbf{14\ R\ 7} \\ 30\overline{)427} \\ \underline{30} \\ 127 \\ \underline{120} \\ 7 \end{array}$$

22.
$$\begin{array}{r} \$5.04 \\ 6\overline{)\$30.24} \\ \underline{30} \\ 0\ 2 \\ \underline{0} \\ 24 \\ \underline{24} \\ 0 \end{array}$$

$m = \$5.04$

23. $\dfrac{2}{3} \times \dfrac{1}{2} = \dfrac{2}{6} = \dfrac{1}{3}$

$\dfrac{5}{1} \times \dfrac{1}{3} = \dfrac{5}{3} = 1\dfrac{2}{3}$

24. $1\dfrac{1}{4} + 2 = 3\dfrac{1}{4}$

$4\dfrac{4}{4} - 3\dfrac{1}{4} = 1\dfrac{3}{4}$

25. $\sqrt{100}\ \overset{\textstyle <}{\bigcirc}\ 5^2$
$\quad\ 10 \qquad\quad 25$

26. **20**

27.
$$\begin{array}{r} \overset{2}{\cancel{3}}{}^{1}1 \\ -\ 1\ 8 \\ \hline \mathbf{1\ 3} \end{array}$$

28. 18, 20, 20, 20, 20, 22, 24, $\boxed{26}$, 27, 28, 28, 30, 30, 31, 31
Median: 26

29. **Isosceles triangle**

30. **Reflection**

LESSON 97, WARM-UP

a. $1\dfrac{1}{6};\ 1\dfrac{1}{3};\ 1\dfrac{1}{2}$

b. 3

c. $33\dfrac{1}{3}$

d. $66\dfrac{2}{3}$

e. $66\dfrac{2}{3}$

f. 7

g. XXXV

h. 34

Problem Solving

Three combinations of two socks can be made: one red, one white; one red, one blue; and one white, one blue. Six permutations of two socks can be made: **(R, W), (W, R), (R, B), (B, R), (W, B), (B, W)**

LESSON 97, LESSON PRACTICE

a. $\dfrac{30}{20} = \dfrac{3}{2}$

b. $\dfrac{20}{30} = \dfrac{2}{3}$

c. $\dfrac{8}{10} = \dfrac{4}{5}$

d. $\dfrac{10}{8} = \dfrac{5}{4}$

LESSON 97, MIXED PRACTICE

1. $\dfrac{15}{10} = \dfrac{3}{2}$

2. $\dfrac{1}{4} + \dfrac{1}{4} = \dfrac{1}{2}$

3.
$$\begin{array}{r} \overset{1\,1}{\$4.00} \\ \$0.75 \\ \$0.20 \\ +\ \$0.05 \\ \hline \$5.00 \end{array} \qquad \begin{array}{r} \mathbf{\$2.50} \\ 2\overline{)\$5.00} \\ \underline{4} \\ 1\,0 \\ \underline{1\,0} \\ 00 \\ \underline{0} \\ 0 \end{array}$$

4. $\frac{1}{2} \div \frac{1}{8}$

$\frac{1}{2} \times \frac{8}{1} = \frac{8}{2} = \mathbf{4}$

5. **2.5**

$2\frac{5}{10} = \mathbf{2\frac{1}{2}}$

6.
$$\begin{array}{r} 1\,\overset{1}{\not2}.\,\overset{10}{\not1}{}^{1}1 \\ -\ 1\,1.\,1\,2 \\ \hline \mathbf{0.\,9\,9} \end{array}$$

7. (a) $\mathbf{\frac{1}{4}}$

(b) **4 quarts**

(c) 4 × 4 = **16 quarts**

8. $\frac{2}{3} = \mathbf{\frac{10}{15}}$; $\frac{2}{5} = \mathbf{\frac{6}{15}}$

$\frac{10}{15} - \frac{6}{15} = \mathbf{\frac{4}{15}}$

9. $\frac{5}{10} = \mathbf{0.5}$

$\frac{5}{10} = \mathbf{\frac{1}{2}}$

10. $\frac{1}{2} \div 2 \;\text{(<)}\; 2 \div \frac{1}{2}$

$\frac{1}{2} \times \frac{1}{2} = \frac{1}{4} \qquad \frac{2}{1} \times \frac{2}{1} = \frac{4}{1} = 4$

11. 30 mm + 40 mm = 70 mm
90 mm − 70 mm = **20 mm**

12. $\frac{3}{1} \times \frac{3}{2} = \frac{9}{2} = \mathbf{4\frac{1}{2}}$

13. $\frac{2}{3} \times \frac{1}{3} = \mathbf{\frac{2}{9}}$

14. $\frac{7}{10} + \frac{7}{10} = \frac{14}{10} = 1\frac{4}{10} = \mathbf{1\frac{2}{5}}$

15.
$$\begin{array}{r} \overset{2\,1\ 1}{43.15} \\ 8.69 \\ 7.2 \\ +\ \ 5.0 \\ \hline \mathbf{64.04} \end{array}$$

16.
$$\begin{array}{r} \$\overset{9}{1\!\!\!/0\!\!\!/}.\,\overset{9}{0\!\!\!/}{}^{1}0 \\ -\ \$\ 0.\,1\,9 \\ \hline \$9.\,8\,1 \end{array} \qquad \begin{array}{r} \mathbf{\$1.09} \\ 9\overline{)\$9.81} \\ \underline{9} \\ 0\,8 \\ \underline{0} \\ 81 \\ \underline{81} \\ 0 \end{array}$$

17.
$$\begin{array}{r} \$0.72 \\ \times\quad\ 6 \\ \hline \mathbf{\$4.32} \end{array}$$

18.
$$\begin{array}{r} 35 \\ \times\ 35 \\ \hline 175 \\ 1050 \\ \hline \mathbf{1225} \end{array}$$

19.
$$\begin{array}{r} \mathbf{20\ R\ 20} \\ 24\overline{)500} \\ \underline{48} \\ 20 \\ \underline{0} \\ 20 \end{array}$$

20. $\frac{50}{100} = \mathbf{\frac{1}{2}}$

21.
$$\begin{array}{r} 102 \\ 12\overline{)1224} \\ \underline{12} \\ 02 \\ \underline{0} \\ 24 \\ \underline{24} \\ 0 \end{array}$$

$y = \mathbf{102}$

22. $2\frac{4}{4} - 1\frac{3}{4} = 1\frac{1}{4}$

$5\frac{3}{4} - 1\frac{1}{4} = 4\frac{2}{4} = \mathbf{4\frac{1}{2}}$

23. $1\frac{1}{4} + 1\frac{1}{4} + 1\frac{1}{4} + 1\frac{1}{4} = 4\frac{4}{4} = \mathbf{5}$

24. $\frac{3}{10} \times \frac{10}{10} = \mathbf{\frac{30}{100}}$

25. (a) $\mathbf{1\frac{1}{4}}$ **inches**

(b) $1\frac{1}{4}$ in. $+ 1\frac{1}{4}$ in. $+ 1\frac{1}{4}$ in. $+ 1\frac{1}{4}$ in.

$= 4\frac{4}{4} =$ **5 inches**

26. $\sqrt{64}$ sq. inches $=$ **8 inches**

27. $\frac{1}{64} \times \frac{640}{1} = \frac{640}{64}$

```
    10
64)640
   64
    00
     0
     0
```

28. The rule is "Add the two previous terms."

13 + 21 = **34**
21 + 34 = **55**
34 + 55 = **89**

29. (a) **1, 2, 4, 8, 16, 32, 64**

(b) **Odd**

(c) **8**

(d) **8**

30. $\mathbf{\frac{13}{50}}$

LESSON 98, WARM-UP

a. $\mathbf{1\frac{1}{2}, 1\frac{3}{4}, 2}$

b. $\mathbf{\frac{4}{3}, 4}$

c. **$10**

d. **$5**

e. **10**

f. **8**

g. **XVIII**

h. **17**

Problem Solving

3 × 3 × 4 = **36 1-inch cubes**

LESSON 98, LESSON PRACTICE

a.

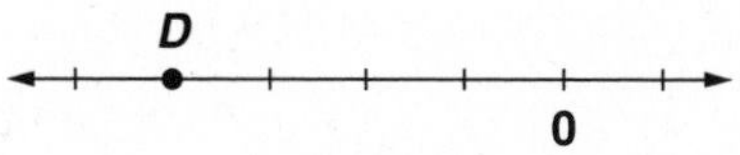

b. **−9**

c. **−12°F**

d. **−5°C**

e. **−20 dollars**

f. **0**

LESSON 98, MIXED PRACTICE

1. $\frac{8}{12} = \mathbf{\frac{2}{3}}$

2. (a) **Pyramid**

(b) **5 faces**

3.

```
   1.2 miles
×    2
   2.4 miles
```

4.

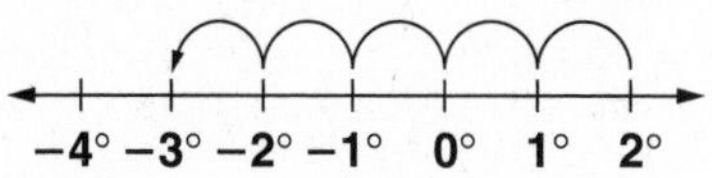

−3°F

5. ***D***

6. ***A***

7. **C.** $\angle KNM$

8. $$\begin{array}{r} \overset{1\,1}{6.5} \\ 2.47 \\ +\ 0.875 \\ \hline \mathbf{9.845} \end{array}$$

9. $$\begin{array}{r} \overset{1}{4}.26 \\ {}_{1}8.0 \\ +\ 15.9 \\ \hline \mathbf{28.16} \end{array}$$

10. $$\begin{array}{r} 23.45 \\ -\ \ 1.20 \\ \hline \mathbf{22.25} \end{array}$$

11. $$\begin{array}{r} 0.367 \\ -\ 0.100 \\ \hline \mathbf{0.267} \end{array}$$

12. $$\begin{array}{r} \$1.25 \\ \times\ \ \ \ 7 \\ \hline \mathbf{\$8.75} \end{array}$$

13. $$\begin{array}{r} 750 \\ \times\ 608 \\ \hline 6\,000 \\ 450\,000 \\ \hline \mathbf{456{,}000} \end{array}$$

14. $$\begin{array}{r} \mathbf{22\ R\ 12} \\ 16\overline{)364} \\ \underline{32} \\ 44 \\ \underline{32} \\ 12 \end{array}$$

15. $$\begin{array}{r} \mathbf{\$0.36} \\ 20\overline{)\$7.20} \\ \underline{6\,0} \\ 1\,20 \\ \underline{1\,20} \\ 0 \end{array}$$

16. $$\begin{array}{r} 3\frac{1}{2} \\ +\ 1\frac{1}{2} \\ \hline 4\frac{2}{2} \end{array} = \mathbf{5}$$

17. $$\begin{array}{r} 5\frac{8}{15} \\ -\ 4\frac{7}{15} \\ \hline \mathbf{1\frac{1}{15}} \end{array}$$

18. $$\begin{array}{r} 5\frac{3}{3} \\ -\ 1\frac{1}{3} \\ \hline \mathbf{4\frac{2}{3}} \end{array}$$

19. $\frac{1}{1} \times \frac{5}{7} = \mathbf{\frac{5}{7}}$

20. $\frac{4}{5} \times \frac{25}{1} = \frac{100}{5} = \mathbf{20}$

21. $\frac{3}{4} \times \frac{3}{2} = \frac{9}{8} = \mathbf{1\frac{1}{8}}$

22. $\frac{7}{10} \times \frac{10}{10} = \mathbf{\frac{70}{100}}$

23. $\frac{30 \div 10}{100 \div 10} = \mathbf{\frac{3}{10}}$

24. **79°F**

25. 81 + 9 = **90**

26. (a) Factors of 70: 1, 2, 5, 7, (10), 14, 35, 70

Factors of 100: 1, 2, 4, 5, (10), 20, 25, 50, 100

GCF is **10**

(b) $\frac{70 \div 10}{100 \div 10} = \mathbf{\frac{7}{10}}$

27. (a) $\underbrace{\frac{1}{2} \times \frac{1}{3}}_{\frac{1}{6}} \ \textcircled{<}\ \frac{1}{2}$

(b) $\underbrace{\frac{1}{2} \times \frac{1}{3}}_{\frac{1}{6}} \ \textcircled{<}\ \frac{1}{3}$

28. **−1°F**

29. **D. 9**

30. **0.037, 0.367, 0.376, 0.38**

Lesson 99, Warm-Up

a. $1\frac{1}{4}, 1\frac{3}{8}, 1\frac{1}{2}$

b. $2; \frac{1}{2}$

c. $12\frac{1}{2}$

d. $37\frac{1}{2}$

e. $37\frac{1}{2}$

f. **8**

g. **XXXVII**

h. **32**

Problem Solving

There are 16 cups in one gallon.
16 − 8 = 8 cups
8 − 4 = 4 cups
4 − 2 = 2 cups
2 − 1 = **1 cup will be left in the gallon container.**

Lesson 99, Lesson Practice

a. $\begin{array}{r} 4.3 \\ +\ 2.0 \\ \hline \mathbf{6.3} \end{array}$

b. $\begin{array}{r} 12.0 \\ +\ \ 1.2 \\ \hline \mathbf{13.2} \end{array}$

c. $\begin{array}{r} {}^{1}6.4 \\ +\ 24.0 \\ \hline \mathbf{30.4} \end{array}$

d. $\begin{array}{r} 4.0 \\ 1.3 \\ +\ 0.6 \\ \hline \mathbf{5.9} \end{array}$

e. $\begin{array}{r} 5.20 \\ 0.75 \\ +\ 2.00 \\ \hline \mathbf{7.95} \end{array}$

f. $\begin{array}{r} \overset{1}{5}6.0 \\ +\ 75.4 \\ \hline \mathbf{131.4} \end{array}$

g. $\begin{array}{r} 8.0 \\ {}^{1}4.7 \\ +\ 12.1 \\ \hline \mathbf{24.8} \end{array}$

h. $\begin{array}{r} 9.0 \\ {}^{1}4.8 \\ +\ 12.0 \\ \hline \mathbf{25.8} \end{array}$

i. $\begin{array}{r} 4.75 \\ -\ 2.00 \\ \hline \mathbf{2.75} \end{array}$

j. $\begin{array}{r} 12.4 \\ -\ \ 5.0 \\ \hline \mathbf{7.4} \end{array}$

k. **4**

l. 12 ⊜ 12.0

Lesson 99, Mixed Practice

1. $\frac{60}{50} = \mathbf{\frac{6}{5}}$

2. $\frac{2}{6} = \mathbf{\frac{1}{3}}$

$\frac{1}{3} \times \frac{100}{1} = \frac{100}{3}$

$$\begin{array}{r} \mathbf{33\tfrac{1}{3}\%} \\ 3\overline{)100} \\ \underline{9} \\ 10 \\ \underline{9} \\ 1 \end{array}$$

3. $\begin{array}{r} \$0.20 \\ 5\overline{)\$1.00} \\ \underline{1\ 0} \\ 00 \\ \underline{0} \\ 0 \end{array}$ $\qquad \begin{array}{r} \$0.20 \\ \times\ \ 12 \\ \hline 40 \\ 2\ 00 \\ \hline \mathbf{\$2.40} \end{array}$

4. $$\begin{array}{r} 13.8 \\ +\ 1.0 \\ \hline \textbf{14.8} \end{array}$$ **seconds**

5. **3.5, $3\frac{1}{2}$**

6. $$\begin{array}{r} 10 \\ 10\overline{)100} \\ \underline{10} \\ 00 \\ \underline{0} \\ 0 \end{array} \qquad \begin{array}{r} 10 \\ \times\ 10 \\ \hline 100 \end{array}$$

$n^2 = \mathbf{100}$

7. **1620.3**

8. 7 ft × 3 ft = **21 sq. ft**

9. $\frac{3}{4} = \mathbf{\frac{6}{8}}$

$\frac{7}{8} - \frac{6}{8} = \mathbf{\frac{1}{8}}$

10. **$7**

11. $$\begin{array}{r} 23 \text{ mm} \\ +\ 23 \text{ mm} \\ \hline 46 \text{ mm} \end{array} \qquad \begin{array}{r} \overset{9}{\cancel{10}}{}^{1}0 \text{ mm} \\ -\ 4\,6 \text{ mm} \\ \hline \mathbf{5\,4 \text{ mm}} \end{array}$$

12. $$\begin{array}{r} 3.4 \\ +\ 5.0 \\ \hline \mathbf{8.4} \end{array}$$

13. $$\begin{array}{r} 7.25 \\ -\ 7.00 \\ \hline \mathbf{0.25} \end{array}$$

14. 5 − 4 = **1**

15. $$\begin{array}{r} 60 \\ \times\ 60 \\ \hline \mathbf{3600} \end{array}$$

16. $$\begin{array}{r} \mathbf{34} \\ 28\overline{)952} \\ \underline{84} \\ 112 \\ \underline{112} \\ 0 \end{array}$$

17. $$\begin{array}{r} \mathbf{\$2.03} \\ 9\overline{)\$18.27} \\ \underline{18} \\ 0\,2 \\ \underline{0} \\ 27 \\ \underline{27} \\ 0 \end{array}$$

18. $4\frac{5}{8} + 1\frac{7}{8} = 5\frac{12}{8} = 6\frac{4}{8} = \mathbf{6\frac{1}{2}}$

19. $2\frac{3}{5} - 1 = 1\frac{3}{5}$

$4\frac{5}{5} - 1\frac{3}{5} = \mathbf{3\frac{2}{5}}$

20. $\frac{3}{4} \times \frac{1}{3} = \frac{3}{12} = \mathbf{\frac{1}{4}}$

21. $\frac{3}{4} \times \frac{1}{3} = \frac{3}{12} = \mathbf{\frac{1}{4}}$

22. $\frac{9}{10} \times \frac{10}{10} = \mathbf{\frac{90}{100}}$

23. $\frac{20 \div 20}{100 \div 20} = \mathbf{\frac{1}{5}}$

24. **C. It subtracts 6.**

25. 6 − 6 = **0**

26. **−10°F**

27. 3, 4, 7, 12, 19, **28**, **39**, **52**

1 3 5 7 9 11 13

28. (a) $\frac{3}{6} = \mathbf{\frac{1}{2}}$

(b) **One possibility: The upturned face is an even number.**

29. **$(2 \times 10^9) + (6 \times 10^8)$**

30. 2 liters ⓧ< 1 gallon

Lesson 100, Warm-Up

a. **$1\frac{1}{2}, 2, 2\frac{1}{2}$**

b. **$\frac{1}{3}; \frac{5}{3}$**

c. **$25**

d. **$2.50**

e. **$0.25**

f. **XXVII**

g. **28**

Problem Solving

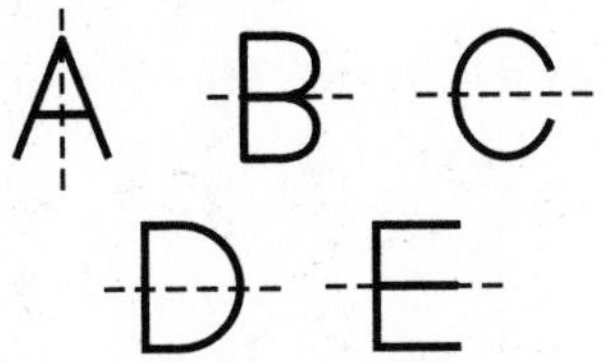

Lesson 100, Lesson Practice

a. **3.2**

b. **0.32**

c. **32**

d. **3.02**

e. $$\begin{array}{r} \overset{1}{3}.\overset{1}{6}5 \\ +\ 6.35 \\ \hline 10.00 \end{array} = \mathbf{10}$$

f. $$\begin{array}{r} \overset{1}{\cancel{2}}\,\overset{12}{\cancel{3}}.{}^{1}1\,6 \\ -\ 1\,9.\,4\,6 \\ \hline 3.\,7\,0 \end{array} = \mathbf{3.7}$$

g. $$\begin{array}{r} 4.\,\overset{1}{\cancel{2}}{}^{1}3 \\ -\ 3.\,1\,8 \\ \hline \mathbf{1.\,0\,5} \end{array}$$

h. **2.50**

i. **6.0**

Lesson 100, Mixed Practice

1. $\frac{20}{60} = \mathbf{\frac{1}{3}}$

2. $$\begin{array}{r} \$1.25 \\ \times\quad 10 \\ \hline \mathbf{\$12.50} \end{array}$$

3. $\frac{3}{4} \times \frac{28}{1} = \frac{84}{4} =$ **21 diners**

   $\frac{3}{4} \times \frac{100}{1} = \frac{300}{4}$

   $$\begin{array}{r} \mathbf{75\%} \\ 4\overline{)300} \\ \underline{28} \\ 20 \\ \underline{20} \\ 0 \end{array}$$

4. (a) **10 pins**

   (b) **6 small squares**

5. **8.0**

6. (a) ***D***

   (b) ***B***

7. $\frac{5}{6} = \mathbf{\frac{10}{12}}; \frac{3}{4} = \mathbf{\frac{9}{12}}$

   $\frac{10}{12} - \frac{9}{12} = \mathbf{\frac{1}{12}}$

8. 5 m × 100 = **500 cm**

9. 40 mm ÷ 2 = 20 mm

   $$\begin{array}{r} 40\text{ mm} \\ 20\text{ mm} \\ +\ 20\text{ mm} \\ \hline \mathbf{80\text{ mm}} \end{array}$$

10. $$\begin{array}{r} 6.2 \\ 3.0 \\ +\ 4.25 \\ \hline \mathbf{13.45} \end{array}$$

11. 1000 − 100 = **900**

12.
$$\begin{array}{r} 6.37 \\ -\ 6.00 \\ \hline \mathbf{0.37} \end{array}$$

13.
$$\begin{array}{r} 234 \\ \times\ 506 \\ \hline 1\,404 \\ 117\,000 \\ \hline \mathbf{118{,}404} \end{array}$$

14.
$$\begin{array}{r} \$1.75 \\ \times\ \ 10 \\ \hline \mathbf{\$17.50} \end{array}$$

15.
$$\begin{array}{r} \mathbf{\$1.75} \\ 10\overline{)\$17.50} \\ 10 \\ \hline 7\,5 \\ 7\,0 \\ \hline 50 \\ 50 \\ \hline 0 \end{array}$$

16. $\frac{1}{50} \times \frac{2}{2} = \mathbf{\frac{2}{100}}$

17. $\frac{40 \div 20}{100 \div 20} = \mathbf{\frac{2}{5}}$

18. **8**

19.
$$\begin{array}{r} 52 \\ 16\overline{)832} \\ 80 \\ \hline 32 \\ 32 \\ \hline 0 \end{array}$$

$w = \mathbf{52}$

20. $\frac{5}{9} + \frac{5}{9} + \frac{5}{9} = \frac{15}{9} = 1\frac{6}{9} = \mathbf{1\frac{2}{3}}$

21. $\frac{9}{10} \times \frac{9}{10} = \mathbf{\frac{81}{100}}$

22. $\frac{2}{3} \times \frac{4}{3} = \mathbf{\frac{8}{9}}$

23. $\frac{3}{1} \times \frac{4}{3} = \frac{12}{3} = \mathbf{4}$

24. 10 × 3 feet = **30 feet**

25. **8 months**

26. (a) **6 faces**

(b) **12 edges**

27. **−5°C**

28. 7 + 6 = **13 children**

29. **9 children**

30. **Some children (8) have no siblings.**

INVESTIGATION 10

1. **origin**
2. **(4, 3)**
3. **(–2, 4)**
4. **(–3, –2)**
5. **(5, –3)**
6. **(4, 1)**
7. **(–4, 1)**
8. **(3, –4)**

Activity

a.

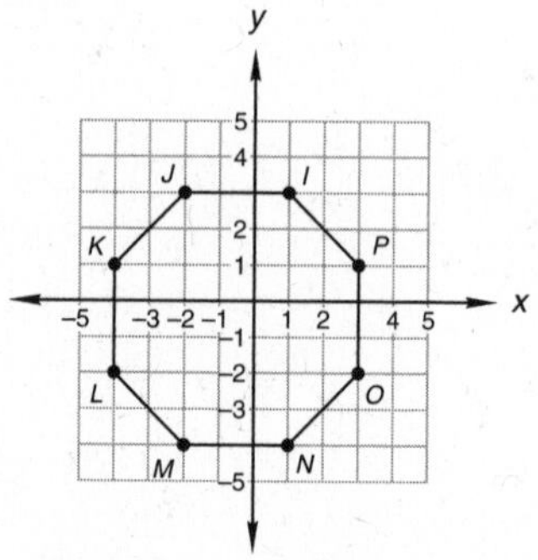

b. **See student work.**

LESSON 101, WARM-UP

a. **Round each number to the nearest 10 and add.**

b. **$3\frac{1}{3}$; $2\frac{1}{2}$; 2**

c. **2**

d. **XXXIII**

e. **<**

Problem Solving

30 days after Monday:
30 ÷ 7 = 4 R 2, so 30 days after Monday is the same as 2 days after Monday: **Wednesday.**

50 days after Saturday:
50 ÷ 7 = 7 R 1, so 50 days after Saturday is the same as 1 day after Saturday: **Sunday.**

78 days after Tuesday:
78 ÷ 7 = 11 R 1, so 78 days after Tuesday is the same as 1 day after Tuesday: **Wednesday.**

LESSON 101, LESSON PRACTICE

a. **4**

b. **7**

c. **7**

d. **6**

e. **13**

f. **25**

g. 10 × 5 = **50**

h.
$$\begin{array}{r} 37 \\ +\ 11 \\ \hline \mathbf{48} \end{array}$$

i. 9 in. + 9 in. + 5 in. + 5 in. = **28 in.**

LESSON 101, MIXED PRACTICE

1. $\frac{60}{40} = \mathbf{\frac{3}{2}}$

2.
$$\begin{array}{r} 25\text{ cm} \\ \times\ 8 \\ \hline 200\text{ cm} \end{array}$$

$$\begin{array}{r} \mathbf{2}\ \textbf{meters} \\ 100\overline{)200} \\ 200 \\ \hline 000 \end{array}$$

3.
$$\begin{array}{r} 1\,\overset{7}{\not 8}\,{}^{1}2\,6 \\ -\quad 5\,0 \\ \hline \mathbf{1\,7\,7\,6} \end{array}$$

4. $\frac{3}{4} \times \frac{100}{1} = \frac{300}{4}$

$$\begin{array}{r} \mathbf{75} \\ 4\overline{)300} \\ 28 \\ \hline 20 \\ 20 \\ \hline 0 \end{array}$$

5. **30 mm; 3 cm**

6.
$$\begin{array}{r} 1\frac{1}{2}\text{ or } \mathbf{1.5\ cm} \\ 2\overline{)3} \\ 2 \\ \hline 1 \end{array}$$

7. **$9**

8. 19 − 8 = **11**

9.
$$\begin{array}{r} \mathbf{80}\ \textbf{yards} \\ 3\overline{)240} \\ 24 \\ \hline 00 \\ 0 \\ \hline 0 \end{array}$$

10.
$$\begin{array}{r} 30\text{ mm} \\ 2\overline{)60} \\ 6 \\ \hline 00 \\ 00 \\ \hline 0 \end{array} \qquad \begin{array}{r} 20\text{ mm} \\ 3\overline{)60} \\ 6 \\ \hline 00 \\ 0 \\ \hline 0 \end{array}$$

60 mm + 30 mm + 20 mm = **110 mm**

11.
$$\begin{array}{r} 4.00 \\ {}_{1}8.57 \\ +\ 12.30 \\ \hline \mathbf{24.87} \end{array}$$

12.
$$\begin{array}{r} 16.37 \\ -\ 12.00 \\ \hline \mathbf{4.37} \end{array}$$

13.
$$\begin{array}{r} \$3.58 \\ \times\ \ \ \ 10 \\ \hline \mathbf{\$35.80} \end{array}$$

14.
$$\begin{array}{r} 24 \\ \times\ 24 \\ \hline 96 \\ 480 \\ \hline \mathbf{576} \end{array}$$

15.
$$\begin{array}{r} \mathbf{172} \\ 25\overline{)4300} \\ 25 \\ \hline 180 \\ 175 \\ \hline 50 \\ 50 \\ \hline 0 \end{array}$$

16.
$$\begin{array}{r} \$1.44 \\ 14\overline{)\$20.16} \\ 14 \\ \hline 6\,1 \\ 5\,6 \\ \hline 56 \\ 56 \\ \hline 0 \end{array}$$

$w = \mathbf{\$1.44}$

17. $3 + 4 = \mathbf{7}$

18. $\frac{5}{6} = \mathbf{\frac{10}{12}}; \frac{1}{4} = \mathbf{\frac{3}{12}}$

$\frac{10}{12} - \frac{3}{12} = \mathbf{\frac{7}{12}}$

19. $6\frac{3}{5} + 1\frac{3}{5} = 7\frac{6}{5} = \mathbf{8\frac{1}{5}}$

20. $8\frac{5}{6} - 1\frac{1}{6} = 7\frac{4}{6} = \mathbf{7\frac{2}{3}}$

21. $\frac{2}{10} \times \frac{5}{10} = \frac{10}{100} = \mathbf{\frac{1}{10}}$

22. $\frac{2}{1} \times \frac{5}{4} = \frac{10}{4} = 2\frac{2}{4} = \mathbf{2\frac{1}{2}}$

23. $\frac{9}{50} \times \frac{2}{2} = \mathbf{\frac{18}{100}}$

24.
$$\begin{array}{r} 29.2 \text{ seconds} \\ -\ \ \ 1.0 \text{ seconds} \\ \hline \mathbf{28.2 \text{ seconds}} \end{array}$$

25. $5 + 3 =$ **8 points**

26. $\frac{50 \div 50}{100 \div 50} = \mathbf{\frac{1}{2}}$

27. (a) **(−3, 2)**

(b) **(1, −2)**

28. 13.7, 13.8, (13.9), 14.0, 14.1
13.9

29.
$$\begin{array}{r} 12 \text{ in.} \\ \times\ 12 \text{ in.} \\ \hline 24 \\ 120 \\ \hline \mathbf{144} \text{ sq. in.} \end{array}$$

30. **Length: 1 in.; width: $\frac{1}{2}$ in.;**

Perimeter: $1 \text{ in.} + 1 \text{ in.} + \frac{1}{2} \text{ in.} + \frac{1}{2} \text{ in.}$

$= 2\frac{2}{2} \text{ in.} =$ **3 in.**

Lesson 102, Warm-Up

a. **Round each to the nearest whole number and add.**

b. **16 ounces; 32 ounces**

c. $\mathbf{\frac{2}{3}, 1\frac{1}{3}, 1\frac{1}{2}}$

d. **1**

e. **XXXVIII**

f. **<**

Problem Solving

(1, 1), (1, 2), (1, 3), (1, 4), (1, 5), (1, 6),
(2, 2), (2, 3), (2, 4), (2, 5), (2, 6),
(3, 3), (3, 4), (3, 5), (3, 6),
(4, 4), (4, 5), (4, 6),
(5, 5), (5, 6),
(6, 6);

21 combinations

LESSON 102, LESSON PRACTICE

a.
$$\begin{array}{r} 0.\overset{2}{\cancel{3}}{}^{1}0 \\ -\ 0.1\,5 \\ \hline \mathbf{0.1\,5} \end{array}$$

b.
$$\begin{array}{r} 0.\overset{2}{\cancel{3}}{}^{1}0 \\ -\ 0.2\,5 \\ \hline \mathbf{0.0\,5} \end{array}$$

c.
$$\begin{array}{r} \overset{3}{\cancel{4}}.\,{}^{1}\overset{1}{\cancel{2}}{}^{1}0 \\ -\ 0.4\,2 \\ \hline \mathbf{3.7\,8} \end{array}$$

d.
$$\begin{array}{r} 3.\overset{4}{\cancel{5}}{}^{1}0 \\ -\ 0.3\,5 \\ \hline \mathbf{3.1\,5} \end{array}$$

e.
$$\begin{array}{r} \overset{9}{\cancel{10}}.{}^{1}0 \\ -\ 6.5 \\ \hline \mathbf{3.5} \end{array}$$

f.
$$\begin{array}{r} 6.5 \\ -\ 4.0 \\ \hline \mathbf{2.5} \end{array}$$

g.
$$\begin{array}{r} \overset{0}{\cancel{1}}.{}^{1}0 \\ -\ 0.9 \\ \hline \mathbf{0.1} \end{array}$$

h.
$$\begin{array}{r} \overset{0}{\cancel{1}}.{}^{1}0 \\ -\ 0.1 \\ \hline \mathbf{0.9} \end{array}$$

i.
$$\begin{array}{r} \overset{0}{\cancel{1}}.\overset{9}{\cancel{0}}{}^{1}0 \\ -\ 0.2\,5 \\ \hline \mathbf{0.7\,5} \end{array}$$

j.
$$\begin{array}{r} 2.5 \\ -\ 1.0 \\ \hline \mathbf{1.5} \end{array}$$

k.
$$\begin{array}{r} \overset{1}{\cancel{2}}.{}^{1}0 \text{ liters} \\ -\ 1.2 \text{ liters} \\ \hline \mathbf{0.8 \text{ liters}} \end{array}$$

LESSON 102, MIXED PRACTICE

1. **One possibility:**

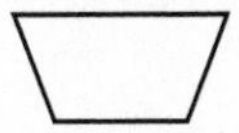

Trapezoid

2. 1 gal = 8 pt

About 8 pounds

3. 7 + 4 = **11**

4.
$$\begin{array}{r} \overset{1}{4}3 \text{ people} \\ +\ 27 \text{ people} \\ \hline 70 \text{ people} \end{array} \qquad \begin{array}{r} \mathbf{35 \text{ people}} \\ 2\overline{)70 \text{ people}} \\ \underline{6}\phantom{0 \text{ people}} \\ 10\phantom{\text{ people}} \\ \underline{10}\phantom{\text{ people}} \\ 0\phantom{\text{ people}} \end{array}$$

5.
$$\begin{array}{r} 4 \\ 25\overline{)100} \\ \underline{100} \\ 0 \end{array}$$

$4 \times 4 = 16$

$m^2 = \mathbf{16}$

6. $\frac{10}{100} = \mathbf{0.10 \text{ or } 0.1}$

$\frac{10}{100} = \mathbf{\frac{1}{10}}$

$\frac{10}{100} = \mathbf{10\%}$

7. $\frac{1}{5} = \mathbf{\frac{8}{40}}; \frac{7}{8} = \mathbf{\frac{35}{40}}$

$\frac{8}{40} + \frac{35}{40} = \frac{43}{40} = \mathbf{1\frac{3}{40}}$

8. One tenth ⊜ Ten hundredths

0.1 0.10

9. **6, 12, 18, 24**

10. (a) **4 inches**

(b) 4 in. + 2 in. + 4 in. + 2 in. = **12 inches**

11. 60 mm ÷ 2 = 30 mm

30 mm ÷ 2 = 15 mm

60 mm + 30 mm + 15 mm = **105 mm**

12.
$$\begin{array}{r} 0.\overset{3}{\cancel{4}}{}^{1}0 \\ -\ 0.1\,2 \\ \hline \mathbf{0.2\,8} \end{array}$$

13.
$$\begin{array}{r} \overset{5}{\cancel{6}}.\overset{1}{\cancel{2}}{}^{1}0 \\ -\ 0.7\,1 \\ \hline \mathbf{5.4\,9} \end{array}$$

14.
$$\begin{array}{r} \overset{3\,2}{315} \\ {}_{2}273 \\ 4197 \\ 586 \\ 92 \\ +\ 3634 \\ \hline \mathbf{9097} \end{array}$$

15.
$$\begin{array}{r} \$4.36 \\ \times\quad 9 \\ \hline \mathbf{\$39.24} \end{array}$$

16.
$$\begin{array}{r} 540 \\ \times\ 780 \\ \hline 43\,200 \\ 378\,000 \\ \hline \mathbf{421{,}200} \end{array}$$

17.
$$\begin{array}{r} \mathbf{72} \\ 6\overline{)432} \\ \underline{42} \\ 12 \\ \underline{12} \\ 0 \end{array}$$

18.
$$\begin{array}{r} \mathbf{72} \\ 12\overline{)864} \\ \underline{84} \\ 24 \\ \underline{24} \\ 0 \end{array}$$

19. $1\frac{2}{3} + 1\frac{2}{3} = 2\frac{4}{3} = 3\frac{1}{3}$

$4\frac{3}{3} - 3\frac{1}{3} = \mathbf{1\frac{2}{3}}$

20. $\frac{3}{1} \times \frac{2}{5} = \frac{6}{5}$

$\frac{5}{6} \times \frac{6}{5} = \frac{30}{30} = \mathbf{1}$

21. $\frac{2}{1} \times \frac{3}{1} = \frac{6}{1} = \mathbf{6}$

22. $\frac{1}{3} \times \frac{1}{2} = \mathbf{\frac{1}{6}}$

23. $\frac{12}{50} \times \frac{2}{2} = \mathbf{\frac{24}{100}}$

24. 7 hr + 9 hr + 2 hr + 2 hr + 4 hr = **24 hours**

25. $\frac{9}{24} = \mathbf{\frac{3}{8}}$

26.
$$\begin{array}{r} \overset{1}{\cancel{2}}.{}^{1}0 \text{ liters} \\ -\ 1.\,4 \text{ liters} \\ \hline \mathbf{0.\,6 \text{ liters}} \end{array}$$

27. The rule is "Count up by 0.3."
3.7, 4.0, 4.3, 4.6

28. 4 blocks × 3 blocks = 12 blocks
12 blocks × 2 blocks = **24 blocks**

29.
$$\begin{array}{r} \overset{1}{23} \\ 16 \\ +\ 41 \\ \hline 80 \end{array} \qquad \mathbf{\frac{23}{80}}$$

30. **−1, 0.1, 1, 1.1**

Lesson 103, Warm-Up

a. **Round each to the nearest whole number and multiply.**

b. **3 ft; 300 ft**

c. $\mathbf{\frac{3}{4}, 1\frac{1}{8}, 1\frac{1}{2}}$

d. **11**

e. **XXXVI**

f. **=**

Problem Solving

$2 \times 2 \times 2 =$ **8 1-inch cubes**

Lesson 103, Lesson Practice

a. 4 in. × 2 in. = 8 sq. in.
8 sq. in. × 4 in. = **32 cubic inches**

b. 3 cm × 3 cm = 9 sq. cm
9 sq. cm × 3 cm = **27 cubic centimeters**

c. 10 in. × 5 in. = 50 sq. in.
50 sq. in. × 2 in. = **100 cubic inches**

d. 2 ft × 2 ft = 4 sq. ft
4 sq. ft × 5 ft = **20 cubic feet**

e. 3 ft × 2 ft = 6 sq. ft
6 sq. ft × 8 ft = 48 cubic feet
Ella could fit **48 boxes** that are 1-foot cubes in her closet.

f. 3 in. × 5 in. = 15 sq. in.
15 sq. in. × 8 in. = **120 cubic inches**

Lesson 103, Mixed Practice

1. $\frac{15}{25} = \mathbf{\frac{3}{5}}$

2. $\frac{1}{3} \times \frac{12}{1} = \frac{12}{3} = 4$
About 4 ounces
$\frac{1}{3} \times \frac{100}{1} = \frac{100}{3}$

$$\begin{array}{r} 33\tfrac{1}{3}\% \\ 3\overline{)100} \\ \underline{9} \\ 10 \\ \underline{9} \\ 1 \end{array}$$

About $33\frac{1}{3}\%$

3. Prime numbers on a number cube: 2, 3, 5
$\frac{3}{6} = \mathbf{\frac{1}{2}}$

4. **1.5; $1\frac{1}{2}$**

5. **1**

6. 30 cm + 30 cm = **60 centimeters**

7. 10 ÷ 5 = **2**

8. **$13**

9. ***D***

10. 8 in. × 4 in. = **32 sq. in.**

11. 4 cm ÷ 2 = 2 cm
4 cm + 2 cm = 6 cm
10 cm − 6 cm = **4 cm**

12.
$$\begin{array}{r} \overset{2}{\not{3}}.\overset{9}{\not{0}}{}^{1}0 \\ -\ 2.\,3\,5 \\ \hline \mathbf{0.\,6\,5} \end{array}$$

13.
$$\begin{array}{r} \overset{9}{\not{1}\not{0}}.\overset{9}{\not{0}}{}^{1}0 \\ -\ 4.\,0\,6 \\ \hline \mathbf{5.\,9\,4} \end{array}$$

14.
$$\begin{array}{r} {}_{1}4.35 \\ 12.60 \\ +\ 15.00 \\ \hline \mathbf{31.95} \end{array}$$

15.
$$\begin{array}{r} 47 \\ \times\ 7 \\ \hline 329 \end{array} \qquad \begin{array}{r} 329 \\ \times\ 360 \\ \hline 19\,740 \\ 98\,700 \\ \hline \mathbf{118{,}440} \end{array}$$

16. 2 × 2 × 2 × 2 × 2 = **32**

17. $\begin{array}{r} \mathbf{\$2.35} \\ 20\overline{)\$47.00} \\ \underline{40} \\ 7\,0 \\ \underline{6\,0} \\ 1\,00 \\ \underline{1\,00} \\ 0 \end{array}$

18. $5 - 3 = \mathbf{2}$

19. $\begin{array}{r} 132 \\ 16\overline{)2112} \\ \underline{16} \\ 51 \\ \underline{48} \\ 32 \\ \underline{32} \\ 0 \end{array}$

$x = \mathbf{132}$

20. $1\frac{3}{3} - \frac{2}{3} = 1\frac{1}{3}$

$3\frac{2}{3} + 1\frac{1}{3} = 4\frac{3}{3} = \mathbf{5}$

21. $\frac{4}{1} \times \frac{1}{4} = \frac{4}{4} = 1$

$\frac{1}{2} \times \frac{1}{1} = \mathbf{\frac{1}{2}}$

22. $\frac{1}{1} \times \frac{5}{7} = \mathbf{\frac{5}{7}}$

23. $\frac{3}{2} \times \frac{3}{2} = \frac{9}{4} = \mathbf{2\frac{1}{4}}$

24. $\frac{4}{10} \times \frac{5}{10} = \frac{20}{100} = \mathbf{\frac{1}{5}}$

25. $\frac{1}{25} \times \frac{4}{4} = \mathbf{\frac{4}{100}}$

26. $\frac{500 \div 500}{1000 \div 500} = \mathbf{\frac{1}{2}}$

27. (a) 5 cm × 3 cm = 15 sq. cm

15 sq. cm × 3 cm = **45 cubic cm**

(b) **6 faces**

28. (a) **(2, 1)**

(b) **(−1, 1)**

(c) **(−1, −2)**

(d) **(2, −2)**

29. (a) $\mathbf{\frac{3}{4}}$ **in.**

(b) $\frac{3}{4}$ in. + $\frac{3}{4}$ in. + $\frac{3}{4}$ in. = $\frac{9}{4}$ in. = $\mathbf{2\frac{1}{4}}$ **in.**

30. **Equilateral (and isosceles), acute**

Lesson 104, Warm-Up

a. **Round each to the nearest whole number and divide.**

b. $\mathbf{1\frac{1}{5}, 1\frac{1}{2}, 2\frac{1}{2}}$

c. **XXIX**

d. **>**

Problem Solving

4 × 4 × 4 = **64**

5 × 5 × 5 = **125**

Lesson 104, Lesson Practice

a. **$6**

b. **$15**

c. **$119**

d. $13 + $7 = **$20**

e. **5**

f. **12**

g. **96**

h. **7**

i. **46**

j. **90**

k. $10 \times 7 =$ **70**

l. **68 seconds**

Lesson 104, Mixed Practice

1. **One possibility:**

2. (a) $18 \div 2 =$ **9 girls**

(b) $18 + 9 =$ **27 players**

(c) $\frac{18}{9} = \mathbf{\frac{2}{1}}$

3. (a) **75, 80, 80, 80, 85, 90, 90**

(b) **80**

(c) **80**

4. $\mathbf{\frac{1}{2} \times \frac{1}{3} = \frac{1}{6}}$

5. **7**

6. $\frac{1}{2} \div \frac{1}{3} \;\textcircled{>}\; \frac{1}{3} \div \frac{1}{2}$

$\frac{1}{2} \times \frac{3}{1} = \frac{3}{2} \qquad \frac{1}{3} \times \frac{2}{1} = \frac{2}{3}$

7.

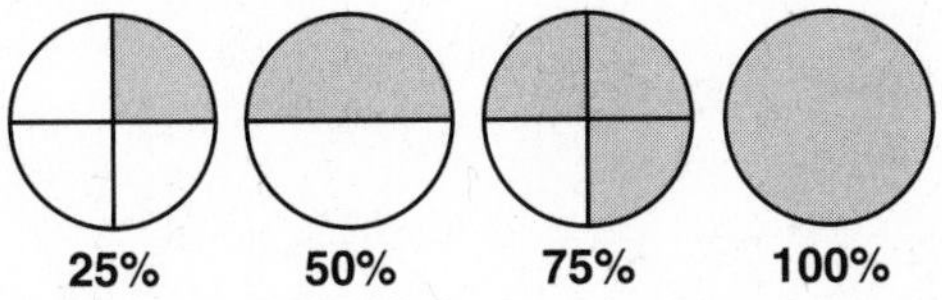

8. **4**

9. (a) **$10**

(b) **$10**

10. **7, 14, 21, 28, 35**

11. ***A***

12. $7 \text{ cm} \times 4 \text{ cm} =$ **28 sq. cm**

13.

$$\begin{array}{r} \overset{1}{6}.40 \\ 2.87 \\ +\ 4.00 \\ \hline \mathbf{13.27} \end{array}$$

14.

$$\begin{array}{r} \$1\overset{5}{\not{6}}.\overset{9}{\not{0}}{}^{1}0 \\ -\ \$5.\ 7\ 4 \\ \hline \$10.\ 2\ 6 \end{array} \qquad \begin{array}{r} \mathbf{\$1.71} \\ 6\overline{)\$10.26} \\ \underline{6} \\ 4\,2 \\ \underline{4\,2} \\ 06 \\ \underline{6} \\ 0 \end{array}$$

15.

$$\begin{array}{r} \$5.64 \\ \times \quad 10 \\ \hline \mathbf{\$56.40} \end{array}$$

16.

$$\begin{array}{r} 976 \\ \times\ 267 \\ \hline 6\,832 \\ 58\,560 \\ 195\,200 \\ \hline \mathbf{260{,}592} \end{array}$$

17. $\frac{640}{32} = 20$

$\frac{320}{16} = 20$

$\frac{160}{8} = 20$

$\frac{80}{4} = 20$

20

18. $\frac{2}{3} = \mathbf{\frac{6}{9}}$

$\frac{6}{9} + \frac{7}{9} = \frac{13}{9} = \mathbf{1\frac{4}{9}}$

19. $2\frac{3}{3} - \frac{1}{3} = 2\frac{2}{3}$

$5\frac{2}{3} + 2\frac{2}{3} = 7\frac{4}{3} = \mathbf{8\frac{1}{3}}$

20. $\frac{1}{2} \times \frac{1}{3} = \frac{1}{6}$

$\frac{2}{1} \times \frac{1}{6} = \frac{2}{6} = \mathbf{\frac{1}{3}}$

21. $\frac{3}{10} \times \frac{30}{1} = \frac{90}{10} = \mathbf{9}$

22. $\frac{4}{25} \times \frac{4}{4} = \mathbf{\frac{16}{100}}$

23. **A. G4**

24. **Grant**

25. **K2**

26. $100 - 10 = \mathbf{90}$

27. (a) **(0, 2)**

(b) **(−2, −1)**

(c) **(2, −1)**

28. **Answers may vary. One possibility:**

29. (a) 10 in. × 4 in. = 40 sq. in.
40 sq. in. × 4 in. = **160 cubic inches**

(b) **8 vertices**

30. $\frac{26}{100} = \mathbf{\frac{13}{50}}$

Lesson 105, Warm-Up

a. **Round the price to the nearest dollar, and multiply by the number of yards needed.**

b. **$4**

c. **$8**

d. **$16**

e. **7**

f. **XXXII**

g. **>**

Problem Solving

No line of symmetry: 

One line of symmetry: 

Two lines of symmetry: 

Lesson 105, Lesson Practice

a. 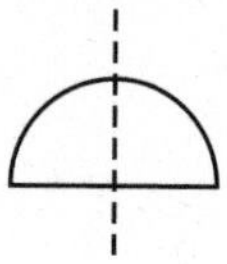

b. 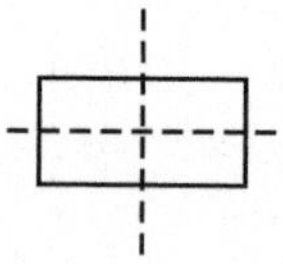

c. 

d. 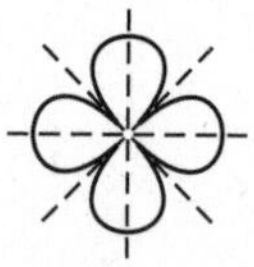

e. **8 lines of symmetry**

f. **S**

Lesson 105, Mixed Practice

1. $\frac{4 \text{ boys}}{5 \text{ girls}} = \frac{40 \text{ boys}}{50 \text{ girls}}$

50 girls

2. **10 tenths**

3.
$$\begin{array}{r} \overset{1}{2}5¢ \\ 10¢ \\ 5¢ \\ +\ \ 3¢ \\ \hline 43¢ \end{array}$$
1 coin

4. **10.23 seconds**

5. $20 \times 2 =$ **40 comic books**

6. $\frac{1}{2} = \mathbf{\frac{5}{10}}$

$\frac{9}{10} - \frac{5}{10} = \frac{4}{10} = \mathbf{\frac{2}{5}}$

7.
$$\begin{array}{r} 40 \\ {}_{1}60 \\ +\ 125 \\ \hline 225 \end{array} \qquad \begin{array}{r} \mathbf{75}\ \textbf{pages} \\ 3\overline{)225} \\ \underline{21} \\ 15 \\ \underline{15} \\ 0 \end{array}$$

8. $12 \times \$2 =$ **$24**

9.
$$\begin{array}{r} \mathbf{3} \\ 7\overline{)21} \\ \underline{21} \\ 0 \end{array}$$

10. (a) **12 sq. units**

(b) 4 units + 3 units + 4 units + 3 units = **14 units**

11. **∠*DAB* (or ∠*BAD*)**

12. **Trapezoid**

13. $\frac{1}{100} + \frac{9}{100} = \frac{10}{100} = \mathbf{\frac{1}{10}}$

14. $\frac{63}{100} - \frac{13}{100} = \frac{50}{100} = \mathbf{\frac{1}{2}}$

15. $\frac{5}{10} \times \frac{5}{10} = \frac{25}{100} = \mathbf{\frac{1}{4}}$

16. $\frac{3}{5} \times \frac{4}{3} = \frac{12}{15} = \mathbf{\frac{4}{5}}$

17.
$$\begin{array}{r} {}_{1}\overset{1}{3}.76 \\ 12.00 \\ +\ \ 6.80 \\ \hline \mathbf{22.56} \end{array}$$

18.
$$\begin{array}{r} 1\overset{1}{\cancel{2}}.\overset{9}{\cancel{0}}{}^{1}0 \\ -\ \ 1.25 \\ \hline \mathbf{10.75} \end{array}$$

19. $\sqrt{64} + \sqrt{36}$

$8 + 6 =$ **14**

20.
$$\begin{array}{r} 31 \\ \times\ 31 \\ \hline 31 \\ \underline{930} \\ \mathbf{961} \end{array}$$

21.
$$\begin{array}{r} \mathbf{213} \\ 28\overline{)5964} \\ \underline{56} \\ 36 \\ \underline{28} \\ 84 \\ \underline{84} \\ 0 \end{array}$$

22.
$$\begin{array}{r} 426 \\ 14\overline{)5964} \\ \underline{56} \\ 36 \\ \underline{28} \\ 84 \\ \underline{84} \\ 0 \end{array}$$
$m =$ **426**

23. $\frac{3}{20} \times \mathbf{\frac{5}{5}} = \frac{15}{100}$

24. $\frac{7}{25} \times \frac{4}{4} = \mathbf{\frac{28}{100}}$

25. **A. It divides by 3.**

26. $30 \div 3 =$ **10**

27.
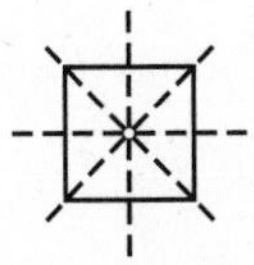

28. (a) 7 in. × 2 in. = 14 sq. in.
14 sq. in. × 10 in. = **140 cubic inches**

(b) **12 edges**

29. **4 units**

30. **N**

Lesson 106, Warm-Up

a. **Round the price to the nearest dollar, and divide by the number of pounds.**

b. **2000 lb; 4000 lb; 1000 lb**

c. $33\frac{1}{3}$

d. $66\frac{2}{3}$

e. **64**

f. **210**

g. **80**

Problem Solving

1 roll of pennies: $0.01 × 50 = $0.50
1 roll of nickels: $0.05 × 40 = $2.00
2 nickel rolls ($4.00) are equal in value to **8 penny rolls.**

Lesson 106, Lesson Practice

a. **Six and eight hundred seventy-five thousandths**

b. **Twenty-five thousandths**

c. **One thousand, six hundred twenty-five ten-thousandths**

d. **4**

e. **3**

f. **1**

g. 0.375 (>) 0.0375

h. **0.1, 0.1025, 0.125, 0.15**

i. **0.125**

Lesson 106, Mixed Practice

1. $100 ÷ $25 = 4
6 games × 4 = **24 games**

2. $\begin{array}{r} \overset{0}{\cancel{1}}.\overset{9}{\cancel{0}}{}^{1}0 \\ -\ 0.3\,7 \\ \hline \mathbf{0.6\,3} \end{array}$ **meter**

3. **1.50;** $\mathbf{1\frac{1}{2}}$

4. 8 × 8 = **64**

5. **8, 16, 24, 32, 40**

6. (a) $\frac{3}{5} \times \frac{30}{1} = \frac{90}{5}$

$5\overline{)90}$ = **18 girls**
5
40
40
0

(b) $\begin{array}{r} \overset{2}{\cancel{3}}{}^{1}0 \\ -\ 1\,8 \\ \hline \mathbf{1\,2} \end{array}$ **boys**

(c) $\frac{12}{18} = \mathbf{\frac{2}{3}}$

7. $\begin{array}{r} \$_{1}9 \\ \$12 \\ +\ \$5 \\ \hline \mathbf{\$26} \end{array}$

8. **Five and three hundred seventy-five thousandths**

9. (a) 4 units + 4 units + 4 units + 4 units
= **16 units**

(b) 4 units × 4 units = **16 sq. units**

10. **0.875, 0.9, 0.96, 1**

11.
$$\begin{array}{r} 4\frac{3}{8} \\ +\ 1\frac{3}{8} \\ \hline 5\frac{6}{8} \end{array} = \mathbf{5\frac{3}{4}}$$

12.
$$\begin{array}{r} 3\frac{7}{10} \\ +\ \frac{3}{10} \\ \hline 3\frac{10}{10} \end{array} = \mathbf{4}$$

13.
$$\begin{array}{r} 3\frac{10}{10} \\ -\ 1\frac{3}{10} \\ \hline \mathbf{2\frac{7}{10}} \end{array}$$

14.
$$\begin{array}{r} \overset{1}{1}.2300 \\ 0.4567 \\ +\ 0.5000 \\ \hline \mathbf{2.1867} \end{array}$$

15.
$$\begin{array}{r} \overset{3}{\not{4}}.{}^{1}0 \\ -\ 1.\ 3 \\ \hline \mathbf{2.\ 7} \end{array}$$

16.
$$\begin{array}{r} 57 \\ \times\ \ 8 \\ \hline 456 \end{array} \qquad \begin{array}{r} 456 \\ \times\ 250 \\ \hline 22\,800 \\ 91\,200 \\ \hline \mathbf{114{,}000} \end{array}$$

17.
$$\begin{array}{r} \$7.25 \\ \times\ \ \ \ 5 \\ \hline \mathbf{\$36.25} \end{array}$$

18.
$$\begin{array}{r} \mathbf{\$3.25} \\ 8\overline{)\$26.00} \\ \underline{24} \\ 2\,0 \\ \underline{1\,6} \\ 40 \\ \underline{40} \\ 0 \end{array}$$

19.
$$\begin{array}{r} \mathbf{20\ R\ 16} \\ 21\overline{)436} \\ \underline{42} \\ 16 \\ \underline{0} \\ 16 \end{array}$$

20.
$$\begin{array}{r} \mathbf{315} \\ 16\overline{)5040} \\ \underline{48} \\ 24 \\ \underline{16} \\ 80 \\ \underline{80} \\ 0 \end{array}$$

21. $\frac{5}{1} \times \frac{3}{10} = \frac{15}{10} = 1\frac{5}{10} = \mathbf{1\frac{1}{2}}$

22. $\frac{5}{1} \times \frac{3}{2} = \frac{15}{2} = \mathbf{7\frac{1}{2}}$

23. $\frac{1}{6} = \mathbf{\frac{4}{24}}; \frac{1}{8} = \mathbf{\frac{3}{24}}$

$\frac{4}{24} + \frac{3}{24} = \mathbf{\frac{7}{24}}$

24. $\frac{1}{3} \times \frac{30}{1} = \frac{30}{3} =$ **10 pens**

$\frac{1}{3} \times \frac{100}{1} = \frac{100}{3}$

$$\begin{array}{r} \mathbf{33\frac{1}{3}\%} \\ 3\overline{)100} \\ \underline{9} \\ 10 \\ \underline{9} \\ 1 \end{array}$$

25. **C. red and black**

26. $\sqrt{81}$ ⊜ 3^2

$9 \qquad 3 \times 3 = 9$

27. (a) 4 in. × 4 in. = 16 sq. in.

16 sq. in × 4 in. = **64 cubic inches**

(b) **Square**

28.

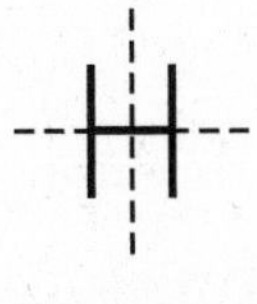

29. $4 + 6 + 7 = 17$

$$\begin{array}{r} 5\tfrac{2}{3} \text{ times} \\ 3\overline{)17} \\ \underline{15} \\ 2 \end{array}$$

30. 1 pt = 16 oz
3 pt = 3 × 16 = 48 oz
B. 48 oz

Lesson 107, Warm-Up

a. **Round the price to the nearest fifty cents, and round gallons to the nearest whole number; then multiply the two numbers.**

b. **4 qt; 16 qt; 2 qt**

c. **$20**

d. **$10**

e. **$4**

f. **190**

g. **65**

Problem Solving
8 possible outcomes:
(H, H, H),
(H, H, T),
(H, T, H),
(H, T, T),
(T, H, H),
(T, H, T),
(T, T, H),
(T, T, T)

Lesson 107, Lesson Practice

a. $\dfrac{120 \div 2}{200 \div 2} = \dfrac{60}{100} = \mathbf{60\%}$

b. $\dfrac{10}{50} \times \dfrac{2}{2} = \dfrac{20}{100} = \mathbf{20\%}$

c. $\dfrac{60 \div 3}{300 \div 3} = \dfrac{20}{100}$
20

d. $\dfrac{48 \div 2}{200 \div 2} = \dfrac{24}{100} = \mathbf{24\%}$

e. $\dfrac{30}{50} \times \dfrac{2}{20} = \dfrac{60}{100} = \mathbf{60\%}$

f. $\dfrac{1}{2} \times 100\% = \dfrac{100\%}{2}$

$$\begin{array}{r} \mathbf{50\%} \\ 2\overline{)100\%} \\ \underline{10} \\ 00 \\ \underline{0} \\ 0 \end{array}$$

g. $\dfrac{1}{12} \times 100\% = \dfrac{100\%}{12}$

$$\begin{array}{r} 8\tfrac{4}{12}\% = \mathbf{8\tfrac{1}{3}\%} \\ 12\overline{)100\%} \\ \underline{96} \\ 4 \end{array}$$

Lesson 107, Mixed Practice

1.
$$\begin{array}{r} 63.8 \text{ seconds} \\ -\ \ 1.0 \text{ seconds} \\ \hline \mathbf{62.8 \text{ seconds}} \end{array}$$

2. 10 in. × 7 in. = **70 sq. in.**

3.
$$\begin{array}{r} \mathbf{16 \text{ bundles}} \\ 15\overline{)245} \\ \underline{15} \\ 95 \\ \underline{90} \\ 5 \end{array}$$

4. 8 × $7 = **$56**

5. $\dfrac{60 \div 2}{200 \div 2} = \dfrac{30}{100} = \mathbf{30\%}$

6. $\dfrac{1}{10} + \dfrac{1}{10}$ ⊜ $0.1 + 0.1$
$\dfrac{2}{10} = 0.2$ $\qquad$ 0.2

7. 20 ÷ 4 = **5**

8. $\dfrac{10}{50} \times \dfrac{2}{2} = \dfrac{20}{100} = \mathbf{20\%}$

9. $\frac{1}{3} = \mathbf{\frac{2}{6}}$

$\frac{2}{6} + \frac{1}{6} = \frac{3}{6} = \mathbf{\frac{1}{2}}$

10. (a) 5 units + 5 units + 3 units + 3 units = **16 units**

(b) 5 units × 3 units = **15 sq. units**

11. 9 cm ÷ 3 = **3 cm**

12. ***C***

13. $\frac{31}{100} + \frac{29}{100} = \frac{60}{100} = \mathbf{\frac{3}{5}}$

14. $4\frac{10}{10} - 3\frac{7}{10} = \mathbf{1\frac{3}{10}}$

15.
```
  4
  5.¹0
- 3. 7
  1. 3
```
(answer: **1.3**)

16.
```
  $3.65
×    10
 $36.50
```
(answer: **$36.50**)

17.
```
     468
   × 579
   4 212
  32 760
 234 000
 270,972
```
(answer: **270,972**)

18.
```
      $3.65
10)$36.50
    30
     6 5
     6 0
       50
       50
        0
```
(answer: **$3.65**)

19.
```
  1753
5)8765
  5
  37
  35
   26
   25
    15
    15
     0
```
(answer: **1753**)

20.
```
     20
32)640
   64
    00
     0
     0
```
(answer: **20**)

21. $\frac{3}{10} \times \frac{7}{10} = \mathbf{\frac{21}{100}}$

22. $\frac{4}{1} \times \frac{5}{3} = \frac{20}{3} = \mathbf{6\frac{2}{3}}$

23. **12 votes**

24. $\frac{8}{32} = \mathbf{\frac{1}{4}}$

25.
```
   ²
   12       4)32
    7         32
    5          0
 +  8
   32
```
8 votes

26. $\frac{25 \div 25}{100 \div 25} = \mathbf{\frac{1}{4}}$

27. $10^3 - \sqrt{100}$

1000 − 10 = **990**

28. **C. Obtuse**

29. ***A* (2, −2)**
***B* (−1, −2)**
***C* (−3, 1)**

30. 4 in. × 2 in. = 8 sq. in.
8 sq. in. × 8 in. = **64 cubic inches**

LESSON 108, WARM-UP

a. **Round the price to the nearest fifty cents, and round gallons to the nearest whole number; then multiply the two numbers.**

b. **100 years; 1000 years**

c. **$20**

d. **$60**

e. $\mathbf{\frac{1}{4}}$

f. **150**

g. **66**

Problem Solving

$3 \times 3 \times 3 =$ **27 1-inch cubes**

Lesson 108, Lesson Practice

a. 45 minutes before 2:00 p.m. is **1:15 p.m.**

b. From 12:30 p.m. to 3:40 p.m. is **3 hr 10 min**

c. **7 days**

d. **B. 4:30 a.m.**

e. From 10:45 a.m. to 3:50 p.m. is **5 hr 5 min**

Lesson 108, Mixed Practice

1. **45,454,500 milligrams**

2.
$$\begin{array}{r} \overset{2\ 4}{\$1.26} \\ \$1.26 \\ \$0.49 \\ \$0.49 \\ \$0.49 \\ +\ \$0.24 \\ \hline \mathbf{\$4.23} \end{array}$$

3.
$$\begin{array}{r} 2.5 \\ +\ 2.5 \\ \hline 5.0 \end{array}$$ or **5 miles**

4. $20 \div 4 = 5$
$2 \times 5 = 10$
$10 - 1 = 9$
$2y - 1 =$ **9**

5. **320**

6. (a) $\frac{15}{25} \times \frac{4}{4} = \frac{60}{100} =$ **60%**

(b) $25 - 15 = 10$
$\frac{15}{10} = \mathbf{\frac{3}{2}}$

7.
$$\begin{array}{r} \overset{1}{1}3 \\ +\ 8 \\ \hline \mathbf{21} \end{array}$$

8. $80\% = \frac{80}{100}; \frac{80 \div 20}{100 \div 20} = \mathbf{\frac{4}{5}}$

9. 50% ⊜ $\frac{1}{2}$
$\frac{50}{100} = \frac{1}{2}$

10.
$$\begin{array}{r} 45 \\ \times\ 45 \\ \hline 225 \\ 1800 \\ \hline \mathbf{2025} \end{array}$$

11. **Seventy-six and three hundred forty-five thousandths; 3**

12. (a) 7 units + 7 units + 6 units + 6 units = **26 units**

(b) 7 units × 6 units = **42 sq. units**

13. 48 mm ÷ 2 = 24 mm
48 mm + 24 mm + 24 mm = **96 mm**

14.
$$\begin{array}{r} \overset{1}{2}.386 \\ 1.200 \\ 16.250 \\ +\ 10.000 \\ \hline \mathbf{29.836} \end{array}$$

15.
$$\begin{array}{r} \overset{2}{\cancel{3}}.\overset{9}{\cancel{0}}{}^{1}0 \\ -\ 0.45 \\ \hline 2.55 \end{array} \qquad \begin{array}{r} \overset{3}{\cancel{4}}.\overset{1}{\cancel{2}}{}^{1}0 \\ -\ 2.55 \\ \hline \mathbf{1.65} \end{array}$$

16.
$$\begin{array}{r} \mathbf{\$2.47} \\ 15\overline{)\$37.05} \\ 30 \\ \hline 7\,0 \\ 6\,0 \\ \hline 1\,05 \\ 1\,05 \\ \hline 0 \end{array}$$

17. $\mathbf{\frac{3}{6}}$

$\frac{3}{6} + \frac{1}{6} = \frac{4}{6} = \mathbf{\frac{2}{3}}$

18. $\frac{1}{2} \times \frac{3}{2} = \mathbf{\frac{3}{4}}$

19. $\frac{3}{10} \times \frac{3}{10} = \mathbf{\frac{9}{100}}$

20. $\frac{4}{11} + \frac{5}{11} = \mathbf{\frac{9}{11}}$

21. $4\frac{5}{7} - \frac{1}{7} = \mathbf{4\frac{4}{7}}$

22. $\frac{5}{6} \times \frac{24}{1} = \frac{120}{6}$

20 juice bars

$$\begin{array}{r} 20 \\ 6\overline{)120} \\ \underline{12} \\ 00 \\ \underline{0} \\ 0 \end{array}$$

23. **17**

24. 4 + 4 + 5 + 6 + 3 + 2 = 24 students
25 students − 24 students = **1 student**

25. $\begin{array}{r} \overset{1}{\cancel{2}}{}^{1}0 \\ -\ 1\ 3 \\ \hline \mathbf{7} \end{array}$

26. 20, 20, 20, 20, 19, 19, 19, 19, 18, 18, 18, 18, (18), 17, 17, 17, 17, 17, 17, 16, 16, 16, 15, 15, 13
18

27. 5 ft × 2 ft = 10 sq. ft
10 sq. ft × 8 ft = **80 cubic feet**

28. $\frac{2}{3} \times 100\% = \frac{200\%}{3}$

$$\begin{array}{r} \mathbf{66\frac{2}{3}\%} \\ 3\overline{)200} \\ \underline{18} \\ 20 \\ \underline{18} \\ 2 \end{array}$$

29. **5 lines of symmetry**

30. **10 sides; decagon**

Lesson 109, Warm-Up

a. **Round miles to the nearest ten and gallons to the nearest whole number; then divide miles by gallons.**

b. **10**

c. **30**

d. **20**

e. **9**

f. **105**

g. **40**

Problem Solving

There are 24 hours in 1 day, 60 minutes in each hour, and 60 seconds in each minute. We multiply: 24 × 60 × 60 = **86,400 seconds**

Lesson 109, Lesson Practice

a. $\begin{array}{r} 0.3 \\ \times\ \ 4 \\ \hline \mathbf{1.2} \end{array}$

b. $\begin{array}{r} 3 \\ \times\ 0.6 \\ \hline \mathbf{1.8} \end{array}$

c. $\begin{array}{r} 0.12 \\ \times\ \ 12 \\ \hline 24 \\ 1\ 20 \\ \hline \mathbf{1.44} \end{array}$

d. $\begin{array}{r} 1.4 \\ \times\ 0.7 \\ \hline \mathbf{0.98} \end{array}$

e. $\begin{array}{r} 0.3 \\ \times\ 0.5 \\ \hline \mathbf{0.15} \end{array}$

f. $\begin{array}{r} 1.2 \\ \times\ 3 \\ \hline \mathbf{3.6} \end{array}$

g. $\begin{array}{r} 1.5 \\ \times\ 0.5 \\ \hline \mathbf{0.75} \end{array}$

h. $\begin{array}{r} 0.25 \\ \times\ 1.1 \\ \hline 25 \\ 250 \\ \hline \mathbf{0.275} \end{array}$

i. $\frac{3}{10} \times \frac{3}{10} \ (=)\ 0.3 \times 0.3$

$\frac{9}{100} = 0.09 \qquad 0.09$

j. $\begin{array}{r} 0.8 \text{ cm} \\ \times\ 0.8 \text{ cm} \\ \hline \mathbf{0.64 \text{ sq. cm}} \end{array}$

LESSON 109, MIXED PRACTICE

1. **See student work.**

2. $\frac{40}{50} \times \frac{2}{2} = \frac{80}{100} = \mathbf{80\%}$

3. $\frac{1}{10} \times \frac{1}{10} \ (=)\ 0.1 \times 0.1$

$\frac{1}{100} = 0.01 \qquad 0.01$

4. **11:25 p.m.**

5. **101.101**

6. $\begin{array}{r} \mathbf{200 \text{ grams}} \\ 3\overline{)600} \\ 6 \\ \hline 00 \\ 0 \\ \hline 00 \\ 0 \\ \hline 0 \end{array}$

7. **10, 20, 30, 40, 50**

8. $\begin{array}{r} \$\overset{1}{\not{2}}{}^{1}3 \\ -\ \$7 \\ \hline \mathbf{\$16} \end{array}$

9. (a) 6 units + 6 units + 5 units + 5 units = **22 units**

(b) 6 units × 5 units = **30 sq. units**

10. (a) $\frac{10 \div 10}{100 \div 10} = \mathbf{\frac{1}{10}}$

(b) $\frac{20 \div 20}{100 \div 20} = \mathbf{\frac{1}{5}}$

11. $\begin{array}{r} \overset{1\,1}{32.30} \\ 4.96 \\ 7.50 \\ +\ 11.00 \\ \hline \mathbf{55.76} \end{array}$

12. $\begin{array}{r} \overset{0}{\not{1}}.{}^{1}36 \\ -\ 0.80 \\ \hline 0.56 \end{array} \qquad \begin{array}{r} \overset{0}{\not{1}}.\overset{9}{\not{0}}{}^{1}0 \\ -\ 0.56 \\ \hline \mathbf{0.44} \end{array}$

13. $\begin{array}{r} 12 \\ \times\ 1.2 \\ \hline 24 \\ 120 \\ \hline \mathbf{14.4} \end{array}$

14. $\begin{array}{r} 0.15 \\ \times\ 0.9 \\ \hline \mathbf{0.135} \end{array}$

15. $\begin{array}{r} 0.16 \\ \times\ 10 \\ \hline 1.60 \end{array}$ or **1.6**

16. $\begin{array}{r} 285 \\ 13\overline{)3705} \\ 26 \\ \hline 110 \\ 104 \\ \hline 65 \\ 65 \\ \hline 0 \end{array}$

m = **285**

17. $\begin{array}{r} \mathbf{\$1.46} \\ 6\overline{)\$8.76} \\ 6 \\ \hline 27 \\ 24 \\ \hline 36 \\ 36 \\ \hline 0 \end{array}$

18. $28\overline{)980}$ = **35**

$$\begin{array}{r} 84 \\ \hline 140 \\ 140 \\ \hline 0 \end{array}$$

19.

$$\begin{array}{r} 1\frac{3}{5} \\ +\ 1\frac{1}{5} \\ \hline \mathbf{2\frac{4}{5}} \end{array}$$

20.

$$\begin{array}{r} 4\frac{3}{10} \\ +\ 1\frac{2}{10} \\ \hline 5\frac{5}{10} \end{array} = \mathbf{5\frac{1}{2}}$$

21.

$$\begin{array}{r} 4\frac{3}{10} \\ -\ 1\frac{2}{10} \\ \hline \mathbf{3\frac{1}{10}} \end{array}$$

22. $\frac{2}{3} = \mathbf{\frac{4}{6}}$; $\frac{1}{2} = \mathbf{\frac{3}{6}}$

$\frac{4}{6} - \frac{3}{6} = \mathbf{\frac{1}{6}}$

23. $\frac{3}{10} \times \frac{1}{3} = \frac{3}{30} = \mathbf{\frac{1}{10}}$

24. $\frac{3}{4} \times \frac{5}{3} = \frac{15}{12} = 1\frac{3}{12} = \mathbf{1\frac{1}{4}}$

25. $\frac{3}{10} \times \frac{1}{3} = \frac{3}{30} = \mathbf{\frac{1}{10}}$

26.

$$\begin{array}{r} 12 \text{ ft} \\ \times\ 15 \text{ ft} \\ \hline 60 \\ 120 \\ \hline 180 \text{ sq. ft} \end{array}$$

180 tiles are needed.

27.

$$\begin{array}{r} \overset{1}{1}2 \text{ ft} \\ 12 \text{ ft} \\ 15 \text{ ft} \\ +\ 15 \text{ ft} \\ \hline \mathbf{54 \text{ feet}} \end{array}$$

28. 3 ft × 2 ft = 6 sq. ft

6 sq. ft × 6 ft = **36 cubic feet**

29. **2 hours**

30. **1 hour**

Lesson 110, Warm-Up

a. **Round each number to the nearest whole number and multiply.**

b. **100 cm; 1000 cm**

c. $\mathbf{\frac{2}{3}, 1\frac{1}{3}, 2\frac{2}{3}}$

d. **6**

e. **1120**

f. **45**

Problem Solving

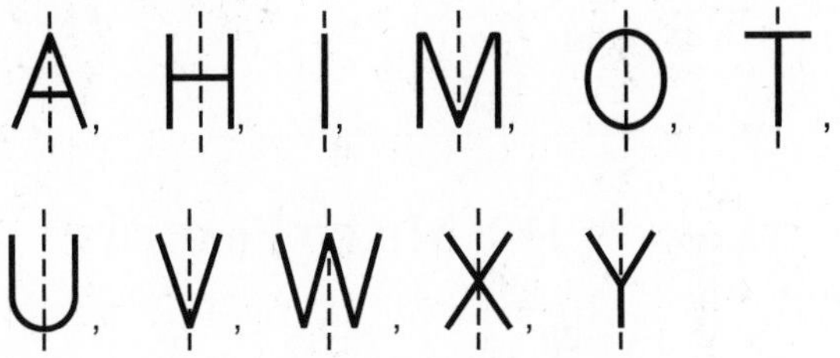

Lesson 110, Lesson Practice

a.

$$\begin{array}{r} 0.25 \\ \times\ 0.3 \\ \hline \mathbf{0.075} \end{array}$$

b.

$$\begin{array}{r} 0.12 \\ \times\ 0.12 \\ \hline 24 \\ 120 \\ \hline \mathbf{0.0144} \end{array}$$

c.

$$\begin{array}{r} 0.125 \\ \times\ 0.3 \\ \hline \mathbf{0.0375} \end{array}$$

d.

$$\begin{array}{r} 0.05 \\ \times\ 0.03 \\ \hline \mathbf{0.0015} \end{array}$$

e. $\begin{array}{r} 0.03 \\ \times\ 0.3 \\ \hline \mathbf{0.009} \end{array}$

f. $\begin{array}{r} 3.2 \\ \times\ 0.03 \\ \hline \mathbf{0.096} \end{array}$

g. $\begin{array}{r} 0.16 \\ \times\ 0.6 \\ \hline \mathbf{0.096} \end{array}$

h. $\begin{array}{r} 0.12 \\ \times\ 0.2 \\ \hline \mathbf{0.024} \end{array}$

i. $\begin{array}{r} 0.01 \\ \times\ 0.1 \\ \hline \mathbf{0.001} \end{array}$

j. $\begin{array}{r} 0.12 \\ \times\ 0.07 \\ \hline \mathbf{0.0084} \end{array}$

k. $\begin{array}{r} 0.4\text{ m} \\ \times\ 0.2\text{ m} \\ \hline \mathbf{0.08\text{ sq. m}} \end{array}$

LESSON 110, MIXED PRACTICE

1. $5 \times 4 = \mathbf{20}$

2. $\frac{5}{10} \times \frac{10}{10} = \frac{50}{100} = \mathbf{50\%}$

3. (a) $\frac{30 \div 10}{100 \div 10} = \mathbf{\frac{3}{10}}$

(b) $\frac{40 \div 10}{100 \div 10} = \frac{4}{10}$ or $\mathbf{\frac{2}{5}}$

4. $\frac{2}{5} \times \frac{100}{1} = \frac{200}{5}$

$\begin{array}{r} 40 \\ 5\overline{)200} \\ \underline{20} \\ 00 \\ \underline{0} \\ 0 \end{array}$

$\begin{array}{r} 100 \\ -\ 40 \\ \hline \mathbf{60\text{ passengers}} \end{array}$

5. **4.2 cm; 42 mm**

6. $\begin{array}{r} \mathbf{1.4\text{ cm}} \\ 3\overline{)4.2\text{ cm}} \\ \underline{3\phantom{.2\text{ cm}}} \\ 1\,2\phantom{\text{ cm}} \\ \underline{1\,2}\phantom{\text{ cm}} \\ 0\phantom{\text{ cm}} \end{array}$

7. $\frac{5}{6} = \mathbf{\frac{10}{12}}; \frac{3}{4} = \mathbf{\frac{9}{12}}$

$\frac{10}{12} + \frac{9}{12} = \frac{19}{12} = \mathbf{1\frac{7}{12}}$

8. (a) 4 units + 2 units + 2 units + 2 units + 6 units + 4 units = **20 units**

(b) **20 sq. units**

9. $\overline{CD}$ **(or** $\overline{DC}$**)**

10. $\overline{AD}$ **or (**$\overline{DA}$**) and** $\overline{BC}$ **(or** $\overline{CB}$**)**

11. $\mathbf{\frac{375}{1000}}$**; Three hundred seventy-five thousandths**

12. $\begin{array}{r} \overset{5}{\not{6}}.\,\overset{9}{\not{0}}{}^{1}0 \\ -\ 4.\,3\,2 \\ \hline \mathbf{1.\,6\,8} \end{array}$

13. $\begin{array}{r} 0.12 \\ \times\ 0.11 \\ \hline 12 \\ 120 \\ \hline \mathbf{0.0132} \end{array}$

14. $\begin{array}{r} 0.28 \\ \times\ 0.04 \\ \hline \mathbf{0.0112} \end{array}$

15. $\begin{array}{r} 0.25 \\ \times\quad 10 \\ \hline 2.50 \end{array}$ or **2.5**

16. $\begin{array}{r} 195 \\ 19\overline{)3705} \\ \underline{19} \\ 180 \\ \underline{171} \\ 95 \\ \underline{95} \\ 0 \end{array}$

$x = \mathbf{195}$

17. $\sqrt{400} = \mathbf{20}$

18. $\begin{array}{r} 30 \\ \times\ 30 \\ \hline \mathbf{900} \end{array}$

19. $\frac{5}{13} + \frac{10}{13} = \frac{15}{13} = \mathbf{1\frac{2}{13}}$

20. $\frac{11}{12} - \frac{7}{12} = \frac{4}{12} = \mathbf{\frac{1}{3}}$

21. $\frac{1}{1} \times \frac{5}{6} = \mathbf{\frac{5}{6}}$

22. $\frac{2}{1} \times \frac{6}{5} = \frac{12}{5} = \mathbf{2\frac{2}{5}}$

23. $\frac{5}{6} \times \frac{1}{2} = \mathbf{\frac{5}{12}}$

24. $\begin{array}{r} {}^{1} \\ 14\% \\ 11\% \\ 25\% \\ +\ 50\% \\ \hline \mathbf{100\%} \end{array}$

25. $\frac{1}{4} \times \frac{100}{1} = \frac{100}{4}$

$\begin{array}{r} 25\% \\ 4\overline{)100} \\ \underline{8} \\ 20 \\ \underline{20} \\ 0 \end{array}$

A

26. $\frac{14}{28} = \mathbf{\frac{1}{2}}$

27. 

28. **Rotation**

29. **Greene, Bolden, Thompson**

30. $\begin{array}{r} \overset{9}{\cancel{1}}\overset{9}{\cancel{0}}.\ \cancel{0}^{1}4 \text{ sec} \\ -\ \ 9.\ 8\ 7 \text{ sec} \\ \hline \mathbf{0.\ 1\ 7 \text{ second}} \end{array}$

INVESTIGATION 11

1. $1 \times 4 = \mathbf{4\ ft}$

2. length: $2 \times 4 = \mathbf{8\ ft}$
width: $1 \times 4 = \mathbf{4\ ft}$

3. $5 \times 4 = 20$ ft (length)
$3 \times 4 = 12$ ft (width)
$20 \times 12 = \mathbf{240\ ft^2}$
$240 - 12 = \mathbf{208\ ft^2}$

4. length: $1\frac{1}{2} \times 4 = \mathbf{6\ ft}$
width: $1 \times 4 = \mathbf{4\ ft}$

5. length: $1 \times 4 = \mathbf{4\ ft}$
width: $\frac{1}{2} \times 4 = \mathbf{2\ ft}$

6. **5 ft; the chest**

7. $3 \times 200 = \mathbf{600\ yd}$

8. $1\frac{1}{2} \times 200 = \mathbf{300\ yd}$

9. $3\frac{1}{2} \times 200 =$ **700 yd; Answers may vary.**

10. $2\frac{3}{4} \times 200 = \mathbf{550\ yd}$

11. about 2.5 in.; **500 yd**

12. $6 \times 3 =$ **18 cm;** $6 \times \frac{1}{2} =$ **3 cm**

13. $\mathbf{\frac{1}{6}}$ **mi;** $\mathbf{\frac{5}{6}}$ **mi**

14. $2 \times 6 = 12$
$12 + 1 =$ **13 cm**

15. $\mathbf{\frac{1}{3}}$ **mi**

Extensions

a. **See student work.**

b. **See student work.**

c. **See student work.** Measurements of the student's scale model should be as follows:

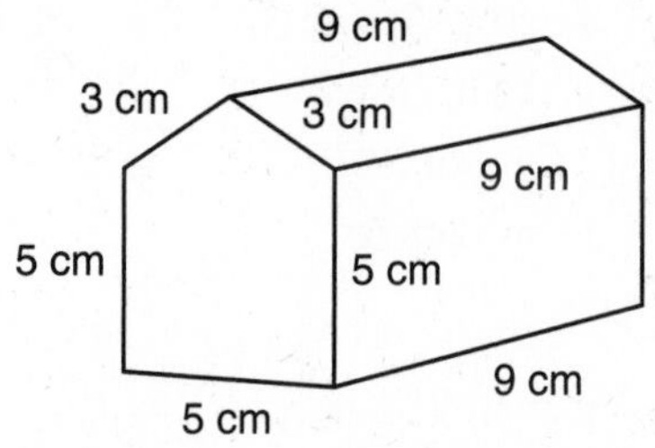

Lesson 111, Warm-Up

a. **Round each number to the nearest whole number and divide.**

b. **1000 m; 100 m**

c. **\$5**

d. **\$2.50**

e. **\$1**

f. **15**

g. **2090**

h. **<**

Problem Solving

1 roll of quarters: \$0.25 × 40 = \$10.00
1 roll of dimes: \$0.10 × 50 = \$5.00
2 rolls of quarters (\$20.00) are equal in value to **4 rolls of dimes.**

Lesson 111, Lesson Practice

a. 1.234 × 10 = **12.34**

b. 1.234 × 1000 = **1234**

c. 0.1234 × 100 = **12.34**

d. 0.345 × 10 = **3.45**

e. 0.345 × 100 = **34.5**

f. 0.345 × 1000 = **345**

g. 5.67 × 10 = **56.7**

h. 5.67 × 1000 = **5670**

i. 5.67 × 100 = **567**

Lesson 111, Mixed Practice

1.
$$\begin{array}{r} 23 \\ 25 \\ +\ 30 \\ \hline 78 \end{array} \qquad \begin{array}{r} \textbf{26 horses} \\ 3\overline{)78} \\ \underline{6} \\ 18 \\ \underline{18} \\ 0 \end{array}$$

2.
$$\begin{array}{r} 1\,\overset{1}{\not2}\,\overset{{}^10}{\not1}\,{}^11 \\ -\ 1\,1\,6\,7 \\ \hline \textbf{4 4 years old} \end{array}$$

3. (a) $\frac{25 \div 25}{100 \div 25} = \mathbf{\frac{1}{4}}$

   (b) $\frac{50 \div 50}{100 \div 50} = \mathbf{\frac{1}{2}}$

4. (a) **6, 12, 18, 24, 30, 36**

   (b) **9, 18, 27, 36**

   (c) **18, 36**

5. $\frac{50}{100} = \mathbf{50\%}$

   $\frac{50}{100} = \mathbf{0.50}$

   $\frac{50}{100} = \mathbf{\frac{1}{2}}$

6. **Sphere**

7. 12 months ÷ 2 = 6 months
   12 months + 6 months = **18 months**

8. (a) 3 units + 3 units + 2 units + 2 units
+ 3 units + 3 units + 2 units
+ 2 units = **20 units**

(b) **17 sq. units**

9.
$$3\overline{)45}^{\ 15 \text{ mm}}$$
$$\underline{3}$$
$$15$$
$$\underline{15}$$
$$0$$

$$\begin{array}{r} 45 \text{ mm} \\ + \ 15 \text{ mm} \\ \hline 60 \text{ mm} \end{array}$$

$$\begin{array}{r} 90 \text{ mm} \\ - \ 60 \text{ mm} \\ \hline \textbf{30 mm} \end{array}$$

10. **12.3**

11. **3420**

12.
$$\begin{array}{r} {\scriptstyle 2\,1\ 1} \\ 15.000 \\ 9.670 \\ 3.292 \\ + \ 5.500 \\ \hline 33.462 \end{array}$$

Thirty-three and four hundred sixty-two thousandths

13.
$$\begin{array}{r} 4.\,\overset{2}{\cancel{3}}{}^{1}0 \\ - \ 1.\,2\ 1 \\ \hline \textbf{3.\ 0 9} \end{array}$$

14.
$$\begin{array}{r} 0.14 \\ \times \ \ 0.6 \\ \hline \textbf{0.084} \end{array}$$

15.
$$\begin{array}{r} 4\ 8 \\ \times \ \ 0.7 \\ \hline \textbf{33.6} \end{array}$$

16. $0.735 \times 100 =$ **73.5**

17. $\dfrac{3}{4} = \dfrac{\mathbf{6}}{\mathbf{8}}$

$\dfrac{6}{8} + \dfrac{3}{8} = \dfrac{9}{8} = 1\dfrac{1}{8}$

18.
$$16\overline{)4000}^{\ 250}$$
$$\underline{32}$$
$$80$$
$$\underline{80}$$
$$00$$
$$\underline{0}$$
$$0$$

19.
$$10\overline{)\$18.00}^{\ \$1.80}$$
$$\underline{10}$$
$$8\ 0$$
$$\underline{8\ 0}$$
$$00$$
$$\underline{0}$$
$$0$$

20.
$$\begin{array}{r} \dfrac{7}{11} \\ + \ \dfrac{8}{11} \\ \hline \dfrac{15}{11} = 1\dfrac{\mathbf{4}}{\mathbf{11}} \end{array}$$

21.
$$\begin{array}{r} 3\dfrac{7}{12} \\ + \ \dfrac{1}{12} \\ \hline 3\dfrac{8}{12} = 3\dfrac{\mathbf{2}}{\mathbf{3}} \end{array}$$

22.
$$\begin{array}{r} 5\dfrac{9}{10} \\ - \ 5\dfrac{3}{10} \\ \hline \dfrac{6}{10} = \dfrac{\mathbf{3}}{\mathbf{5}} \end{array}$$

23. $\dfrac{7}{2} \times \dfrac{1}{2} = \dfrac{7}{4} = 1\dfrac{\mathbf{3}}{\mathbf{4}}$

24. $\dfrac{2}{3} \times \dfrac{4}{1} = \dfrac{8}{3} = 2\dfrac{\mathbf{2}}{\mathbf{3}}$

25. $\dfrac{3}{1} \times \dfrac{4}{3} = \dfrac{12}{3} = \mathbf{4}$

26. $\sqrt{9} + \sqrt{16} \; \bigcirc\!\!> \; \sqrt{9 + 16}$
$\quad 3 + 4 \qquad\qquad \sqrt{25}$
$\qquad 7 \qquad\qquad\quad 5$

27. $\dfrac{2}{12} \times \dfrac{100}{1} = \dfrac{200}{12}$

$$12\overline{)200}^{\ 16\frac{8}{12}} = \mathbf{16\dfrac{2}{3}\%}$$
$$\underline{12}$$
$$80$$
$$\underline{72}$$

28. **10:10 a.m.**

29. **5:40 p.m.**

30. **Departure and arrival times are local times, so six hours of the difference is attributable to the fact that Philadelphia is in a time zone that is 3 hours ahead of Los Angeles. About half an hour of difference, and often more, is due to the west-to-east direction of the jet stream.**

Lesson 112, Warm-Up

a. **Round tickets to the nearest ten, and multiply by the cost per ticket.**

b. **16 oz; 32 oz; 8 oz**

c. **3**

d. **9**

e. **27**

f. **12**

g. **1150**

h. **>**

Problem Solving

Heights of Bounces

First	4 ft
Second	2 ft
Third	1 ft
Fourth	$\frac{1}{2}$ ft
Fifth	$\frac{1}{4}$ ft

Lesson 112, Lesson Practice

a. Multiples of 2: 2, 4, (6), 8, 10, . . .
Multiples of 3: 3, (6), 9, 12, 15, . . .
LCM is **6**

b. Multiples of 3: 3, 6, 9, 12, (15), 18, . . .
Multiples of 5: 5, 10, (15), 20, 25, . . .
LCM is **15**

c. Multiples of 5: 5, (10), 15, 20, . . .
Multiples of 10: (10), 20, 30, 40, . . .
LCM is **10**

d. Multiples of 2: 2, (4), 6, 8, 10, . . .
Multiples of 4: (4), 8, 12, 16, 20, . . .
LCM is **4**

e. Multiples of 3: 3, (6), 9, 12, . . .
Multiples of 6: (6), 12, 18, 24, . . .
LCM is **6**

f. Multiples of 6: 6, 12, 18, 24, (30), . . .
Multiples of 10: 10, 20, (30), 40, 50, . . .
LCM is **30**

g. Multiples of 8: 8, 16, 24, 32, (40), . . .
Multiples of 10: 10, 20, 30, (40), 50, . . .
LCM is **40**

Lesson 112, Mixed Practice

1. 1 ton = 2000 lbs

$$\begin{array}{r} 2000 \text{ lbs} \\ \times \quad 4 \\ \hline \end{array}$$
About 8000 lbs

2.
$$\begin{array}{r} 12 \\ \times \ 10 \\ \hline \end{array}$$
About 120 inches

3. $5.25 × 10 = **$52.50**

4. **4**

5. **One possibility:**

6. $12\frac{5}{10} = \mathbf{12\frac{1}{2}}$

7. $\frac{25}{100} = \mathbf{25\%}$

$\frac{25}{100} = \mathbf{0.25}$

$\frac{25}{100} = \mathbf{\frac{1}{4}}$

8. **Cylinder**

9. $\frac{1}{2}$ in. + $\frac{1}{2}$ in. + $\frac{1}{2}$ in. = $\frac{3}{2}$ in. = $\mathbf{1\frac{1}{2}}$ **in.**

10. Multiples of 6: 6, 12, (18), 24, . . .
Multiples of 9: 9, (18), 27, 36, . . .
LCM is **18**

11.

$$\begin{array}{r} 15\text{ mm} \\ \times\quad 2 \\ \hline \mathbf{30\text{ mm}} \end{array}$$

12.

$$\begin{array}{r} 4.2\text{ cm} \\ +\ 3.0\text{ cm} \\ \hline 7.2\text{ cm} \end{array} \qquad \begin{array}{r} 9.2\text{ cm} \\ -\ 7.2\text{ cm} \\ \hline \mathbf{2.0\text{ cm}} \end{array}$$

13.

$$\begin{array}{r} {}^{1\,1} \\ 4.380 \\ {}_{2}7.525 \\ 23.700 \\ +\ 9.000 \\ \hline \mathbf{44.605} \end{array}$$

14.

$$\begin{array}{r} 4.\,\cancel{3}^{2}\,{}^{1}0 \\ -\ 0.\,2\,1 \\ \hline 4.\,0\,9 \end{array} \qquad \begin{array}{r} \cancel{5}^{4}.\,\cancel{0}^{9}\,{}^{1}0 \\ -\ 4.\,0\,9 \\ \hline \mathbf{0.\,9\,1} \end{array}$$

15.

$$\begin{array}{r} 3.6 \\ \times\ 40 \\ \hline 144.0 \end{array}$$ or **144**

16.

$$\begin{array}{r} 0.15 \\ \times\ 0.5 \\ \hline \mathbf{0.075} \end{array}$$

17. **1.25**

18.

$$\begin{array}{r} 75 \\ 4\overline{)300} \\ 28 \\ \hline 20 \\ 20 \\ \hline 0 \end{array}$$

w = **75**

19.

$$\begin{array}{r} \mathbf{75} \\ 40\overline{)3000} \\ 280 \\ \hline 200 \\ 200 \\ \hline 0 \end{array}$$

20.

$$\begin{array}{r} \mathbf{132} \\ 25\overline{)3300} \\ 25 \\ \hline 80 \\ 75 \\ \hline 50 \\ 50 \\ \hline 0 \end{array}$$

21. $4\frac{7}{7} - 1\frac{2}{7} = 3\frac{5}{7}$

$3\frac{3}{7} + 3\frac{5}{7} = 6\frac{8}{7} = \mathbf{7\frac{1}{7}}$

22. $\frac{3}{1} \times \frac{1}{2} = \frac{3}{2} = 1\frac{1}{2}$

$1\frac{1}{2} - 1\frac{1}{2} = \mathbf{0}$

23. $\frac{1}{4} = \mathbf{\frac{3}{12}}$; $\frac{2}{3} = \mathbf{\frac{8}{12}}$

$\frac{8}{12} - \frac{3}{12} = \mathbf{\frac{5}{12}}$

24. **Triangle**

25. **F2**

26. $3^2 + 4^2$ (=) 5^2

$9 + 16 \qquad 25$

25

27. $\frac{1}{8} \times 100\% = \frac{100}{8}$

$$\begin{array}{r} 12\frac{4}{8} \\ 8\overline{)100} \\ 8 \\ \hline 20 \\ 16 \\ \hline 4 \end{array}$$

$12\frac{4}{8}$ = $\mathbf{12\frac{1}{2}\%}$

28. Arrive at 8:09 a.m.
Leave at 9:43 a.m.
1 hr 34 min

29. 6:11 a.m. to 8:09 a.m. ⟶ 1 hr 58 min
9:43 a.m. to 10:38 a.m. ⟶ + 55 min
2 hr 53 min

30. 9:58 a.m. to 11:03 a.m. ⟶ 1 hr 5 min
12:04 a.m. to 1:33 a.m. ⟶ + 1 hr 29 min
2 hr 34 min

2 hr 53 min − 2 hr 34 min = **19 min**

The route taken through St. Louis is shorter than the route through Chicago.

Lesson 113, Warm-Up

a. **Round gallons to the nearest whole number, round the price to the nearest ten cents, and multiply the numbers.**

b. $\mathbf{\frac{2}{3}, \frac{3}{4}, 1\frac{1}{4}}$

c. **3**

d. $\mathbf{9\frac{1}{2}}$

e. **9**

f. **6**

g. **145**

h. **<**

Problem Solving

$5 \times 4 \times 3 =$ **60 1-inch cubes**

Lesson 113, Lesson Practice

a. $\mathbf{\frac{7}{4}; 1\frac{3}{4}}$

b. $\mathbf{\frac{7}{2}; 3\frac{1}{2}}$

c. $\mathbf{\frac{8}{3}; 2\frac{2}{3}}$

d. $\mathbf{\frac{9}{2}}$

e. $\mathbf{\frac{5}{3}}$

f. $\mathbf{\frac{11}{4}}$

g. $\mathbf{\frac{25}{8}}$

Lesson 113, Mixed Practice

1. $5\overline{)1400}$ = **280 miles**

$$\begin{array}{r} 10 \\ \hline 40 \\ 40 \\ \hline 00 \\ 0 \\ \hline 0 \end{array}$$

2. $$\begin{array}{r} 600 \\ \times\ 200 \\ \hline \mathbf{120{,}000} \end{array}$$

3. (a) $\frac{1}{10} \times \frac{10}{10} = \mathbf{\frac{10}{100}}$

(b) $\frac{10}{100} =$ **10%**

4. $\frac{1}{6} \times \frac{108}{1} = \frac{108}{6}$

$6\overline{)108}$ = 18

$$\begin{array}{r} 6 \\ \hline 48 \\ 48 \\ \hline 0 \end{array}$$

About 18 pounds

5. $\mathbf{\frac{3}{2}; 1\frac{1}{2}}$

6. (a) 1 in. + 1 in. + 1 in. + 1 in. = **4 inches**

(b) 1 in. × 1 in. = **1 sq. inch**

7. $\frac{3}{12} = \mathbf{\frac{1}{4}}$

$\frac{1}{4} \times \frac{100}{1} = \frac{100}{4} =$ **25%**

8. (a) **Rectangular solid**

(b) **6 faces**

9. Multiples of 4: 4, 8, (12), 16, . . .

Multiples of 6: 6, (12), 18, 24, . . .

LCM is **12**

10. $5\frac{2}{6} = \mathbf{5\frac{1}{3}}$

11.
$$\begin{array}{r} {}^{1\,1}\\ {}_{1}4.239 \\ 25.000 \\ 6.790 \\ +\ 12.500 \\ \hline \mathbf{48.529} \end{array}$$

12.
$$\begin{array}{r} {}^{3\ \ 9}\ \ \\ \not{4}.\,\not{0}{}^{1}0 \\ -\ 3.\,7\,5 \\ \hline 0.\,2\,5 \end{array} \qquad \begin{array}{r} 6.875 \\ -\ 0.250 \\ \hline \mathbf{6.625} \end{array}$$

13.
$$\begin{array}{r} 3.7 \\ \times\ 0.8 \\ \hline \mathbf{2.96} \end{array}$$

14.
$$\begin{array}{r} 0.125 \\ \times\ \ \ \ 100 \\ \hline 12.500 \end{array} \text{ or } \mathbf{12.5}$$

15.
$$\begin{array}{r} 0.32 \\ \times\ 0.04 \\ \hline \mathbf{0.0128} \end{array}$$

16.
$$\begin{array}{r} \mathbf{24} \\ 17\overline{)408} \\ \underline{34}\ \\ 68 \\ \underline{68} \\ 0 \end{array}$$

17.
$$\begin{array}{r} \mathbf{26\ R\ 3} \\ 27\overline{)705} \\ \underline{54}\ \\ 165 \\ \underline{162} \\ 3 \end{array}$$

18.
$$\begin{array}{r} \mathbf{\$3.54} \\ 5\overline{)\$17.70} \\ \underline{15}\ \ \ \ \\ 2\,7\ \ \\ \underline{2\,5}\ \ \\ 20 \\ \underline{20} \\ 0 \end{array}$$

19.
$$\begin{array}{r} 3\frac{7}{10} \\ +\ 4\ \ \ \\ \hline \mathbf{7\frac{7}{10}} \end{array}$$

20.
$$\begin{array}{r} 5\frac{5}{8} \\ +\ \frac{1}{8} \\ \hline 5\frac{6}{8} \end{array} = \mathbf{5\frac{3}{4}}$$

21.
$$\begin{array}{r} 6\frac{10}{10} \\ -\ 4\frac{3}{10} \\ \hline \mathbf{2\frac{7}{10}} \end{array}$$

22. $\frac{5}{6} \times \frac{4}{1} = \frac{20}{6} = 3\frac{2}{6} = \mathbf{3\frac{1}{3}}$

23. $\frac{3}{8} \times \frac{1}{2} = \mathbf{\frac{3}{16}}$

24. $\frac{3}{8} \times \frac{2}{1} = \frac{6}{8} = \mathbf{\frac{3}{4}}$

25. $\frac{1}{6} = \mathbf{\frac{2}{12}}$; $\frac{1}{4} = \mathbf{\frac{3}{12}}$

$\frac{2}{12} + \frac{3}{12} = \mathbf{\frac{5}{12}}$

26. 5 ft × 2 ft = 10 sq. ft

10 sq. ft × 3 ft = **30 cubic feet**

27. 5 ft × 2 ft = **10 sq. ft**

28. Leave at 9:30 a.m.

Arrive at 5:30 p.m.

8 hours

29. 1:40 p.m. + 15 min

1:55 p.m.

30.
$$\begin{array}{r} 40 \\ 8\overline{)320} \\ \underline{32}\ \\ 00 \\ \underline{0} \\ 0 \end{array}$$

B. 40 miles

Lesson 114, Warm-Up

a. **Round each number to the nearest whole number and multiply.**

b. **12 in.; 30 in.**

c. **3**

d. **9**

e. **15**

f. **0**

g. **1600**

h. **=**

Problem Solving

Any two of the following are acceptable:

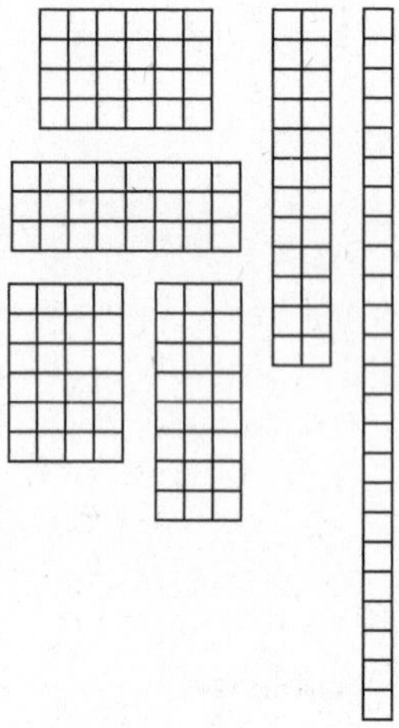

Lesson 114, Lesson Practice

a. **4, 5**

b. **No solution**

c. **6 by 4; 24 by 1**

d. **Only three combinations need to be tried. The largest number that can play game A is 4, the smallest is 0. Try each number between 0 and 4 for game A, and see whether game B can be played by all the remaining party guests.**

Number of Game A	Number of Game B
4	0
2	3
0	6

Lesson 114, Mixed Practice

1.

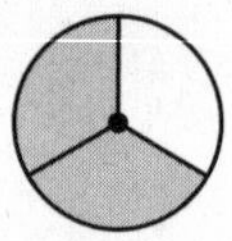

$\frac{2}{3} \times \frac{100}{1} = \frac{200}{3}$

$3\overline{)200}$ = $\mathbf{66\frac{2}{3}\%}$

18
20
18
2

2. **B. feet**

3. **D.**

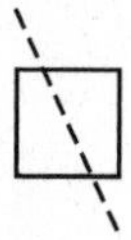

4. 28×16

168
280
448 miles

5. $1\frac{3}{4} = \frac{4}{4} + \frac{3}{4} = \mathbf{\frac{7}{4}}$

6. $\frac{1}{2} = \mathbf{\frac{3}{6}}$

$\frac{5}{6} - \frac{3}{6} = \frac{2}{6} = \mathbf{\frac{1}{3}}$

7. Multiples of 8: 8, 16, (24), 32, . . .
Multiples of 6: 6, 12, 18, (24), . . .
LCM is **24**

8. $\mathbf{\frac{1}{3}}$

9. $\mathbf{\frac{1}{3}}$

10. 6 cm − 2 cm = 4 cm
6 cm + 4 cm = **10 cm**

11.
$$\begin{array}{r} \overset{2}{4}\overset{1}{5}.\overset{1}{0}00 \\ 16.700 \\ 8.290 \\ +\ \ 4.325 \\ \hline \mathbf{74.315} \end{array}$$

12.
$$\begin{array}{r} 3.2 \\ -\ 1.0 \\ \hline 2.2 \end{array} \qquad \begin{array}{r} 4.2 \\ -\ 2.2 \\ \hline 2.0 \text{ or } \mathbf{2} \end{array}$$

13.
$$\begin{array}{r} 0.75 \\ \times\ 0.05 \\ \hline \mathbf{0.0375} \end{array}$$

14.
$$\begin{array}{r} 38 \\ \times\ 0.6 \\ \hline \mathbf{22.8} \end{array}$$

15. **750**

16.
$$\begin{array}{r} \mathbf{\$2.03} \\ 12\overline{)\$24.36} \\ \underline{24} \\ 0\,3 \\ \underline{0} \\ 36 \\ \underline{36} \\ 0 \end{array}$$

17.
$$\begin{array}{r} \mathbf{184} \\ 25\overline{)4600} \\ \underline{25} \\ 210 \\ \underline{200} \\ 100 \\ \underline{100} \\ 0 \end{array}$$

18. $6\frac{9}{10} - \frac{1}{10} = 6\frac{8}{10} = \mathbf{6\frac{4}{5}}$

19. $5\frac{4}{9} + 3\frac{5}{9} = 8\frac{9}{9} = \mathbf{9}$

20. $\frac{4}{1} \times \frac{8}{1} = \frac{32}{1} = \mathbf{32}$

21. $\frac{4}{1} \times \frac{1}{8} = \frac{4}{8} = \mathbf{\frac{1}{2}}$

22. $\frac{18}{30} = \mathbf{\frac{3}{5}}$

23. (a) $\frac{3}{10} \times \frac{10}{10} = \frac{30}{100} = \mathbf{30\%}$

(b) $100\% - 30\% = \mathbf{70\%}$

24. (a) $\frac{60}{100} = \mathbf{\frac{3}{5}}$

(b) $\frac{70}{100} = \mathbf{\frac{7}{10}}$

25.
$$\begin{array}{r} \overset{1}{1}2 \text{ in.} \\ 6 \text{ in.} \\ 12 \text{ in.} \\ +\ \ 6 \text{ in.} \\ \hline 36 \text{ in.} \end{array} \qquad \begin{array}{r} \mathbf{9\ in.} \\ 4\overline{)36 \text{ in.}} \end{array}$$

26. (a)
$$\begin{array}{r} 12 \text{ in.} \\ \times\ \ 6 \text{ in.} \\ \hline \mathbf{72\ sq.\ in.} \end{array}$$

(b) 9 in. × 9 in. = **81 sq. in.**

27. $\frac{1}{6} \times 100\% = \frac{100\%}{6}$

$$\begin{array}{r} 16\frac{4}{6} \\ 6\overline{)100} \\ \underline{6} \\ 40 \\ \underline{36} \\ 4 \end{array} \quad = \mathbf{16\frac{2}{3}\%}$$

28. 6 cm × 6 cm = **36 sq. cm**
3 cm × 3 cm = **9 sq. cm**

29.
$$\begin{array}{r} \overset{1}{3}6 \text{ sq. cm} \\ +\ \ 9 \text{ sq. cm} \\ \hline \mathbf{45\ sq.\ cm} \end{array}$$

30.

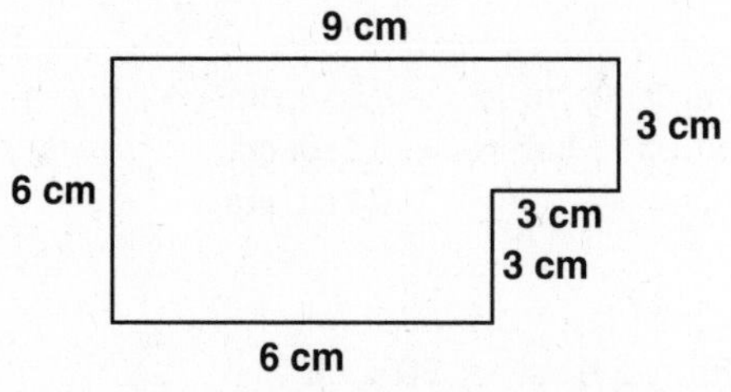

Perimeter = 9 cm + 3 cm + 3 cm + 3 cm + 6 cm + 6 cm = **30 cm**

Lesson 115, Warm-Up

a. **20 inches; 25 square inches**

b. **16 ounces; 40 ounces**

c. **20**

d. **40**

e. **60**

f. **2004**

g. **>**

Problem Solving

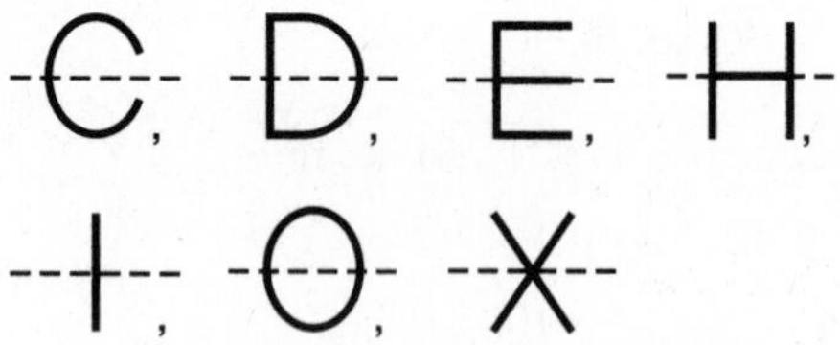

Depending on how it is written, the letter K may also have a horizontal line of symmetry.

Lesson 115, Lesson Practice

a.
$$\begin{array}{lr} 7\text{ m} \times 4\text{ m} = & 28\text{ sq. m} \\ 3\text{ m} \times 4\text{ m} = & +\ 12\text{ sq. m} \\ \hline & \mathbf{40\text{ sq. m}} \end{array}$$

b.
$$\begin{array}{lr} 5\text{ in.} \times 10\text{ in.} = & 50\text{ sq. in.} \\ 3\text{ in.} \times 6\text{ in.} = & +\ 18\text{ sq. in.} \\ \hline & \mathbf{68\text{ sq. in.}} \end{array}$$

c.
$$\begin{array}{lr} 6\text{ cm} \times 2\text{ cm} = & 12\text{ sq. cm} \\ 6\text{ cm} \times 2\text{ cm} = & +\ 12\text{ sq. cm} \\ \hline & \mathbf{24\text{ sq. cm}} \end{array}$$

d.
$$\begin{array}{lr} 6\text{ ft} \times 5\text{ ft} = & 30\text{ sq. ft} \\ 5\text{ ft} \times 1\text{ ft} = & +\ 5\text{ sq. ft} \\ \hline & \mathbf{35\text{ sq. ft}} \end{array}$$

Lesson 115, Mixed Practice

1.
$$\begin{array}{r} 4\text{ feet} \\ 12\overline{)48} \\ 48 \\ \hline 0 \end{array} \qquad \begin{array}{r} 24\text{ ft} \\ -\ 4\text{ ft} \\ \hline \mathbf{20\text{ ft}} \end{array}$$

2. $\mathbf{\frac{5}{2}; 2\frac{1}{2}}$

3. $\mathbf{\frac{1}{2}}$

4. $\mathbf{\frac{1}{4}}$

5. $\mathbf{7\frac{3}{5}}$

6. In $1\frac{1}{2}$ hours the time will be after noon, so the "a.m." will switch to "p.m." The time will be **1:10 p.m.**

7. **C.** $\mathbf{\frac{1}{4}, \frac{3}{4}}$

8. Multiples of 5: 5, 10, (15), 20, . . .
Multiples of 3: 3, 6, 9, 12, (15), . . .
LCM is **15**

9. 4 m + 3 m + 4 m + 3 m = **14 m**

10. 4 m × 3 m = **12 sq. m**

11.
$$\begin{array}{r} \overset{1\,1}{42.980} \\ 50.000 \\ 23.500 \\ +\ \ 0.025 \\ \hline \mathbf{116.505} \end{array}$$

12.
$$\begin{array}{r} \overset{5}{\cancel{6}}.\,\overset{9}{\cancel{0}}{}^{1}0 \\ -\ 5.\,1\,8 \\ \hline 0.\,8\,2 \end{array}$$
Eighty-two hundredths

13.
$$\begin{array}{r} 0.375 \\ \times\ \ \ \ 10 \\ \hline 3.750 \end{array}$$ or **3.75**

14. $\begin{array}{r} 0.14 \\ \times\ 0.06 \\ \hline \mathbf{0.0084} \end{array}$

15. $\begin{array}{r} 7.8 \\ \times\ 19 \\ \hline 702 \\ 780 \\ \hline \mathbf{148.2} \end{array}$

16. $30\overline{)2340}$ = **78**
$\begin{array}{r} 210 \\ \hline 240 \\ 240 \\ \hline 0 \end{array}$

17. $18\overline{)2340}$ = **130**
$\begin{array}{r} 18 \\ \hline 54 \\ 54 \\ \hline 00 \\ 0 \\ \hline 0 \end{array}$

18. $7\overline{)8765}$ = **1252 R 1**
$\begin{array}{r} 7 \\ \hline 17 \\ 14 \\ \hline 36 \\ 35 \\ \hline 15 \\ 14 \\ \hline 1 \end{array}$

19. $\frac{5}{6} + 1\frac{5}{6} = 1\frac{10}{6} = 2\frac{4}{6} = \mathbf{2\frac{2}{3}}$

20. $7\frac{5}{8} - 7\frac{1}{8} = \frac{4}{8} = \mathbf{\frac{1}{2}}$

21. $\frac{4}{5} \times \frac{2}{3} = \mathbf{\frac{8}{15}}$

22. $\frac{4}{5} \times \frac{3}{2} = \frac{12}{10} = 1\frac{2}{10} = \mathbf{1\frac{1}{5}}$

23. $\frac{2}{5} \times \frac{3}{3} = \mathbf{\frac{6}{15}}$

24. $\frac{2}{3} \times \frac{5}{5} = \mathbf{\frac{10}{15}}$

25. $\frac{6}{15} + \frac{10}{15} = \frac{16}{15} = \mathbf{1\frac{1}{15}}$

26. $\frac{1}{2}\text{ in.} + \frac{1}{2}\text{ in.} + \frac{1}{2}\text{ in.} + \frac{1}{2}\text{ in.}$
$+ \frac{1}{2}\text{ in.} = \frac{5}{2}\text{ in.} = \mathbf{2\frac{1}{2}\text{ in.}}$

27. **5 lines of symmetry**

28. $\begin{array}{rr} 1\text{ ft} \times 2\text{ ft} = & 2\text{ sq. ft} \\ 1\text{ ft} \times 3\text{ ft} = & +\ 3\text{ sq. ft} \\ \hline & \mathbf{5\text{ sq. ft}} \end{array}$

29. 1 ft + 3 ft + 3 ft + 1 ft
+ 2 ft + 2 ft = **12 ft**

30. $\frac{1}{4}\text{ mi} \times \frac{1}{2}\text{ mi} = \frac{1}{8}\text{ sq. mi}$

$\mathbf{\frac{1}{8}}$ **of a sq. mi**

Lesson 116, Warm-Up

a. **20 in.; 24 sq. in.**

b. **60 seconds; 150 seconds**

c. **$30**

d. **$3**

e. **$0.30**

f. **9**

g. **1776**

h. **<**

Problem Solving

3 × 8 = 24 column inches
24 × $20 = $480 per day
$480 × 2 = **$960**

LESSON 116, LESSON PRACTICE

a. $\frac{1}{2} \times \frac{4}{4} = \frac{4}{8}$

$\frac{1}{8} + \frac{4}{8} = \mathbf{\frac{5}{8}}$

b. $\frac{1}{2} \times \frac{2}{2} = \frac{2}{4}$

$\frac{2}{4} - \frac{1}{4} = \mathbf{\frac{1}{4}}$

c. $\frac{3}{4} \times \frac{2}{2} = \frac{6}{8}$

$\frac{6}{8} + \frac{1}{8} = \mathbf{\frac{7}{8}}$

d. $\frac{2}{3} \times \frac{3}{3} = \frac{6}{9}$

$\frac{6}{9} - \frac{1}{9} = \mathbf{\frac{5}{9}}$

e. $\frac{1}{3} \times \frac{4}{4} = \frac{4}{12}$

$\frac{1}{4} \times \frac{3}{3} = \frac{3}{12}$

$\frac{4}{12} + \frac{3}{12} = \mathbf{\frac{7}{12}}$

f. $\frac{1}{2} \times \frac{3}{3} = \frac{3}{6}$

$\frac{1}{3} \times \frac{2}{2} = \frac{2}{6}$

$\frac{3}{6} - \frac{2}{6} = \mathbf{\frac{1}{6}}$

g. $\frac{1}{2} \times \frac{2}{2} = \frac{2}{4}$

$$\begin{array}{r} 3\frac{1}{4} \\ + \ 2\frac{2}{4} \\ \hline \mathbf{5\frac{3}{4}} \end{array}$$

h. $\frac{1}{2} \times \frac{4}{4} = \frac{4}{8}$

$$\begin{array}{r} 2\frac{1}{8} \\ + \ 5\frac{4}{8} \\ \hline \mathbf{7\frac{5}{8}} \end{array}$$

i. $\frac{1}{2} \times \frac{3}{3} = \frac{3}{6}$

$$\begin{array}{r} 3\frac{3}{6} \\ - \ 1\frac{1}{6} \\ \hline 2\frac{2}{6} \end{array} = \mathbf{2\frac{1}{3}}$$

j. $\frac{1}{2} \times \frac{2}{2} = \frac{2}{4}$

$$\begin{array}{r} 2\frac{3}{4} \\ - \ 2\frac{2}{4} \\ \hline \mathbf{\frac{1}{4}} \end{array}$$

k. $\frac{1}{4} \times \frac{2}{2} = \frac{2}{8}$

$$\begin{array}{r} 5\frac{5}{8} \\ + \ 1\frac{2}{8} \\ \hline \mathbf{6\frac{7}{8}} \end{array}$$

l. $\frac{1}{2} \times \frac{3}{3} = \frac{3}{6}$

$\frac{1}{3} \times \frac{2}{2} = \frac{2}{6}$

$$\begin{array}{r} 3\frac{3}{6} \\ + \ 1\frac{2}{6} \\ \hline \mathbf{4\frac{5}{6}} \end{array}$$

m. $\frac{3}{4} \times \frac{3}{3} = \frac{9}{12}$

$\frac{2}{3} \times \frac{4}{4} = \frac{8}{12}$

$$\begin{array}{r} 4\frac{9}{12} \\ - \ 1\frac{8}{12} \\ \hline \mathbf{3\frac{1}{12}} \end{array}$$

n. $\frac{1}{2} \times \frac{5}{5} = \frac{5}{10}$

$\frac{1}{5} \times \frac{2}{2} = \frac{2}{10}$

$$\begin{array}{r} 4\frac{5}{10} \\ -\ 1\frac{2}{10} \\ \hline \mathbf{3\frac{3}{10}} \end{array}$$

Lesson 116, Mixed Practice

1.

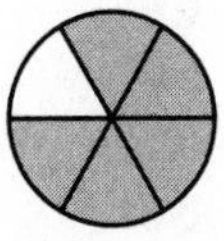

$\frac{5}{6} \times \frac{100}{1} = \frac{500}{6}$

$6\overline{)500}$ = $83\frac{2}{6}$ = $\mathbf{83\frac{1}{3}\%}$

$$\begin{array}{r} 48 \\ \hline 20 \\ 18 \\ \hline 2 \end{array}$$

2.

$$\begin{array}{r} 1\,8\,\overset{6}{\cancel{7}}\,{}^{1}5 \\ -\ 1\,8\,4\,9 \\ \hline \mathbf{2\,6\ years} \end{array}$$

3. (a) **25%**

(b) $\mathbf{\frac{3}{4}}$

4. **D.**

5. $\mathbf{\frac{4}{6} < \frac{5}{6}}$

$\frac{2}{3}$ Ⓛ $\frac{5}{6}$

6. $\mathbf{\frac{9}{4}; 2\frac{1}{4}}$

7. $\frac{60}{100} = \mathbf{\frac{3}{5}}$

8.

$$\begin{array}{r} 0.5\text{ cm} \\ \times\ \ 4 \\ \hline 2.0\text{ cm} \end{array}$$ or **2 cm**

9.

$$\begin{array}{r} 0.5 \\ \times\ 0.5 \\ \hline \mathbf{0.25\ sq.\ cm} \end{array}$$

10. 60 mm − 40 mm = 20 mm

70 mm + 20 mm = **90 mm**

11. $\frac{1}{4} \times \frac{2}{2} = \frac{2}{8}$

$\frac{2}{8} + \frac{1}{8} = \mathbf{\frac{3}{8}}$

12. $\frac{1}{2} \times \frac{2}{2} = \frac{2}{4}$

$\frac{3}{4} - \frac{2}{4} = \mathbf{\frac{1}{4}}$

13. $\frac{3}{4} \times \frac{2}{2} = \frac{6}{8}$

$\frac{7}{8} - \frac{6}{8} = \mathbf{\frac{1}{8}}$

14. $\frac{1}{2} \times \frac{4}{4} = \frac{4}{8}$

$$\begin{array}{r} 2\frac{5}{8} \\ -\ 1\frac{4}{8} \\ \hline \mathbf{1\frac{1}{8}} \end{array}$$

15. $\frac{1}{2} \times \frac{4}{4} = \frac{4}{8}$

$$\begin{array}{r} 3\frac{4}{8} \\ -\ 2\frac{1}{8} \\ \hline \mathbf{1\frac{3}{8}} \end{array}$$

16. $\frac{1}{3} \times \frac{2}{2} = \frac{2}{6}$

$$\begin{array}{r} 5\frac{1}{6} \\ +\ 1\frac{2}{6} \\ \hline 6\frac{3}{6} \end{array} = \mathbf{6\frac{1}{2}}$$

17. $\frac{3}{5} \times \frac{3}{1} = \frac{9}{5} = \mathbf{1\frac{4}{5}}$

18. $\frac{3}{1} \times \frac{5}{3} = \frac{15}{3} = \mathbf{5}$

19. **650**

20. $$\begin{array}{r} 4.6 \\ \times\ \ 80 \\ \hline 368.0 \end{array}$$ or **368**

21. $$\begin{array}{r} 0.18 \\ \times\ \ 0.4 \\ \hline \mathbf{0.072} \end{array}$$

22. $$\begin{array}{r} \mathbf{\$1.32} \\ 10\overline{)\$13.20} \\ \underline{10} \\ 3\,2 \\ \underline{3\,0} \\ 20 \\ \underline{20} \\ 0 \end{array}$$

23. $$\begin{array}{r} \mathbf{\$1.10} \\ 12\overline{)\$13.20} \\ \underline{12} \\ 1\,2 \\ \underline{1\,2} \\ 00 \\ \underline{0} \\ 0 \end{array}$$

24. $$\begin{array}{r} \mathbf{35} \\ 42\overline{)1470} \\ \underline{126} \\ 210 \\ \underline{210} \\ 0 \end{array}$$

25. $\angle ADC$ (or $\angle CDA$)

26. $\frac{1}{4} \times \frac{3}{3} = \frac{3}{12}$

$\frac{2}{3} \times \frac{4}{4} = \frac{8}{12}$

$\frac{3}{12} + \frac{8}{12} = \mathbf{\frac{11}{12}}$

27. $$\begin{array}{rr} 3 \text{ ft} \times 3 \text{ ft} = & 9 \text{ sq. ft} \\ 2 \text{ ft} \times 4 \text{ ft} = & +\ 8 \text{ sq. ft} \\ \hline & \mathbf{17\ sq.\ ft} \end{array}$$

28. Leave at 12:20 p.m.
Arrive at 2:05 p.m.
1 hr 45 min

29. 1 hour before 9:00 p.m. is 8:00 p.m.
20 min before 8:00 p.m. is **7:40 p.m.**

30. **Yes, if the train arrives in Fort Collins as scheduled (at 11:40 p.m.), Tim can be back on campus before midnight. He should get to campus at about 11:45 p.m.**

Lesson 117, Warm-Up

a. **Round liters to the nearest ten and the cost to the nearest 10 cents; then multiply.**

b. **60 minutes; 210 minutes**

c. **$4**

d. **$8**

e. **$12**

f. **28**

g. **1900**

h. **>**

Problem Solving

25% of $40 = $10
$40 − $10 = **$30**

Lesson 117, Lesson Practice

a. $$\begin{array}{r} \mathbf{0.13} \\ 4\overline{)0.52} \\ \underline{4} \\ 12 \\ \underline{12} \\ 0 \end{array}$$

b. $$\begin{array}{r} \mathbf{0.6} \\ 6\overline{)3.6} \\ \underline{3\,6} \\ 0 \end{array}$$

c.
$$\begin{array}{r} \mathbf{0.17} \\ 5\overline{)0.85} \\ \underline{5} \\ 35 \\ \underline{35} \\ 0 \end{array}$$

d.
$$\begin{array}{r} \mathbf{1.5} \\ 5\overline{)7.5} \\ \underline{5} \\ 2\,5 \\ \underline{2\,5} \\ 0 \end{array}$$

e.
$$\begin{array}{r} \mathbf{0.13} \\ 5\overline{)0.65} \\ \underline{5} \\ 15 \\ \underline{15} \\ 0 \end{array}$$

f.
$$\begin{array}{r} \mathbf{0.7} \\ 3\overline{)2.1} \\ \underline{2\,1} \\ 0 \end{array}$$

g.
$$\begin{array}{r} \mathbf{0.04} \\ 4\overline{)0.16} \\ \underline{16} \\ 0 \end{array}$$

h.
$$\begin{array}{r} \mathbf{0.05} \\ 7\overline{)0.35} \\ \underline{35} \\ 0 \end{array}$$

i.
$$\begin{array}{r} \mathbf{0.0005} \\ 5\overline{)0.0025} \\ \underline{25} \\ 0 \end{array}$$

j.
$$\begin{array}{r} \mathbf{0.02} \\ 4\overline{)0.08} \\ \underline{8} \\ 0 \end{array}$$

k.
$$\begin{array}{r} \mathbf{0.04} \\ 6\overline{)0.24} \\ \underline{24} \\ 0 \end{array}$$

l.
$$\begin{array}{r} \mathbf{0.0048} \\ 3\overline{)0.0144} \\ \underline{12} \\ 24 \\ \underline{24} \\ 0 \end{array}$$

m.
$$\begin{array}{r} 1.89\text{ L} \\ 2\overline{)3.78} \\ \underline{2} \\ 1\,7 \\ \underline{1\,6} \\ 18 \\ \underline{18} \\ 0 \end{array}$$

About 1.89 liters

Lesson 117, Mixed Practice

1. $\mathbf{\frac{1}{6} + \frac{1}{3} = \frac{1}{2}}$

2. $$\begin{array}{r} 18 \\ 2\overline{)36} \\ \underline{2} \\ 16 \\ \underline{16} \\ 0 \end{array}$$

   $18 - 8 =$ **10 points**

3. **June 21**

4. (a) $\mathbf{\frac{2}{3}}$

   (b) $\mathbf{33\frac{1}{3}\%}$

5. $\mathbf{\frac{1}{10}}$**; 0.1;**

   $\frac{1}{10} \times \frac{100\%}{1} = \frac{100\%}{10} =$ **10%**

6. $$\begin{array}{r} 6\text{ in.} \\ \underline{\times\ 8\phantom{\text{ in.}}} \\ 48\text{ in.} \end{array} \qquad \begin{array}{r} \mathbf{4\ feet} \\ 12\overline{)48} \\ \underline{48} \\ 0 \end{array}$$

7. $\mathbf{\frac{4}{3}; 1\frac{1}{3}}$

8. **9807**

9. $\overline{DC}$ **(or $\overline{CD}$)**

10. (a) 3 cm + 3 cm + 4 cm + 4 cm = **14 cm**

(b) 3 cm × 4 cm = **12 sq. cm**

11.
$$\begin{array}{r} 28 \text{ mm} \\ 2\overline{)56} \\ \underline{4} \\ 16 \\ \underline{16} \\ 0 \end{array} \qquad \begin{array}{r} 14 \text{ mm} \\ 2\overline{)28} \\ \underline{2} \\ 08 \\ \underline{8} \\ 0 \end{array}$$

$$\begin{array}{r} \overset{1}{5}6 \text{ mm} \\ 28 \text{ mm} \\ + \ 14 \text{ mm} \\ \hline \mathbf{98 \text{ mm}} \end{array}$$

12.
$$\begin{array}{r} \overset{1}{1}6.000 \\ 3.170 \\ 49.000 \\ + \ 1.125 \\ \hline \mathbf{69.295} \end{array}$$

13.
$$\begin{array}{r} 3.\overset{3}{\cancel{4}}\overset{1}{\cancel{2}}{}^{1}0 \\ - \ 1.242 \\ \hline \mathbf{2.178} \end{array}$$

14. **430**

15.
$$\begin{array}{r} 6.4 \\ \times \ 3.7 \\ \hline 4\,48 \\ 19\,20 \\ \hline \mathbf{23.68} \end{array}$$

16.
$$\begin{array}{r} 0.36 \\ \times \ 0.04 \\ \hline \mathbf{0.0144} \end{array}$$

17.
$$\begin{array}{r} \mathbf{1.8} \\ 2\overline{)3.6} \\ \underline{2} \\ 1\,6 \\ \underline{1\,6} \\ 0 \end{array}$$

18.
$$\begin{array}{r} \mathbf{0.0007} \\ 7\overline{)0.0049} \\ \underline{49} \\ 0 \end{array}$$

19.
$$\begin{array}{r} 1.35 \\ \times \quad 90 \\ \hline 121.50 \end{array}$$ or **121.5**

20. $\frac{3}{4} \times \frac{2}{2} = \frac{6}{8}$

$$\begin{array}{r} 2\frac{1}{8} \\ + \ 1\frac{6}{8} \\ \hline \mathbf{3\frac{7}{8}} \end{array}$$

21. $\frac{1}{3} \times \frac{2}{2} = \frac{2}{6}$

$$\begin{array}{r} \frac{2}{6} \\ + \ \frac{1}{6} \\ \hline \frac{3}{6} = \mathbf{\frac{1}{2}} \end{array}$$

22. $\frac{1}{2} \times \frac{5}{5} = \frac{5}{10}$

$$\begin{array}{r} \frac{7}{10} \\ - \ \frac{5}{10} \\ \hline \frac{2}{10} = \mathbf{\frac{1}{5}} \end{array}$$

23. $\frac{1}{5} \times \frac{2}{2} = \frac{2}{10}$

$$\begin{array}{r} 3\frac{9}{10} \\ - \ \frac{2}{10} \\ \hline \mathbf{3\frac{7}{10}} \end{array}$$

24. $\frac{4}{1} \times \frac{3}{2} = \frac{12}{2} = \mathbf{6}$

25. $\frac{3}{4} \times \frac{4}{1} = \frac{12}{4} = \mathbf{3}$

26. $\frac{18 \div 18}{144 \div 18} = \mathbf{\frac{1}{8}}$

27. $\frac{1}{5} \times \frac{2}{2} = \frac{2}{10}$

$\frac{1}{2} \times \frac{5}{5} = \frac{5}{10}$

$\mathbf{3\frac{2}{10} + 2\frac{5}{10} = 5\frac{7}{10}}$

28. (a) 6 in. × 3 in. = **18 sq. in.**

(b) $\frac{1}{2}$ ft × $\frac{1}{4}$ ft = $\mathbf{\frac{1}{8}}$ **sq. ft**

29.

$$\begin{array}{rcr} 2 \text{ in.} \times 2 \text{ in.} & = & 4 \text{ sq. in.} \\ 5 \text{ in.} \times 5 \text{ in.} & = & \underline{+\ 25 \text{ sq. in.}} \\ & & \mathbf{29 \text{ sq. in.}} \end{array}$$

30.

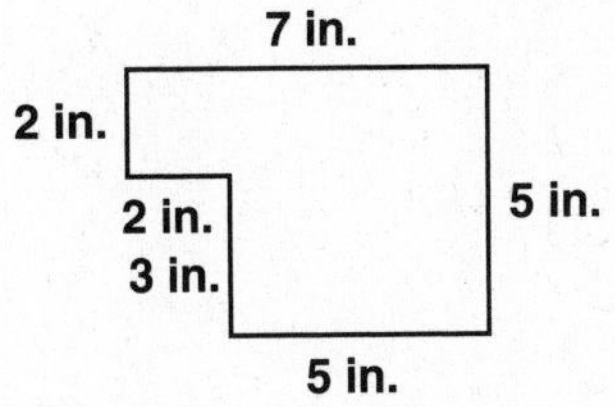

Perimeter = 2 in. + 2 in. + 7 in. + 5 in. + 5 in. + 3 in. = **24 in.**

Lesson 118, Warm-Up

Symmetry Activity

The uppercase letters B, C, D, E, H, I, K, O, and X, exclusively, can be used to create words with horizontal line symmetry. Line symmetry is a form of reflective symmetry. A mirror placed along a line of symmetry reflects half the figure, creating the appearance of a whole figure.

Lesson 118, Lesson Practice

a.

$$\begin{array}{r} \mathbf{0.15} \\ 4\overline{)0.60} \\ \underline{4} \\ 20 \\ \underline{20} \\ 0 \end{array}$$

b.

$$\begin{array}{r} \mathbf{0.024} \\ 5\overline{)0.120} \\ \underline{10} \\ 20 \\ \underline{20} \\ 0 \end{array}$$

c.

$$\begin{array}{r} \mathbf{0.025} \\ 4\overline{)0.100} \\ \underline{8} \\ 20 \\ \underline{20} \\ 0 \end{array}$$

d.

$$\begin{array}{r} \mathbf{0.05} \\ 2\overline{)0.10} \\ \underline{10} \\ 0 \end{array}$$

e.

$$\begin{array}{r} \mathbf{0.08} \\ 5\overline{)0.40} \\ \underline{40} \\ 0 \end{array}$$

f.

$$\begin{array}{r} \mathbf{0.175} \\ 8\overline{)1.400} \\ \underline{8} \\ 60 \\ \underline{56} \\ 40 \\ \underline{40} \\ 0 \end{array}$$

g.

$$\begin{array}{r} \mathbf{0.125} \\ 4\overline{)0.500} \\ \underline{4} \\ 10 \\ \underline{8} \\ 20 \\ \underline{20} \\ 0 \end{array}$$

h.

$$\begin{array}{r} \mathbf{0.075} \\ 8\overline{)0.600} \\ \underline{56} \\ 40 \\ \underline{40} \\ 0 \end{array}$$

i.

$$\begin{array}{r} \mathbf{0.075} \\ 4\overline{)0.300} \\ \underline{28} \\ 20 \\ \underline{20} \\ 0 \end{array}$$

j. **0.25**

k. **3.24**

l. **0.025**

m. **0.324**

n. **0.0025**

o. **0.0324**

p. **1.2**

q. **0.12**

r. **0.012**

Lesson 118, Mixed Practice

1. **C.**

2. $6 \times 5 =$ **30**

3. **8 pencils**
$$\begin{array}{r} 8 \\ 12\overline{)100} \\ \underline{96} \\ 4 \end{array}$$

4. **C.**

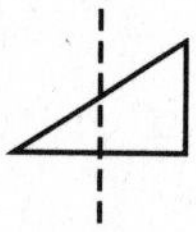

5. $\frac{7}{10} \times \frac{10}{10} = \frac{70}{100} =$ **70%**

6. (a) $\frac{4}{100} = \mathbf{\frac{1}{25}}$

(b) $\frac{5}{100} = \mathbf{\frac{1}{20}}$

7. $\mathbf{\frac{11}{4}; 2\frac{3}{4}}$

8. $\mathbf{\frac{11}{8}}$

$\frac{11}{8} \times \frac{1}{2} = \mathbf{\frac{11}{16}}$

9. **Octagon**

10. $\mathbf{\frac{5}{6}, \frac{5}{5}, \frac{5}{3}}$

11. (a) **0.3 meter**
$$\begin{array}{r} 0.3 \\ 4\overline{)1.2} \\ \underline{1\ 2} \\ 0 \end{array}$$

(b)
$$\begin{array}{r} 0.3 \\ \times\ 0.3 \\ \hline \mathbf{0.09} \end{array}$$
0.09 sq. meter

12.
$$\begin{array}{r} \overset{1\,1}{49.35} \\ 25.00 \\ +\ \ 3.70 \\ \hline \mathbf{78.05} \end{array}$$

13.
$$\begin{array}{r} \overset{1}{\not 2}.\overset{9}{\not 0}\overset{9}{\not 0}{}^{1}0 \\ -\ 1.2\,3\,4 \\ \hline 0.7\,6\,6 \end{array}$$
Seven hundred sixty-six thousandths

14.
$$\begin{array}{r} \mathbf{0.0025} \\ 5\overline{)0.0125} \\ \underline{10} \\ 25 \\ \underline{25} \\ 0 \end{array}$$

15. **420**

16.
$$\begin{array}{r} 0.17 \\ \times\ 0.5 \\ \hline \mathbf{0.085} \end{array}$$

17.
$$\begin{array}{r} \mathbf{0.15} \\ 4\overline{)0.60} \\ \underline{4} \\ 20 \\ \underline{20} \\ 0 \end{array}$$

18. **0.06**

19.
$$\begin{array}{r} \mathbf{0.45} \\ 4\overline{)1.80} \\ \underline{1\ 6} \\ 20 \\ \underline{20} \\ 0 \end{array}$$

20. $\frac{1}{3} \times \frac{3}{3} = \frac{3}{9}$

$$\begin{array}{r} 3\frac{1}{9} \\ + \frac{3}{9} \\ \hline \mathbf{3\frac{4}{9}} \end{array}$$

21. $\frac{1}{3} \times \frac{2}{2} = \frac{2}{6}$

$$\begin{array}{r} \frac{2}{6} \\ + \frac{5}{6} \\ \hline \frac{7}{6} \end{array} = \mathbf{1\frac{1}{6}}$$

22. $\frac{1}{4} \times \frac{2}{2} = \frac{2}{8}$

$$\begin{array}{r} \frac{7}{8} \\ - \frac{2}{8} \\ \hline \mathbf{\frac{5}{8}} \end{array}$$

23. $\frac{1}{2} \times \frac{5}{5} = \frac{5}{10}$

$$\begin{array}{r} 4\frac{5}{10} \\ - 1\frac{3}{10} \\ \hline 3\frac{2}{10} \end{array} = \mathbf{3\frac{1}{5}}$$

24. $\frac{6}{1} \times \frac{2}{3} = \frac{12}{3} = \mathbf{4}$

25. $\frac{6}{1} \times \frac{3}{2} = \frac{18}{2} = \mathbf{9}$

26. $\frac{1}{3} \times \frac{4}{4} = \frac{4}{12}$

$\frac{3}{4} \times \frac{3}{3} = \frac{9}{12}$

$\mathbf{2\frac{9}{12} - 1\frac{4}{12} = 1\frac{5}{12}}$

27. (a) **0.035**

(b) **8.75**

28. $\sqrt{81} + \sqrt{100}$ ⓧ $9^2 + 10^2$ — circled: $<$

$9 + 10 \qquad 81 + 100$

$19 \qquad 181$

29. $6 \text{ cm} \times 4 \text{ cm} = 24 \text{ sq. cm}$

$2 \text{ cm} \times 3 \text{ cm} = + \ 6 \text{ sq. cm}$

30 sq. cm

30.

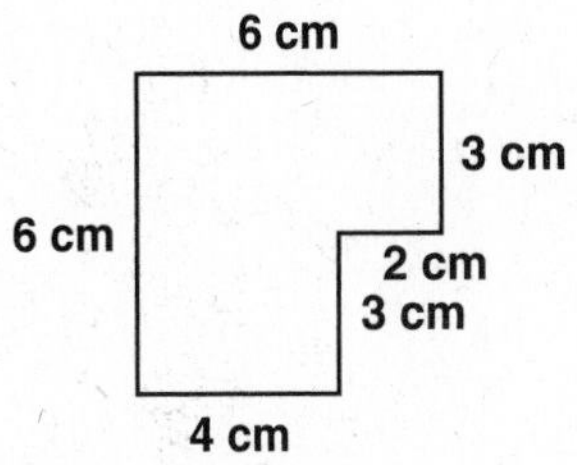

6 cm + 6 cm + 3 cm + 2 cm + 3 cm + 4 cm = 24 cm

Perimeter is 24 cm.

LESSON 119, WARM-UP

a. **To find the perimeter, double the length, double the width, and then add the two numbers. To calculate the area, find 8 × 11 and $\frac{1}{2}$ of 11 and then add the two numbers.**

b. **$20**

c. **$40**

d. **$80**

e. **1969**

f. **<**

Problem Solving

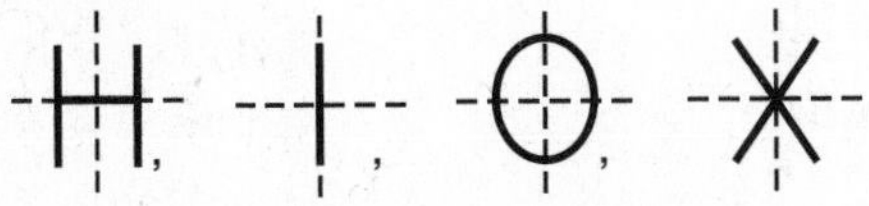

LESSON 119, LESSON PRACTICE

a.
$$
\begin{array}{r}
4 \\
0.3)\overline{1.2} \\
\underline{1\,2} \\
0
\end{array}
$$

b.
$$
\begin{array}{r}
1.4 \\
0.3)\overline{0.42} \\
\underline{3} \\
12 \\
\underline{12} \\
0
\end{array}
$$

c.
$$
\begin{array}{r}
0.2 \\
1.2)\overline{0.24} \\
\underline{24} \\
0
\end{array}
$$

d.
$$
\begin{array}{r}
0.6 \\
0.4)\overline{0.24} \\
\underline{24} \\
0
\end{array}
$$

e.
$$
\begin{array}{r}
14 \\
0.4)\overline{5.6} \\
\underline{4} \\
16 \\
\underline{16} \\
0
\end{array}
$$

f.
$$
\begin{array}{r}
3 \\
1.2)\overline{3.6} \\
\underline{3\,6} \\
0
\end{array}
$$

g.
$$
\begin{array}{r}
4 \\
0.6)\overline{2.4} \\
\underline{2\,4} \\
0
\end{array}
$$

h.
$$
\begin{array}{r}
0.25 \\
0.5)\overline{0.125} \\
\underline{10} \\
25 \\
\underline{25} \\
0
\end{array}
$$

i.
$$
\begin{array}{r}
1.9 \\
1.2)\overline{2.28} \\
\underline{1\,2} \\
108 \\
\underline{108} \\
0
\end{array}
$$

LESSON 119, MIXED PRACTICE

1. See student work.

2. $5 + 6 + 7 + 8 + 9 = 35$

 $35 \div 5 = \mathbf{7}$

3.
$$
\begin{array}{r}
24 \\
\times\ \ 2 \\
\hline
48
\end{array}
\qquad
\begin{array}{r}
48 \\
\times\ \ 2 \\
\hline
\mathbf{96}\ \text{lions}
\end{array}
$$

4.
$$
\begin{array}{r}
\overset{1\ 1\ 1}{\$18.35} \\
+\ \$22.65 \\
\hline
\$41.00
\end{array}
\qquad
\begin{array}{r}
\$1\overset{5}{6},\overset{9}{\cancel{0}}\overset{{}^{1}3}{\cancel{4}}0 \\
-\qquad\ \ \$4\ 1 \\
\hline
\mathbf{\$1\,5,9\,9\,9}
\end{array}
$$

5. C. --⌐--

6. $3\dfrac{1}{3} = \dfrac{\mathbf{10}}{\mathbf{3}}$

 $\dfrac{10}{3} \times \dfrac{3}{4} = \dfrac{30}{12} = 2\dfrac{6}{12} = \mathbf{2\dfrac{1}{2}}$

7. $\angle ADC$ (or $\angle CDA$)

8. Trapezoid

9. $\dfrac{1}{2} \times \dfrac{3}{3} = \dfrac{3}{6}$

 $\dfrac{1}{3} \times \dfrac{2}{2} = \dfrac{2}{6}$

$$
\begin{array}{r}
3\dfrac{3}{6} \\
+\ 1\dfrac{2}{6} \\
\hline
4\dfrac{5}{6}
\end{array}
$$

10. $\dfrac{1}{2} \times \dfrac{3}{3} = \dfrac{3}{6}$

$$
\begin{array}{r}
2\dfrac{1}{6} \\
+\ 1\dfrac{3}{6} \\
\hline
3\dfrac{4}{6} = 3\dfrac{2}{3}
\end{array}
$$

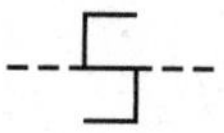

11. $\frac{1}{2} \times \frac{3}{3} = \frac{3}{6}$

$$\begin{array}{r} 5\frac{5}{6} \\ -\ 1\frac{3}{6} \\ \hline 4\frac{2}{6} \end{array} = \mathbf{4\frac{1}{3}}$$

12. $\frac{2}{3} \times \frac{4}{4} = \frac{8}{12}$

$\frac{1}{4} \times \frac{3}{3} = \frac{3}{12}$

$$\begin{array}{r} 4\frac{8}{12} \\ -\ 1\frac{3}{12} \\ \hline \mathbf{3\frac{5}{12}} \end{array}$$

13.
$$\begin{array}{r} \mathbf{0.0024} \\ 6\overline{)0.0144} \\ \underline{12} \\ 24 \\ \underline{24} \\ 0 \end{array}$$

14.
$$\begin{array}{r} \mathbf{0.24} \\ 5\overline{)1.20} \\ \underline{1\ 0} \\ 20 \\ \underline{20} \\ 0 \end{array}$$

15.
$$\begin{array}{r} \mathbf{0.015} \\ 12\overline{)0.180} \\ \underline{12} \\ 60 \\ \underline{60} \\ 0 \end{array}$$

16.
$$\begin{array}{r} \mathbf{0.8} \\ 0.3\overline{)0.24} \\ \underline{24} \\ 0 \end{array}$$

17.
$$\begin{array}{r} \mathbf{2} \\ 0.5\overline{)1.0} \\ \underline{1\ 0} \\ 0 \end{array}$$

18.
$$\begin{array}{r} \mathbf{0.15} \\ 1.2\overline{)0.180} \\ \underline{12} \\ 60 \\ \underline{60} \\ 0 \end{array}$$

19. (a) **0.05**

(b) **0.005**

20.
$$\begin{array}{r} \overset{2}{\cancel{3}}.{}^{1}0 \\ -\ 1.\ 6 \\ \hline 1.\ 4 \end{array} \qquad \begin{array}{r} 1.\ \overset{3}{\cancel{4}}{}^{1}0 \\ -\ 0.\ 1\ 6 \\ \hline \mathbf{1.\ 2\ 4} \end{array}$$

21.
$$\begin{array}{r} 0.12 \\ \times\ 0.30 \\ \hline 0.0360 \end{array} \text{ or } \mathbf{0.036}$$

22.
$$\begin{array}{r} 0.12 \\ \times\quad 10 \\ \hline 1.20 \end{array} \text{ or } \mathbf{1.2}$$

23.
$$\begin{array}{r} 7.6 \\ \times\ 3.9 \\ \hline 6\ 84 \\ \underline{22\ 80} \\ \mathbf{29.64} \end{array}$$

24. $\frac{4}{1} \times \frac{3}{8} = \frac{12}{8} = 1\frac{4}{8} = \mathbf{1\frac{1}{2}}$

25. $\frac{4}{1} \times \frac{8}{3} = \frac{32}{3} = \mathbf{10\frac{2}{3}}$

26. $\frac{1}{2} \times \frac{3}{3} = \frac{3}{6}$

$\frac{1}{3} \times \frac{2}{2} = \frac{2}{6}$

$\frac{3}{6}\text{ ft} + \frac{3}{6}\text{ ft} + \frac{2}{6}\text{ ft} + \frac{2}{6}\text{ ft}$

$= \frac{10}{6}\text{ ft} = 1\frac{4}{6}\text{ ft} = \mathbf{1\frac{2}{3}\ ft}$

27. $\frac{1}{2}\text{ ft} \times \frac{1}{3}\text{ ft} = \mathbf{\frac{1}{6}\ sq.\ ft}$

28. 10 ft × 12 ft = 120 sq. ft

120 sq. ft × 8 ft = **960 cubic feet**

29.
$$\begin{aligned} 10\text{ ft} \times 10\text{ ft} &= \quad 100\text{ sq. ft} \\ 5\text{ ft} \times 5\text{ ft} &= + \;\; 25\text{ sq. ft} \\ &\quad \mathbf{125\text{ sq. ft}} \end{aligned}$$

30.

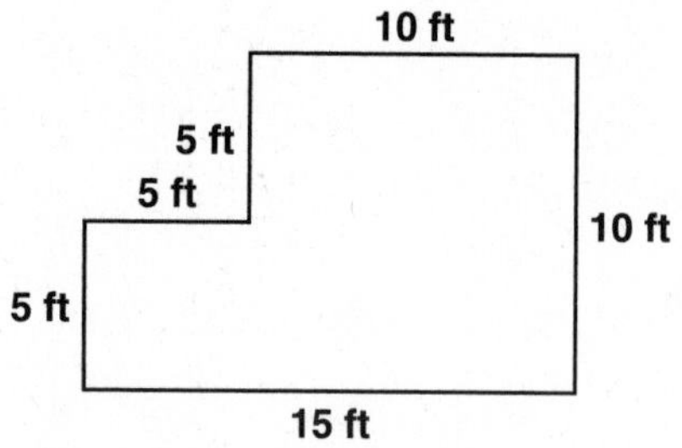

5 ft + 5 ft + 5 ft + 10 ft
+ 10 ft + 15 ft = 50 ft

Perimeter is 50 feet.

Lesson 120, Warm-Up

a. **Add the lengths and divide by 3.**

b. **2**

c. **22**

d. **18**

e. **9**

f. **See student work.**

g. **604**

Problem Solving

$4 \times 4 \times 4 =$ **64 small cubes**

Lesson 120, Lesson Practice

a. $\frac{3}{2} \times \frac{7}{4} = \frac{21}{8} = \mathbf{2\frac{5}{8}}$

b. $\frac{7}{2} \times \frac{5}{3} = \frac{35}{6} = \mathbf{5\frac{5}{6}}$

c. $\frac{3}{1} \times \frac{5}{2} = \frac{15}{2} = \mathbf{7\frac{1}{2}}$

d. $\frac{4}{1} \times \frac{11}{3} = \frac{44}{3} = \mathbf{14\frac{2}{3}}$

e. $\frac{1}{3} \times \frac{7}{3} = \mathbf{\frac{7}{9}}$

f. $\frac{1}{6} \times \frac{17}{6} = \mathbf{\frac{17}{36}}$

Lesson 120, Mixed Practice

1. **See student work.**

2. **Cylinder**

3. **2 + 2 = 2 × 2**

4. **D. 0.05**

5. 3 + 8 = **11**

6. 100 ÷ 2 = **50 hours**

7. (a) 3 in. × 2 = 6 in.
3 in. + 6 in. + 3 in. + 6 in. = **18 in.**

(b) 3 in. × 6 in. = **18 sq. in.**

8. (a) $\frac{3}{6} = \mathbf{\frac{1}{2}}$

(b) $\frac{1}{2} \times \frac{100\%}{1} = \frac{100\%}{2} = \mathbf{50\%}$

9. 8 − 5 = **3 more sides**

10. 2 + 4 + 6 + 8 = 20
20 ÷ 4 = **5**

11. 7 cm − 5 cm = 2 cm
2 cm + 2 cm + 5 cm = **9 cm**

12.
$$\begin{array}{r} \overset{2}{3}8.248 \\ 7.500 \\ 37.230 \\ +\;15.000 \\ \hline \mathbf{97.978} \end{array}$$

13.
$$\begin{array}{r} \overset{0}{\$\not{1}}.\,^{1}4\,9 \\ -\ \$0.\,7\,5 \\ \hline 0.\,7\,4 \end{array} \qquad \begin{array}{r} \$\overset{5}{\not{6}}.\,\overset{9}{\not{0}}\,^{1}0 \\ -\ \$0.\,7\,4 \\ \hline \mathbf{\$5.\,2\,6} \end{array}$$

14. **240**

15.
$$\begin{array}{r} 0.24 \\ \times\ 0.12 \\ \hline 48 \\ 240 \\ \hline \mathbf{0.0288} \end{array}$$

16.
$$\begin{array}{r} 2.4 \\ \times\ 5.7 \\ \hline 1\,68 \\ 12\,00 \\ \hline \mathbf{13.68} \end{array}$$

17.
$$\begin{array}{r} \mathbf{0.0125} \\ 8\overline{)0.1000} \\ \underline{8} \\ 20 \\ \underline{16} \\ 40 \\ \underline{40} \\ 0 \end{array}$$

18.
$$\begin{array}{r} \mathbf{8.7} \\ 0.5\overline{)4.35} \\ \underline{4\,0} \\ 35 \\ \underline{35} \\ 0 \end{array}$$

19.
$$\begin{array}{r} \mathbf{1.2} \\ 1.2\overline{)1.44} \\ \underline{1\,2} \\ 24 \\ \underline{24} \\ 0 \end{array}$$

20. $\frac{1}{3} \times \frac{4}{4} = \frac{4}{12}$

$\frac{3}{4} \times \frac{3}{3} = \frac{9}{12}$

$$\begin{array}{r} 3\frac{4}{12} \\ +\ 7\frac{9}{12} \\ \hline 10\frac{13}{12} \end{array} = \mathbf{11\frac{1}{2}}$$

21. $\frac{3}{7} \times \frac{2}{2} = \frac{6}{14}$

$\frac{1}{2} \times \frac{7}{7} = \frac{7}{14}$

$$\begin{array}{r} \frac{6}{14} \\ +\ \frac{7}{14} \\ \hline \mathbf{\frac{13}{14}} \end{array}$$

22. $\frac{1}{5} \times \frac{3}{3} = \frac{3}{15}$

$$\begin{array}{r} 6\frac{14}{15} \\ -\ 1\frac{3}{15} \\ \hline \mathbf{5\frac{11}{15}} \end{array}$$

23. $\frac{4}{5} \times \frac{3}{3} = \frac{12}{15}$

$\frac{1}{3} \times \frac{5}{5} = \frac{5}{15}$

$$\begin{array}{r} \frac{12}{15} \\ -\ \frac{5}{15} \\ \hline \mathbf{\frac{7}{15}} \end{array}$$

24. $\frac{1}{2} \times \frac{10}{3} = \frac{10}{6} = 1\frac{4}{6} = \mathbf{1\frac{2}{3}}$

25. $\frac{4}{1} \times \frac{5}{2} = \frac{20}{2} = \mathbf{10}$

26.
$$\begin{array}{r} 4.5\text{ m} \\ \times\ \ 3\text{ m} \\ \hline \mathbf{13.5\text{ sq. m}} \end{array}$$

27.
$$\begin{array}{r} 1.5\text{ ft} \\ \times\ \ 2\text{ ft} \\ \hline 3.0\text{ sq. ft.} \end{array} \qquad \begin{array}{r} 0.5\text{ ft} \\ \times\ \ 3\text{ sq. ft} \\ \hline \mathbf{1.5\text{ cubic feet}} \end{array}$$

28. $6 \div 3 = 2$ inches (side of small rectangle)

2 in. $\times$ 3 = 6 in. (side of large rectangle)

6 in. + 6 in. + 6 in. = **18 in.**

29. $\frac{1}{9} \times \frac{100\%}{1} = \frac{100\%}{9}$

$$\begin{array}{r} \mathbf{11\tfrac{1}{9}\%} \\ 9\overline{)100} \\ \underline{9} \\ 10 \\ \underline{9} \\ 1 \end{array}$$

30. 16 small triangles

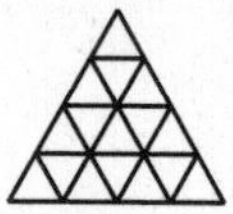

INVESTIGATION 12

1. **6 triangles**

2. **4 squares**

3. **A. ⬡;**

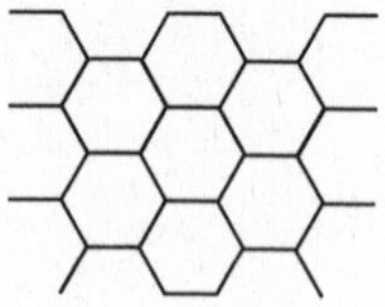

4. **A. and C.**

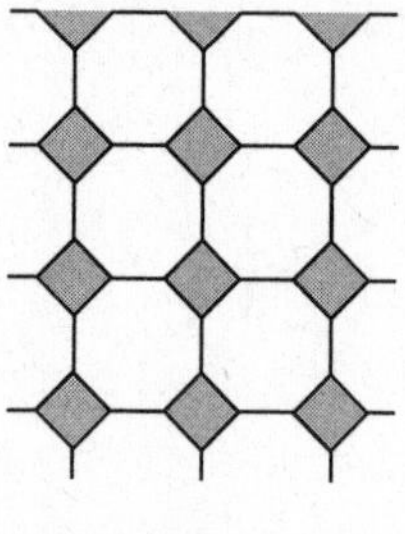

Extensions

a. **See student work.**

b. **See student work.**

Solutions for

Appendix Topics

Topic A, Lesson Practice

1 = I	14 = XIV	27 = XXVII
2 = II	15 = XV	28 = XXVIII
3 = III	16 = XVI	29 = XXIX
4 = IV	17 = XVII	30 = XXX
5 = V	18 = XVIII	31 = XXXI
6 = VI	19 = XIX	32 = XXXII
7 = VII	20 = XX	33 = XXXIII
8 = VIII	21 = XXI	34 = XXXIV
9 = IX	22 = XXII	35 = XXXV
10 = X	23 = XXIII	36 = XXXVI
11 = XI	24 = XXIV	37 = XXXVII
12 = XII	25 = XXV	38 = XXXVIII
13 = XIII	26 = XXVI	39 = XXXIX

Topic B, Lesson Practice

a. 362

b. 285

c. 400

d. 47

e. 3256

f. 1999

Topic C, Lesson Practice

a. 111 (base 5)

b. 201 (base 5)

c. 20 (base 5)

d. 400 (base 5)

e. 123 (base 5)

f. 321 (base 5)

Solutions for

Supplemental Practice

SUPPLEMENTAL PRACTICE, LESSON 5

1. forty-four
2. fifty-five
3. one hundred ten
4. three hundred twelve
5. four hundred twenty-six
6. five dollars and thirty-seven cents
7. two hundred eleven dollars and twenty-five cents
8. six hundred eight dollars
9. seventy-six dollars and twenty-seven cents
10. nine dollars and one cent
11. 114
12. 240
13. 732
14. 607
15. 816
16. \$384
17. \$418
18. \$180.50
19. \$508.15
20. \$650

SUPPLEMENTAL PRACTICE, LESSON 6

1. 29
2. 33
3. 64
4. 114
5. 121
6. 796
7. 1312
8. 1193
9. 370
10. 1023
11. 386
12. 893
13. 1215
14. 1191
15. 1771
16. 344
17. 634
18. 426
19. 495
20. 581

SUPPLEMENTAL PRACTICE, LESSON 7

1. 7254
2. 12,625
3. 11,580
4. 21,300

5. 56,208

6. 18,700

7. 175,000

8. 210,500

9. 356,200

10. 980,000

11. six thousand, five hundred

12. four thousand, two hundred ten

13. one thousand, seven hundred sixty

14. eight thousand, one hundred twelve

15. twenty-one thousand

16. twelve thousand, five hundred

17. forty thousand, eight hundred

18. one hundred eighteen thousand

19. two hundred ten thousand, six hundred

20. one hundred twenty-five thousand, two hundred

Supplemental Practice, Lesson 9

1. 19

2. 14

3. $8

4. 171

5. 481

6. $93

7. 208

8. 478

9. $98

10. 44

11. 506

12. $377

13. 372

14. 642

15. $57

Supplemental Practice, Lesson 13

1. $5.45

2. $10.58

3. $10.00

4. $19.15

5. $24.03

6. $33.80

7. $1.24

8. $19.99

9. $20.20

10. $1.00

11. $5.05

12. $7.00

13. $0.61

14. $4.79

15. $1.75

16. $0.64

17. $7.57

18. $1.46

19. $0.08

20. $0.05

Supplemental Practice, Lesson 17

1. 161

2. 258

3. 228

4. 288

5. 420

6. 312

7. 768

8. 6335

9. 534

10. 5672

11. $228

12. $170

13. $16.68

14. $26.10

15. $31.05

16. 5664

Supplemental Practice, Lesson 18

1. 180

2. 120

3. 210

4. 216

5. 280

6. 120

7. 81

8. 100

9. 0

10. 360

11. 700

12. 900

13. 900

14. 2000

15. 420

16. 480

17. 0

18. 600

19. 100

20. 2100

Supplemental Practice, Lesson 22

1. 3 R 1

2. 8 R 1

3. 4 R 2
4. 6 R 3
5. 8 R 5
6. 6 R 2
7. 8 R 1
8. 8 R 2
9. 3 R 5
10. 5 R 3
11. 6 R 7
12. 7 R 7
13. 5 R 5
14. 4 R 4
15. 3 R 5
16. 5 R 3
17. 2 R 5
18. 3 R 5

Supplemental Practice, Lesson 24

1. 4
2. 0
3. 5
4. 2
5. 1
6. 2
7. 8
8. 2
9. 6
10. 6
11. 3
12. 8
13. 2
14. 2
15. 18
16. 90
17. 18
18. 3
19. 45
20. 39

Supplemental Practice, Lesson 26

1. 68
2. 178
3. 78
4. $1.52
5. $1.89
6. $0.31
7. 62
8. 89

9. 24

10. $0.46

11. $0.72

12. $1.46

13. 73

14. 121

15. 129

16. $0.55

17. $0.81

18. $0.63

19. 63 R 1

20. 86 R 2

21. 87 R 2

SUPPLEMENTAL PRACTICE, LESSON 28

1. 10:35 a.m.

2. 8:35 a.m.

3. 12:35 p.m.

4. 10:05 a.m.

5. 11:05 a.m.

6. 12:20 p.m.

7. 12:20 a.m.

8. 10:20 a.m.

9. 11:50 a.m.

10. 40 minutes

11. 11:10 a.m.

12. 11:10 a.m.

13. 10:40 a.m.

14. 12:40 p.m.

15. 50 minutes

16. 11:50 a.m.

17. 1:30 a.m.

18. 2:05 p.m.

19. 5:55 a.m.

20. 3:15 p.m.

SUPPLEMENTAL PRACTICE, LESSON 29

1. 360

2. 1410

3. 3900

4. 2380

5. 3330

6. 450

7. 700

8. 2920

9. 2280

10. 5920

11. 2710

12. 27,960

13. 47,180

14. 9300

15. 47,580

16. 8100

17. 18,000

18. 57,600

19. 33,600

20. 76,800

Supplemental Practice, Lesson 33

1. 50

2. 40

3. 60

4. 60

5. 40

6. 80

7. 90

8. 100

9. 400

10. 200

11. 900

12. 800

13. 500

14. 300

15. 200

16. 700

17. 120

18. 130

19. 360

20. 340

21. 770

22. 530

23. 480

24. 270

Supplemental Practice, Lesson 34

1. 10 R 1

2. 20 R 3

3. 30 R 1

4. 40 R 2

5. 60 R 3

6. 80 R 4

7. 60 R 5

8. 50 R 5

9. 50 R 7

10. $1.05

11. $2.06

12. **$1.08**

13. **207 R 1**

14. **306 R 2**

15. **108 R 1**

16. **$1.06**

17. **$2.02**

18. **$2.40**

Supplemental Practice, Lesson 37

Note: **Answers to problems 1–20 may vary.**

1. 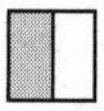

2. 

3. 

4. 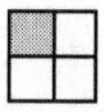

5. 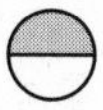

6. 

7. 

8. 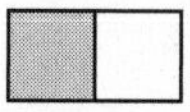

9. 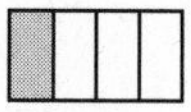

10. 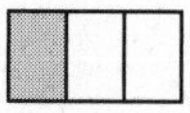

11. 

12. 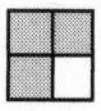

13. 

14. 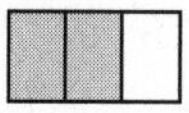

15. 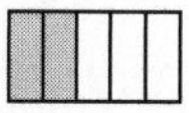

16. 

17. 

18. 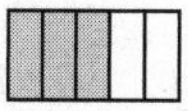

19. 

20. 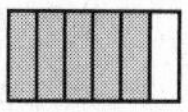

Supplemental Practice, Lesson 38

A. $\frac{1}{5}$

B. $\frac{4}{5}$

C. $1\frac{3}{5}$

D. $2\frac{1}{5}$

E. $3\frac{2}{5}$

F. $\frac{1}{4}$

G. $\frac{3}{4}$

H. $1\frac{2}{4} = 1\frac{1}{2}$

I. $2\frac{1}{4}$

J. $3\frac{3}{4}$

K. $\frac{1}{3}$

L. $\frac{2}{3}$

M. $1\frac{2}{3}$

N. $2\frac{1}{3}$

O. $3\frac{2}{3}$

P. $\frac{1}{10}$

Q. $\frac{7}{10}$

R. $1\frac{3}{10}$

S. $2\frac{9}{10}$

T. $3\frac{5}{10} = 3\frac{1}{2}$

U. $\frac{1}{8}$

V. $\frac{5}{8}$

W. $1\frac{3}{8}$

X. $2\frac{7}{8}$

Y. $3\frac{4}{8} = 3\frac{1}{2}$

Supplemental Practice, Lesson 43

1. $4\frac{1}{3}$
2. $5\frac{2}{3}$
3. $6\frac{2}{5}$
4. $1\frac{3}{4}$
5. $7\frac{7}{8}$
6. $12\frac{2}{3}$
7. $4\frac{3}{10}$
8. $4\frac{9}{10}$
9. $16\frac{1}{2}$
10. $1\frac{5}{6}$
11. $2\frac{2}{3}$
12. $2\frac{1}{4}$
13. 7
14. $\frac{5}{8}$
15. 6

16. 0

17. $9\frac{3}{10}$

18. $4\frac{2}{5}$

19. $5\frac{4}{5}$

20. 1

Supplemental Practice, Lesson 48

1. 5280
2. 642
3. 45,067
4. 5492
5. 71,400
6. 6403
7. 7089
8. 10,407
9. 6010
10. 16,005
11. (6 × 10) + (5 × 1)
12. (7 × 100) + (4 × 10) + (2 × 1)
13. (3 × 100) + (2 × 10)
14. (5 × 100) + (6 × 1)
15. (7 × 1000) + (5 × 100)
16. (2 × 1000) + (1 × 1)
17. (1 × 1000) + (4 × 10)
18. (1 × 1000) + (7 × 100) + (6 × 10)
19. (1 × 1000) + (4 × 100) + (9 × 10) + (2 × 1)
20. (2 × 10,000) + (5 × 1000)

Supplemental Practice, Lesson 50

1. 4
2. 6
3. 7
4. 17
5. 22
6. 3
7. 6
8. 40
9. 48
10. 8
11. 118
12. 51
13. 7
14. 13
15. 34
16. 45
17. 25
18. 180

19. 103

20. 56

Supplemental Practice, Lesson 51

1. 1862
2. 9312
3. $59.28
4. $24.44
5. 5355
6. 5589
7. $34.22
8. $11.84
9. 3456
10. 1971
11. $81.60
12. $33.84
13. 1332
14. 2988
15. $42.92
16. $42.88
17. 4324
18. 3087
19. $15.30
20. $40.94

Supplemental Practice, Lesson 52

1. 10,000,000
2. 1,000,000
3. 100,000
4. 1000
5. 100,000,000
6. 10,000
7. millions
8. thousands
9. hundred thousands
10. hundreds
11. ten millions
12. ten thousands
13. 4
14. 7
15. 5
16. 4
17. 5,000,000
18. 500,000
19. 50,000,000
20. 50,000
21. 1,250,000
22. 5,312,000
23. 10,125,200

24. 13,210,500

25. 25,196,100

26. 327,000,000

27. 645,600,200

28. 716,911,000

29. 120,615,000

30. 984,200,00

31. one million, five hundred thousand

32. ten million, two hundred thousand

33. fifteen million, three hundred fifty-two thousand

34. twenty-five million, seven hundred forty thousand

35. forty-two million, one hundred sixty-four thousand

36. seventy-eight million, three hundred forty-five thousand, two hundred

37. one hundred twenty million

38. two hundred fifty-three million

39. four hundred twelve million, five hundred twenty thousand

40. six hundred thirty-five million, one hundred fifty-four thousand

Supplemental Practice, Lesson 54

1. 21

2. 15

3. $0.12

4. 14

5. 13 R 20

6. $0.10

7. 11 R 20

8. 28

9. $0.19

10. 16 R 10

11. 17

12. $0.15

13. 11 R 30

14. 30 R 14

15. $0.26

16. 21 R 36

17. 19 R 37

18. $0.16

Supplemental Practice, Lesson 56

1. 33,210

2. 307,450

3. $2415.36

4. $920.40

5. 251,875

6. 275,210

7. $687.06

8. **\$5052.40**

9. **470,808**

10. **271,700**

11. **\$3979.57**

12. **\$544.44**

13. **299,440**

14. **475,566**

15. **\$1630.53**

16. **\$8291.10**

17. **516,520**

18. **98,658**

19. **\$1818.90**

20. **\$2132.82**

Supplemental Practice, Lesson 58

1. $3\frac{1}{6}$

2. $7\frac{1}{5}$

3. $6\frac{3}{4}$

4. $2\frac{2}{7}$

5. $3\frac{1}{8}$

6. $6\frac{2}{9}$

7. $3\frac{1}{3}$

8. $7\frac{1}{7}$

9. $8\frac{1}{10}$

10. $22\frac{1}{2}$

11. $15\frac{1}{3}$

12. $11\frac{3}{4}$

13. $11\frac{1}{5}$

14. $13\frac{1}{6}$

15. $6\frac{1}{10}$

16. $4\frac{1}{8}$

17. $41\frac{2}{3}$

18. $15\frac{5}{6}$

19. $14\frac{2}{7}$

20. $11\frac{1}{9}$

21. $33\frac{1}{3}$

Supplemental Practice, Lesson 62

1. **80**

2. **90**

3. **40**

4. 30

5. 2000

6. 3200

7. 3

8. 4

9. 600

10. 1000

11. 400

12. 300

13. 250,000

14. 210,000

15. 20

16. 20

17. 180

18. 1300

19. 400

20. 800

Supplemental Practice, Lesson 63

1. $\frac{2}{3}$

2. $1\frac{1}{3}$

3. $2\frac{3}{4}$

4. $3\frac{1}{4}$

5. $\frac{4}{5}$

6. $1\frac{5}{6}$

7. $1\frac{1}{6}$

8. $1\frac{7}{8}$

9. $4\frac{5}{8}$

10. $2\frac{3}{8}$

11. $\frac{1}{8}$

12. $9\frac{1}{2}$

13. $1\frac{9}{10}$

14. $2\frac{7}{10}$

15. $\frac{1}{2}$

16. $3\frac{11}{12}$

17. $9\frac{9}{10}$

18. $3\frac{3}{5}$

19. $\frac{1}{12}$

20. $\frac{2}{5}$

Supplemental Practice, Lesson 68

1. three and four tenths

2. twenty-three hundredths

3. twelve and nine tenths

4. seven and fourteen hundredths

5. twenty and five tenths

6. fifteen and fifteen hundredths

7. ten and one tenth

8. one and ten hundredths

9. one hundred twenty and eight tenths

10. twenty-one and four hundredths

11. 23.4

12. 0.32

13. 10.5

14. 2.25

15. 52.1

16. 0.05

17. 135.9

18. 76.12

19. 1.06

20. 96.5

Supplemental Practice, Lesson 75

1. $2\frac{2}{3}$

2. $3\frac{1}{2}$

3. 3

4. $1\frac{3}{4}$

5. 2

6. 1

7. $6\frac{1}{2}$

8. 8

9. $5\frac{3}{5}$

10. 10

11. $4\frac{3}{8}$

12. $6\frac{1}{4}$

13. $7\frac{2}{3}$

14. $8\frac{4}{5}$

15. $10\frac{1}{3}$

16. $1\frac{1}{3}$

17. $2\frac{1}{4}$

18. 2

19. 3

20. $5\frac{1}{3}$

21. 3

Supplemental Practice, Lesson 76

1. $\frac{1}{4}$

2. $\frac{3}{8}$

3. $\frac{4}{9}$

4. $\frac{1}{9}$

5. $\frac{5}{12}$

6. $\frac{3}{8}$

7. $\frac{1}{6}$

8. $\frac{2}{15}$

9. $\frac{6}{35}$

10. $\frac{1}{16}$

11. $\frac{2}{9}$

12. $\frac{1}{10}$

13. $\frac{1}{8}$

14. $\frac{9}{16}$

15. $\frac{5}{16}$

16. $\frac{1}{12}$

17. $\frac{3}{16}$

18. $\frac{9}{20}$

19. $\frac{1}{100}$

20. $\frac{15}{32}$

Supplemental Practice, Lesson 79

1. $\frac{2}{2}$

2. $\frac{6}{6}$

3. $\frac{2}{2}$

4. $\frac{4}{4}$

5. $\frac{2}{2}$

6. $\frac{3}{3}$

7. $\frac{5}{5}$

8. $\frac{2}{2}$

9. 4

10. 3

11. 12

12. 6

13. 10

14. 2

15. 6

16. 10

17. 6

18. 15

19. 16

20. 10

Supplemental Practice, Lesson 82

1. 2
2. 4
3. 2
4. 3
5. 2
6. 6
7. 4
8. 3
9. 2
10. 5
11. 1
12. 8
13. $\frac{3}{4}$
14. $\frac{2}{3}$
15. $\frac{3}{5}$
16. $\frac{1}{2}$
17. $\frac{3}{5}$
18. $\frac{2}{3}$

Supplemental Practice, Lesson 86

1. $\frac{2}{3}$
2. $1\frac{1}{2}$
3. $1\frac{1}{3}$
4. 2
5. $1\frac{1}{4}$
6. $2\frac{1}{4}$
7. $1\frac{3}{5}$
8. $1\frac{4}{5}$
9. $2\frac{2}{3}$
10. 3
11. 6
12. 2
13. 6
14. 2
15. 6
16. 2
17. 10
18. 3
19. 12
20. 2
21. 10

Supplemental Practice, Lesson 90

1. $\frac{1}{4}$

2. $\frac{1}{3}$

3. $\frac{2}{3}$

4. $\frac{2}{5}$

5. $\frac{1}{2}$

6. $\frac{1}{4}$

7. $1\frac{1}{2}$

8. $3\frac{3}{4}$

9. $2\frac{1}{2}$

10. $4\frac{4}{5}$

11. $1\frac{1}{3}$

12. $5\frac{2}{3}$

13. $\frac{1}{2}$

14. $\frac{1}{2}$

15. $\frac{2}{3}$

16. $\frac{1}{2}$

17. $\frac{1}{5}$

18. $\frac{1}{2}$

19. $\frac{2}{3}$

20. $\frac{1}{10}$

21. $\frac{3}{4}$

22. $\frac{1}{2}$

23. $\frac{4}{5}$

24. $\frac{3}{5}$

25. $\frac{3}{4}$

26. $\frac{3}{5}$

27. $\frac{1}{6}$

28. $\frac{1}{2}$

29. $3\frac{1}{2}$

30. $2\frac{2}{3}$

31. $6\frac{2}{3}$

32. $\frac{5}{6}$

33. $\frac{2}{3}$

34. $\frac{3}{4}$

35. $\frac{1}{5}$

36. $\frac{1}{4}$

37. $\frac{2}{3}$

38. $\frac{1}{3}$

39. $\frac{3}{10}$

40. $\frac{1}{2}$

41. $\frac{2}{5}$

42. $\frac{1}{4}$

43. $\frac{1}{10}$

44. $\frac{1}{50}$

45. $\frac{3}{5}$

46. $\frac{4}{5}$

47. $\frac{9}{10}$

48. $\frac{3}{10}$

49. $\frac{1}{100}$

50. $\frac{1}{2}$

51. $\frac{1}{5}$

52. $\frac{1}{20}$

53. $\frac{7}{10}$

54. $\frac{99}{100}$

55. $\frac{1}{25}$

56. $\frac{2}{5}$

57. $\frac{3}{4}$

Supplemental Practice, Lesson 91

1. $1\frac{1}{3}$

2. $1\frac{1}{2}$

3. $1\frac{2}{3}$

4. $1\frac{1}{4}$

5. $1\frac{1}{2}$

6. $1\frac{1}{5}$

7. $3\frac{1}{2}$

8. $4\frac{1}{2}$

9. $2\frac{1}{2}$

10. $2\frac{1}{2}$

11. $1\frac{1}{2}$

12. $1\frac{3}{4}$

13. $4\frac{1}{2}$

14. $5\frac{3}{5}$

15. $7\frac{1}{2}$

16. $2\frac{2}{3}$

17. $8\frac{3}{4}$

18. $7\frac{1}{2}$

19. $5\frac{2}{5}$

Supplemental Practice, Lesson 94

1. 36
2. 18
3. 24
4. 16
5. 18 R 1
6. 13 R 24
7. 16 R 4
8. 21
9. 21 R 7
10. 35 R 10
11. 28
12. 30 R 10
13. 11 R 13
14. 20 R 20
15. 24 R 14
16. 29 R 19
17. 14 R 8
18. 30 R 13

Supplemental Practice, Lesson 96

1. $1\frac{1}{3}$
2. $\frac{3}{4}$
3. $\frac{4}{9}$
4. $2\frac{1}{4}$
5. 3
6. $\frac{1}{3}$
7. 4
8. $\frac{1}{4}$
9. 6
10. $\frac{1}{6}$
11. $\frac{1}{2}$
12. 2
13. $1\frac{1}{2}$
14. $\frac{2}{3}$

15. $4\frac{1}{2}$

16. $\frac{2}{9}$

17. 4

18. $\frac{1}{4}$

Supplemental Practice, Lesson 99

1. 9.87
2. 6.51
3. 25.721
4. 5.77
5. 5.442
6. 8.45
7. 8.3
8. 4.023
9. 7.88
10. 0.16
11. 19.833
12. 0.375
13. 7.87
14. 24.27
15. 25.6
16. 2.427
17. 8.67
18. 26.67
19. 17.36
20. 1.25

Supplemental Practice, Lesson 102

1. 0.25
2. 0.07
3. 3.15
4. 2.95
5. 0.08
6. 4.29
7. 3.6
8. 0.375
9. 0.05
10. 4.175
11. 0.001
12. 0.625
13. 2.75
14. 3.96
15. 0.125
16. 5.4
17. 0.075
18. 1.51
19. 5.4
20. 0.049

21. 0.9

22. 0.79

23. 0.8

24. 2.45

25. 1.3

26. 0.99

27. 1.4

28. 0.77

29. 5.6

30. 13.5

31. 3.3

32. 0.1

33. 0.1

34. 0.01

35. 12.5

36. 8.7

37. 8.4

38. 6.65

39. 1.23

40. 0.789

Supplemental Practice, Lesson 104

1. 7

2. 4

3. 4

4. 6

5. 5

6. 5

7. 12

8. 17

9. 14

10. 4

11. 7

12. 5

13. 16

14. 24

15. 13

16. 10

17. 10

18. 10

Supplemental Practice, Lesson 109

1. 1.5

2. 2.4

3. 0.56

4. 3.6

5. 0.16

6. 7.2

7. 0.75

8. 12.5

9. 1.75

10. 0.72

11. 0.96

12. 0.75

13. 0.54

14. 2.35

15. 0.24

16. 0.861

17. 50

18. 0.225

19. 0.135

20. 0.48

Supplemental Practice, Lesson 110

1. 0.09

2. 0.08

3. 0.036

4. 0.0035

5. 0.056

6. 0.0144

7. 0.0096

8. 0.084

9. 0.075

10. 0.092

11. 0.0021

12. 0.0492

13. 0.08

14. 0.025

15. 0.0175

16. 0.065

17. 0.0009

18. 0.001

19. 0.072

20. 0.0072

Supplemental Practice, Lesson 112

1. 12

2. 20

3. 12

4. 6

5. 8

6. 24

7. 18

8. 30

9. 12

10. 40

11. 24

12. 16

13. 30

14. 15

15. 10

16. 30

17. 20

18. 50

19. 60

20. 40

Supplemental Practice, Lesson 113

1. $2\frac{3}{4}, \frac{11}{4}$

2. $3\frac{1}{3}, \frac{10}{3}$

3. $1\frac{3}{8}, \frac{11}{8}$

4. $2\frac{1}{6}, \frac{13}{6}$

5. $\frac{7}{2}$

6. $\frac{7}{3}$

7. $\frac{11}{3}$

8. $\frac{9}{2}$

9. $\frac{9}{8}$

10. $\frac{11}{5}$

11. $\frac{11}{2}$

12. $\frac{13}{3}$

13. $\frac{13}{4}$

14. $\frac{15}{2}$

15. $\frac{10}{3}$

16. $\frac{21}{5}$

Supplemental Practice, Lesson 116

1. $\frac{3}{4}$

2. $\frac{1}{4}$

3. $\frac{7}{8}$

4. $\frac{1}{8}$

5. $\frac{3}{8}$

6. $\frac{5}{8}$

7. $\frac{7}{8}$

8. $\frac{2}{9}$

9. $\frac{3}{5}$

10. $\frac{2}{9}$

11. $\frac{3}{10}$

12. $\frac{2}{5}$

13. $\frac{7}{10}$

14. $\frac{1}{10}$

15. $\frac{3}{4}$

16. $4\frac{3}{4}$

17. $4\frac{7}{8}$

18. $6\frac{7}{8}$

19. $6\frac{1}{2}$

20. $5\frac{2}{3}$

21. $4\frac{5}{6}$

22. $5\frac{5}{8}$

23. $7\frac{4}{5}$

24. $5\frac{9}{10}$

25. $6\frac{11}{12}$

26. $5\frac{3}{4}$

27. $7\frac{5}{6}$

28. $3\frac{3}{8}$

29. $2\frac{3}{8}$

30. $7\frac{2}{3}$

31. $4\frac{1}{5}$

32. $7\frac{1}{2}$

33. $3\frac{1}{3}$

34. $5\frac{1}{8}$

35. $4\frac{1}{12}$

36. $\frac{5}{6}$

37. $\frac{1}{6}$

38. $\frac{7}{12}$

39. $\frac{1}{12}$

40. $\frac{7}{10}$

41. $\frac{3}{10}$

42. $\frac{9}{20}$

43. $\frac{1}{20}$

44. $\frac{11}{12}$

45. $\frac{5}{12}$

46. $1\frac{1}{12}$

47. $\frac{5}{12}$

48. $\frac{5}{12}$

49. $\frac{1}{12}$

50. $1\frac{7}{12}$

51. $\frac{1}{12}$

52. $1\frac{5}{12}$

53. $\frac{1}{12}$

54. $4\frac{7}{12}$

55. $7\frac{9}{10}$

56. $7\frac{11}{12}$

57. $3\frac{1}{4}$

58. $4\frac{7}{12}$

59. $3\frac{1}{8}$

60. $12\frac{11}{15}$

61. $5\frac{17}{20}$

62. $7\frac{5}{6}$

63. $3\frac{1}{3}$

64. $7\frac{1}{12}$

65. $3\frac{7}{24}$

66. $4\frac{9}{10}$

67. $4\frac{1}{6}$

68. $8\frac{5}{6}$

69. $6\frac{1}{15}$

70. $7\frac{17}{20}$

71. $\frac{1}{12}$

72. $5\frac{7}{8}$

73. $2\frac{3}{20}$

Supplemental Practice, Lesson 117

1. 1.14
2. 1.3
3. 0.17
4. 0.7
5. 0.12
6. 1.2
7. 0.18
8. 1.8
9. 1.5
10. 0.22
11. 1.8
12. 0.43
13. 1.6

14. 0.32

15. 1.3

16. 0.21

17. 0.54

18. 1.6

19. 0.16

20. 3.4

21. 0.05

22. 0.07

23. 0.27

24. 0.024

25. 0.09

26. 0.018

27. 0.05

28. 0.004

29. 0.027

30. 0.09

31. 0.06

32. 0.023

33. 0.06

34. 0.22

35. 0.04

36. 0.024

37. 0.04

38. 0.019

39. 0.08

40. 0.009

Supplemental Practice, Lesson 118

1. 0.85

2. 0.024

3. 0.45

4. 0.065

5. 1.55

6. 0.135

7. 0.14

8. 0.25

9. 0.45

10. 0.25

11. 0.375

12. 0.024

13. 0.125

14. 0.12

15. 0.15

16. 0.55

17. 0.45

18. 0.18

19. 0.045

20. 0.0225

Supplemental Practice, Lesson 119

1. 0.5
2. 6
3. 0.3
4. 1.6
5. 4.1
6. 1.4
7. 1.2
8. 1.4
9. 1.9
10. 13
11. 0.33
12. 2.1
13. 23
14. 1.6
15. 15
16. 0.9
17. 10.7
18. 4.1
19. 1.9
20. 2.9

Supplemental Practice, Lesson 120

1. 1
2. $\frac{15}{16}$
3. $7\frac{1}{2}$
4. 10
5. $1\frac{7}{9}$
6. $1\frac{7}{8}$
7. $\frac{5}{6}$
8. $1\frac{1}{6}$
9. 7
10. 10
11. $4\frac{1}{6}$
12. $6\frac{1}{8}$
13. $\frac{8}{9}$
14. $1\frac{3}{8}$
15. 18
16. 5
17. $3\frac{3}{8}$
18. $2\frac{11}{12}$

Solutions for

Facts Practice Tests

A

100 Addition Facts

Add.

3 + 2 **5**	8 + 3 **11**	2 + 1 **3**	5 + 6 **11**	2 + 9 **11**	4 + 8 **12**	8 + 0 **8**	3 + 9 **12**	1 + 0 **1**	6 + 3 **9**
7 + 3 **10**	1 + 6 **7**	4 + 7 **11**	0 + 3 **3**	6 + 4 **10**	5 + 5 **10**	3 + 1 **4**	7 + 2 **9**	8 + 5 **13**	2 + 5 **7**
4 + 0 **4**	5 + 7 **12**	1 + 1 **2**	5 + 4 **9**	2 + 8 **10**	7 + 1 **8**	4 + 6 **10**	0 + 2 **2**	6 + 5 **11**	4 + 9 **13**
8 + 6 **14**	0 + 4 **4**	5 + 8 **13**	7 + 4 **11**	1 + 7 **8**	6 + 6 **12**	4 + 1 **5**	8 + 2 **10**	2 + 4 **6**	6 + 0 **6**
9 + 1 **10**	8 + 8 **16**	2 + 2 **4**	4 + 5 **9**	6 + 2 **8**	0 + 0 **0**	5 + 9 **14**	3 + 3 **6**	8 + 1 **9**	2 + 7 **9**
4 + 4 **8**	7 + 5 **12**	0 + 1 **1**	8 + 7 **15**	3 + 4 **7**	7 + 9 **16**	1 + 2 **3**	6 + 7 **13**	0 + 8 **8**	9 + 2 **11**
0 + 9 **9**	8 + 9 **17**	7 + 6 **13**	1 + 3 **4**	6 + 8 **14**	2 + 0 **2**	8 + 4 **12**	3 + 5 **8**	9 + 8 **17**	5 + 0 **5**
9 + 3 **12**	2 + 6 **8**	3 + 0 **3**	6 + 1 **7**	3 + 6 **9**	5 + 2 **7**	0 + 5 **5**	6 + 9 **15**	1 + 8 **9**	9 + 6 **15**
4 + 3 **7**	9 + 9 **18**	0 + 7 **7**	9 + 4 **13**	7 + 7 **14**	1 + 4 **5**	3 + 7 **10**	7 + 0 **7**	2 + 3 **5**	5 + 1 **6**
9 + 5 **14**	1 + 5 **6**	9 + 0 **9**	3 + 8 **11**	1 + 9 **10**	5 + 3 **8**	4 + 2 **6**	9 + 7 **16**	0 + 6 **6**	7 + 8 **15**

© Saxon Publishers, Inc., and Stephen Hake. Reproduction prohibited.

B

100 Subtraction Facts

Subtract.

16 − 9 **7**	7 − 1 **6**	18 − 9 **9**	11 − 3 **8**	13 − 7 **6**	8 − 2 **6**	11 − 5 **6**	5 − 0 **5**	17 − 9 **8**	6 − 1 **5**
10 − 9 **1**	6 − 2 **4**	13 − 4 **9**	4 − 0 **4**	10 − 5 **5**	5 − 1 **4**	10 − 3 **7**	12 − 6 **6**	10 − 1 **9**	6 − 4 **2**
7 − 2 **5**	14 − 7 **7**	8 − 1 **7**	11 − 6 **5**	3 − 3 **0**	16 − 7 **9**	5 − 2 **3**	12 − 4 **8**	3 − 0 **3**	11 − 7 **4**
17 − 8 **9**	6 − 0 **6**	10 − 6 **4**	4 − 1 **3**	9 − 5 **4**	9 − 0 **9**	5 − 4 **1**	12 − 5 **7**	4 − 2 **2**	9 − 3 **6**
12 − 3 **9**	16 − 8 **8**	9 − 1 **8**	15 − 6 **9**	11 − 4 **7**	13 − 5 **8**	1 − 0 **1**	8 − 5 **3**	9 − 6 **3**	11 − 2 **9**
7 − 0 **7**	10 − 8 **2**	6 − 3 **3**	14 − 5 **9**	3 − 1 **2**	8 − 6 **2**	4 − 4 **0**	11 − 8 **3**	3 − 2 **1**	15 − 9 **6**
13 − 8 **5**	7 − 4 **3**	10 − 7 **3**	0 − 0 **0**	12 − 8 **4**	5 − 5 **0**	4 − 3 **1**	8 − 7 **1**	7 − 3 **4**	7 − 6 **1**
5 − 3 **2**	7 − 5 **2**	2 − 1 **1**	6 − 6 **0**	8 − 4 **4**	2 − 2 **0**	13 − 6 **7**	15 − 8 **7**	2 − 0 **2**	13 − 9 **4**
1 − 1 **0**	11 − 9 **2**	10 − 4 **6**	9 − 2 **7**	14 − 6 **8**	8 − 0 **8**	9 − 4 **5**	10 − 2 **8**	6 − 5 **1**	8 − 3 **5**
7 − 7 **0**	14 − 8 **6**	12 − 9 **3**	9 − 8 **1**	12 − 7 **5**	9 − 9 **0**	15 − 7 **8**	8 − 8 **0**	14 − 9 **5**	9 − 7 **2**

© Saxon Publishers, Inc., and Stephen Hake. Reproduction prohibited.

C

100 Multiplication Facts

Multiply.

9 × 9 **81**	3 × 5 **15**	8 × 5 **40**	2 × 6 **12**	4 × 7 **28**	0 × 3 **0**	7 × 2 **14**	1 × 5 **5**	7 × 8 **56**	4 × 0 **0**
3 × 4 **12**	5 × 9 **45**	0 × 2 **0**	7 × 3 **21**	4 × 1 **4**	2 × 7 **14**	6 × 3 **18**	5 × 4 **20**	1 × 0 **0**	9 × 2 **18**
1 × 1 **1**	9 × 0 **0**	2 × 8 **16**	6 × 4 **24**	0 × 7 **0**	8 × 1 **8**	3 × 3 **9**	4 × 8 **32**	9 × 3 **27**	2 × 0 **0**
4 × 9 **36**	7 × 0 **0**	1 × 2 **2**	8 × 4 **32**	6 × 5 **30**	2 × 9 **18**	9 × 4 **36**	0 × 1 **0**	7 × 4 **28**	5 × 8 **40**
0 × 8 **0**	4 × 2 **8**	9 × 8 **72**	3 × 6 **18**	5 × 5 **25**	1 × 6 **6**	5 × 0 **0**	6 × 6 **36**	2 × 1 **2**	7 × 9 **63**
9 × 1 **9**	2 × 2 **4**	5 × 1 **5**	4 × 3 **12**	0 × 0 **0**	8 × 9 **72**	3 × 7 **21**	9 × 7 **63**	1 × 7 **7**	6 × 0 **0**
5 × 6 **30**	7 × 5 **35**	3 × 0 **0**	8 × 8 **64**	1 × 3 **3**	8 × 3 **24**	5 × 2 **10**	0 × 4 **0**	9 × 5 **45**	6 × 7 **42**
2 × 3 **6**	8 × 6 **48**	0 × 5 **0**	6 × 1 **6**	3 × 8 **24**	7 × 6 **42**	1 × 8 **8**	9 × 6 **54**	4 × 4 **16**	5 × 3 **15**
7 × 7 **49**	1 × 4 **4**	6 × 2 **12**	4 × 5 **20**	2 × 4 **8**	8 × 0 **0**	3 × 1 **3**	6 × 8 **48**	0 × 9 **0**	8 × 7 **56**
3 × 2 **6**	4 × 6 **24**	1 × 9 **9**	5 × 7 **35**	8 × 2 **16**	0 × 6 **0**	7 × 1 **7**	2 × 5 **10**	6 × 9 **54**	3 × 9 **27**

© Saxon Publishers, Inc., and Stephen Hake. Reproduction prohibited.

D

90 Division Facts

Divide.

3 7)21	**5** 2)10	**7** 6)42	**3** 1)3	**6** 4)24	**2** 3)6	**6** 9)54	**3** 6)18	**0** 4)0	**6** 5)30
8 4)32	**7** 8)56	**0** 1)0	**2** 6)12	**6** 3)18	**8** 9)72	**3** 5)15	**4** 2)8	**6** 7)42	**6** 6)36
0 6)0	**2** 5)10	**1** 9)9	**3** 2)6	**9** 7)63	**4** 4)16	**6** 8)48	**2** 1)2	**7** 5)35	**7** 3)21
9 2)18	**1** 6)6	**5** 3)15	**5** 8)40	**0** 2)0	**4** 5)20	**3** 9)27	**8** 1)8	**1** 4)4	**5** 7)35
5 4)20	**7** 9)63	**4** 1)4	**2** 7)14	**1** 3)3	**3** 8)24	**0** 5)0	**4** 6)24	**1** 8)8	**8** 2)16
1 5)5	**8** 8)64	**0** 3)0	**7** 4)28	**7** 7)49	**2** 2)4	**9** 9)81	**4** 3)12	**5** 6)30	**5** 1)5
4 8)32	**1** 1)1	**4** 9)36	**9** 3)27	**7** 2)14	**5** 5)25	**8** 6)48	**0** 8)0	**4** 7)28	**9** 4)36
6 2)12	**9** 5)45	**7** 1)7	**2** 4)8	**0** 7)0	**2** 8)16	**8** 3)24	**5** 9)45	**9** 1)9	**9** 6)54
8 7)56	**0** 9)0	**9** 8)72	**1** 2)2	**8** 5)40	**3** 3)9	**2** 9)18	**6** 1)6	**3** 4)12	**1** 7)7

© Saxon Publishers, Inc., and Stephen Hake. Reproduction prohibited.

E

90 Division Facts

Divide.

20 ÷ 4 = **5**	21 ÷ 7 = **3**	0 ÷ 2 = **0**	27 ÷ 3 = **9**	8 ÷ 1 = **8**	54 ÷ 6 = **9**
15 ÷ 5 = **3**	6 ÷ 3 = **2**	28 ÷ 4 = **7**	18 ÷ 2 = **9**	24 ÷ 6 = **4**	9 ÷ 9 = **1**
56 ÷ 8 = **7**	0 ÷ 6 = **0**	21 ÷ 3 = **7**	1 ÷ 1 = **1**	25 ÷ 5 = **5**	12 ÷ 2 = **6**
5 ÷ 1 = **5**	45 ÷ 9 = **5**	16 ÷ 4 = **4**	30 ÷ 6 = **5**	9 ÷ 3 = **3**	14 ÷ 7 = **2**
0 ÷ 8 = **0**	6 ÷ 2 = **3**	24 ÷ 8 = **3**	10 ÷ 5 = **2**	81 ÷ 9 = **9**	24 ÷ 4 = **6**
16 ÷ 2 = **8**	30 ÷ 5 = **6**	0 ÷ 1 = **0**	28 ÷ 7 = **4**	4 ÷ 4 = **1**	40 ÷ 8 = **5**
3 ÷ 3 = **1**	18 ÷ 6 = **3**	63 ÷ 9 = **7**	40 ÷ 5 = **8**	10 ÷ 2 = **5**	36 ÷ 6 = **6**
32 ÷ 8 = **4**	12 ÷ 4 = **3**	18 ÷ 3 = **6**	35 ÷ 7 = **5**	8 ÷ 8 = **1**	2 ÷ 1 = **2**
45 ÷ 5 = **9**	7 ÷ 7 = **1**	27 ÷ 9 = **3**	9 ÷ 1 = **9**	48 ÷ 6 = **8**	0 ÷ 7 = **0**
4 ÷ 1 = **4**	0 ÷ 9 = **0**	24 ÷ 3 = **8**	32 ÷ 4 = **8**	5 ÷ 5 = **1**	72 ÷ 9 = **8**
56 ÷ 7 = **8**	15 ÷ 3 = **5**	12 ÷ 6 = **2**	8 ÷ 2 = **4**	63 ÷ 7 = **9**	0 ÷ 4 = **0**
14 ÷ 2 = **7**	42 ÷ 6 = **7**	6 ÷ 1 = **6**	16 ÷ 8 = **2**	20 ÷ 5 = **4**	49 ÷ 7 = **7**
36 ÷ 4 = **9**	64 ÷ 8 = **8**	0 ÷ 3 = **0**	54 ÷ 9 = **6**	4 ÷ 2 = **2**	48 ÷ 8 = **6**
18 ÷ 9 = **2**	3 ÷ 1 = **3**	35 ÷ 5 = **7**	8 ÷ 4 = **2**	72 ÷ 8 = **9**	6 ÷ 6 = **1**
0 ÷ 5 = **0**	42 ÷ 7 = **6**	2 ÷ 2 = **1**	36 ÷ 9 = **4**	7 ÷ 1 = **7**	12 ÷ 3 = **4**

© Saxon Publishers, Inc., and Stephen Hake. Reproduction prohibited.

F

64 Multiplication Facts

Multiply.

5 × 6 **30**	4 × 3 **12**	9 × 8 **72**	7 × 5 **35**	2 × 9 **18**	8 × 4 **32**	9 × 3 **27**	6 × 9 **54**
9 × 4 **36**	2 × 5 **10**	7 × 6 **42**	4 × 8 **32**	7 × 9 **63**	5 × 4 **20**	3 × 2 **6**	9 × 7 **63**
3 × 7 **21**	8 × 5 **40**	6 × 2 **12**	5 × 5 **25**	3 × 5 **15**	2 × 4 **8**	7 × 7 **49**	8 × 9 **72**
6 × 4 **24**	2 × 8 **16**	4 × 4 **16**	8 × 2 **16**	3 × 9 **27**	6 × 6 **36**	9 × 9 **81**	5 × 3 **15**
4 × 6 **24**	8 × 8 **64**	5 × 7 **35**	6 × 3 **18**	2 × 2 **4**	7 × 4 **28**	3 × 8 **24**	8 × 6 **48**
2 × 6 **12**	5 × 9 **45**	3 × 3 **9**	9 × 2 **18**	6 × 7 **42**	4 × 5 **20**	7 × 2 **14**	9 × 6 **54**
5 × 2 **10**	7 × 8 **56**	2 × 3 **6**	6 × 8 **48**	4 × 7 **28**	9 × 5 **45**	3 × 6 **18**	8 × 7 **56**
3 × 4 **12**	7 × 3 **21**	5 × 8 **40**	4 × 2 **8**	8 × 3 **24**	2 × 7 **14**	6 × 5 **30**	4 × 9 **36**

© Saxon Publishers, Inc., and Stephen Hake. Reproduction prohibited.

G

48 Uneven Divisions

Divide. Write each answer with a remainder.

3 R 3 4)15	**1 R 5** 9)14	**6 R 3** 7)45	**5 R 1** 3)16	**6 R 2** 6)38	**3 R 1** 2)7
6 R 2 8)50	**5 R 3** 5)28	**5 R 1** 4)21	**2 R 3** 6)15	**1 R 4** 7)11	**2 R 4** 8)20
6 R 2 3)20	**4 R 4** 7)32	**3 R 6** 8)30	**7 R 1** 2)15	**8 R 3** 5)43	**5 R 5** 6)35
6 R 8 9)62	**2 R 2** 4)10	**4 R 3** 6)27	**2 R 3** 9)21	**4 R 3** 4)19	**8 R 1** 3)25
9 R 2 6)56	**8 R 1** 2)17	**3 R 1** 3)10	**1 R 3** 5)8	**4 R 4** 9)40	**4 R 2** 7)30
2 R 1 2)5	**3 R 1** 8)25	**3 R 2** 5)17	**2 R 3** 7)17	**2 R 2** 3)8	**2 R 1** 4)9
2 R 6 7)20	**1 R 4** 6)10	**4 R 1** 2)9	**7 R 2** 4)30	**1 R 7** 8)15	**3 R 2** 9)29
6 R 2 5)32	**4 R 2** 3)14	**5 R 5** 9)50	**8 R 1** 8)65	**5 R 1** 2)11	**3 R 4** 5)19

© Saxon Publishers, Inc., and Stephen Hake. Reproduction prohibited.

H

60 Improper Fractions to Simplify

Simplify.

$\frac{15}{2} = 7\frac{1}{2}$	$\frac{9}{8} = 1\frac{1}{8}$	$\frac{10}{2} = 5$	$\frac{18}{6} = 3$	$\frac{8}{3} = 2\frac{2}{3}$	$\frac{12}{4} = 3$
$\frac{10}{10} = 1$	$\frac{3}{2} = 1\frac{1}{2}$	$\frac{11}{4} = 2\frac{3}{4}$	$\frac{4}{3} = 1\frac{1}{3}$	$\frac{12}{5} = 2\frac{2}{5}$	$\frac{5}{4} = 1\frac{1}{4}$
$\frac{12}{6} = 2$	$\frac{9}{3} = 3$	$\frac{5}{5} = 1$	$\frac{15}{4} = 3\frac{3}{4}$	$\frac{6}{2} = 3$	$\frac{9}{9} = 1$
$\frac{3}{3} = 1$	$\frac{7}{4} = 1\frac{3}{4}$	$\frac{21}{10} = 2\frac{1}{10}$	$\frac{11}{2} = 5\frac{1}{2}$	$\frac{7}{6} = 1\frac{1}{6}$	$\frac{24}{8} = 3$
$\frac{11}{3} = 3\frac{2}{3}$	$\frac{9}{5} = 1\frac{4}{5}$	$\frac{4}{2} = 2$	$\frac{21}{8} = 2\frac{5}{8}$	$\frac{6}{5} = 1\frac{1}{5}$	$\frac{12}{3} = 4$
$\frac{7}{2} = 3\frac{1}{2}$	$\frac{25}{6} = 4\frac{1}{6}$	$\frac{10}{9} = 1\frac{1}{9}$	$\frac{4}{4} = 1$	$\frac{12}{2} = 6$	$\frac{16}{15} = 1\frac{1}{15}$
$\frac{10}{5} = 2$	$\frac{5}{2} = 2\frac{1}{2}$	$\frac{7}{3} = 2\frac{1}{3}$	$\frac{8}{4} = 2$	$\frac{8}{8} = 1$	$\frac{27}{10} = 2\frac{7}{10}$
$\frac{16}{4} = 4$	$\frac{6}{6} = 1$	$\frac{25}{12} = 2\frac{1}{12}$	$\frac{5}{3} = 1\frac{2}{3}$	$\frac{7}{5} = 1\frac{2}{5}$	$\frac{16}{9} = 1\frac{7}{9}$
$\frac{15}{8} = 1\frac{7}{8}$	$\frac{10}{3} = 3\frac{1}{3}$	$\frac{33}{10} = 3\frac{3}{10}$	$\frac{2}{2} = 1$	$\frac{35}{6} = 5\frac{5}{6}$	$\frac{25}{8} = 3\frac{1}{8}$
$\frac{6}{3} = 2$	$\frac{8}{5} = 1\frac{3}{5}$	$\frac{9}{4} = 2\frac{1}{4}$	$\frac{12}{12} = 1$	$\frac{25}{2} = 12\frac{1}{2}$	$\frac{9}{2} = 4\frac{1}{2}$

© Saxon Publishers, Inc., and Stephen Hake. Reproduction prohibited.

I

40 Fractions to Reduce

Reduce each fraction to lowest terms.

$\frac{2}{10} = \mathbf{\frac{1}{5}}$	$\frac{8}{16} = \mathbf{\frac{1}{2}}$	$\frac{2}{6} = \mathbf{\frac{1}{3}}$	$\frac{10}{100} = \mathbf{\frac{1}{10}}$	$\frac{6}{8} = \mathbf{\frac{3}{4}}$
$\frac{10}{15} = \mathbf{\frac{2}{3}}$	$\frac{5}{10} = \mathbf{\frac{1}{2}}$	$\frac{8}{12} = \mathbf{\frac{2}{3}}$	$\frac{9}{15} = \mathbf{\frac{3}{5}}$	$\frac{4}{16} = \mathbf{\frac{1}{4}}$
$\frac{2}{8} = \mathbf{\frac{1}{4}}$	$\frac{4}{10} = \mathbf{\frac{2}{5}}$	$\frac{15}{20} = \mathbf{\frac{3}{4}}$	$\frac{4}{8} = \mathbf{\frac{1}{2}}$	$\frac{4}{6} = \mathbf{\frac{2}{3}}$
$\frac{6}{15} = \mathbf{\frac{2}{5}}$	$\frac{4}{12} = \mathbf{\frac{1}{3}}$	$\frac{25}{100} = \mathbf{\frac{1}{4}}$	$\frac{10}{25} = \mathbf{\frac{2}{5}}$	$\frac{12}{20} = \mathbf{\frac{3}{5}}$
$\frac{20}{100} = \mathbf{\frac{1}{5}}$	$\frac{6}{9} = \mathbf{\frac{2}{3}}$	$\frac{2}{4} = \mathbf{\frac{1}{2}}$	$\frac{3}{12} = \mathbf{\frac{1}{4}}$	$\frac{3}{15} = \mathbf{\frac{1}{5}}$
$\frac{3}{9} = \mathbf{\frac{1}{3}}$	$\frac{2}{12} = \mathbf{\frac{1}{6}}$	$\frac{6}{10} = \mathbf{\frac{3}{5}}$	$\frac{12}{16} = \mathbf{\frac{3}{4}}$	$\frac{50}{100} = \mathbf{\frac{1}{2}}$
$\frac{9}{12} = \mathbf{\frac{3}{4}}$	$\frac{3}{6} = \mathbf{\frac{1}{2}}$	$\frac{5}{15} = \mathbf{\frac{1}{3}}$	$\frac{10}{12} = \mathbf{\frac{5}{6}}$	$\frac{8}{24} = \mathbf{\frac{1}{3}}$
$\frac{12}{15} = \mathbf{\frac{4}{5}}$	$\frac{8}{10} = \mathbf{\frac{4}{5}}$	$\frac{75}{100} = \mathbf{\frac{3}{4}}$	$\frac{6}{12} = \mathbf{\frac{1}{2}}$	$\frac{12}{24} = \mathbf{\frac{1}{2}}$

© Saxon Publishers, Inc., and Stephen Hake. Reproduction prohibited.

J

50 Fractions to Simplify

Simplify.

$\frac{16}{20} = \frac{4}{5}$	$\frac{6}{4} = 1\frac{1}{2}$	$\frac{4}{6} = \frac{2}{3}$	$\frac{10}{8} = 1\frac{1}{4}$	$\frac{3}{12} = \frac{1}{4}$
$\frac{12}{9} = 1\frac{1}{3}$	$\frac{2}{4} = \frac{1}{2}$	$\frac{12}{10} = 1\frac{1}{5}$	$\frac{12}{4} = 3$	$\frac{12}{8} = 1\frac{1}{2}$
$\frac{8}{3} = 2\frac{2}{3}$	$\frac{8}{6} = 1\frac{1}{3}$	$\frac{4}{12} = \frac{1}{3}$	$\frac{10}{4} = 2\frac{1}{2}$	$\frac{4}{10} = \frac{2}{5}$
$\frac{20}{8} = 2\frac{1}{2}$	$\frac{4}{8} = \frac{1}{2}$	$\frac{20}{9} = 2\frac{2}{9}$	$\frac{24}{6} = 4$	$\frac{9}{6} = 1\frac{1}{2}$
$\frac{15}{10} = 1\frac{1}{2}$	$\frac{5}{2} = 2\frac{1}{2}$	$\frac{12}{20} = \frac{3}{5}$	$\frac{15}{9} = 1\frac{2}{3}$	$\frac{8}{12} = \frac{2}{3}$
$\frac{4}{20} = \frac{1}{5}$	$\frac{8}{24} = \frac{1}{3}$	$\frac{10}{6} = 1\frac{2}{3}$	$\frac{3}{6} = \frac{1}{2}$	$\frac{16}{10} = 1\frac{3}{5}$
$\frac{2}{8} = \frac{1}{4}$	$\frac{20}{6} = 3\frac{1}{3}$	$\frac{6}{3} = 2$	$\frac{25}{12} = 2\frac{1}{12}$	$\frac{9}{12} = \frac{3}{4}$
$\frac{10}{2} = 5$	$\frac{8}{8} = 1$	$\frac{50}{100} = \frac{1}{2}$	$\frac{6}{12} = \frac{1}{2}$	$\frac{15}{6} = 2\frac{1}{2}$
$\frac{10}{3} = 3\frac{1}{3}$	$\frac{10}{20} = \frac{1}{2}$	$\frac{24}{9} = 2\frac{2}{3}$	$\frac{6}{8} = \frac{3}{4}$	$\frac{16}{5} = 3\frac{1}{5}$
$\frac{5}{10} = \frac{1}{2}$	$\frac{14}{8} = 1\frac{3}{4}$	$\frac{15}{2} = 7\frac{1}{2}$	$\frac{21}{6} = 3\frac{1}{2}$	$\frac{16}{24} = \frac{2}{3}$

© Saxon Publishers, Inc., and Stephen Hake. Reproduction prohibited.

K

30 Percents to Write as Fractions

Write each percent as a reduced fraction.

$1\% = \frac{1}{100}$	$20\% = \frac{1}{5}$	$55\% = \frac{11}{20}$	$90\% = \frac{9}{10}$	$75\% = \frac{3}{4}$
$99\% = \frac{99}{100}$	$5\% = \frac{1}{20}$	$95\% = \frac{19}{20}$	$80\% = \frac{4}{5}$	$12\% = \frac{3}{25}$
$70\% = \frac{7}{10}$	$65\% = \frac{13}{20}$	$50\% = \frac{1}{2}$	$2\% = \frac{1}{50}$	$48\% = \frac{12}{25}$
$24\% = \frac{6}{25}$	$25\% = \frac{1}{4}$	$98\% = \frac{49}{50}$	$40\% = \frac{2}{5}$	$15\% = \frac{3}{20}$
$60\% = \frac{3}{5}$	$30\% = \frac{3}{10}$	$4\% = \frac{1}{25}$	$35\% = \frac{7}{20}$	$36\% = \frac{9}{25}$
$45\% = \frac{9}{20}$	$8\% = \frac{2}{25}$	$10\% = \frac{1}{10}$	$21\% = \frac{21}{100}$	$85\% = \frac{17}{20}$

© Saxon Publishers, Inc., and Stephen Hake. Reproduction prohibited.

Solutions for

Tests

TEST 1

1. The pattern is "Count up by eights": **40, 48, 56**

2. The pattern is "Count down by sevens": **21, 14, 7**

3. The pattern is "Count up by fives": **45, 50, 55**

4. The last digit is in the farthest place to the right, so the last digit is **9.**

5. If a whole number is not even it must be **odd.**

6. The only number that ends with an even digit is **C. 5732.**

7. The only number that ends with an odd digit is **A. 8431.**

8. An even number of objects can be paired. Since there is one more pen than there are pencils, the pens and pencils cannot be paired completely. This means that the total number of pens and pencils could not be an even number. The only choice with an even number is **B. 14.**

9. 2 hundreds + 4 tens + 9 ones = **249**

10. The digit showing the number of tens is in the second place from the right, so the digit is **4.**

11. 8 hundreds + 3 tens = **830**

12. 1 hundred = **10 tens**

13. **13 < 30**

14. **17 > 6**

15. 102 **<** 111

16. 23 **<** 32

17. $329.72 = **three hundred twenty-nine dollars and seventy-two cents**

18. 115 = **one hundred fifteen**

19. Seven hundred three dollars and forty cents = **$703.40**

20. Three hundred nine = **309**

TEST 2

1.
$$\begin{array}{r} 23 \text{ girls} \\ +\ 18 \text{ boys} \\ \hline \mathbf{41 \text{ boys and girls}} \end{array}$$

2. Half of 50 is 25. There are **25 girls** on Siew's team.

3. **1 + 4 = 5   5 − 1 = 4**
   4 + 1 = 5   5 − 4 = 1

4. Three hundred seven thousand, eight hundred thirteen = **307,813**

5. The ones digit must be 2 because the number is even. The number is greater than 400, so the hundreds digit must be 4. Thus, the tens digit is 3, and the number is **432.**

6. $14 + n = 22$
$$\begin{array}{r} 22 \\ -\ 14 \\ \hline \mathbf{8} \end{array}$$

7. $b + 11 = 50$
$$\begin{array}{r} 50 \\ -\ 11 \\ \hline \mathbf{39} \end{array}$$

8. The digit showing the number of hundreds is in the third place from the right, so the digit is **3.**

9. 37 − 7 ◯ 32 − 2
   30 **=** 30

10. The sum of two even numbers is always **even.**

11. Student may draw a diagram (not required).

We count **11 people** between Kelly and Bobby.

12.
$$\begin{array}{r} 143 \\ 87 \\ +\ 623 \\ \hline \mathbf{853} \end{array}$$

13.
$$\begin{array}{r} 327 \\ -\ 239 \\ \hline \mathbf{88} \end{array}$$

14.
$$\begin{array}{r} 900 \\ -\ 238 \\ \hline \mathbf{662} \end{array}$$

15.
$$\begin{array}{r} 5 \\ 6 \\ 2 \\ 9 \\ +\ 7 \\ \hline \mathbf{29} \end{array}$$

16.
$$\begin{array}{r} \$229 \\ -\ \ \$40 \\ \hline \mathbf{\$189} \end{array}$$

17.
$$\begin{array}{r} \$276 \\ \$23 \\ +\ \ \ \$3 \\ \hline \mathbf{\$302} \end{array}$$

18. The pattern is "Count up by nines": **36**

19. The pattern is "Count up by fives": **50**

20. The pattern is "Count down by sevens": **42**

Test 3

1. **5 + 6 = 11** **11 − 5 = 6**
6 + 5 = 11 **11 − 6 = 5**

2. 4 × 9 = **36**

3. 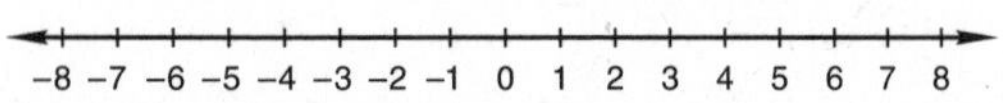

4. **B.** ———

5.
$$\begin{array}{r} 17 \\ -\ \ n \\ \hline 8 \end{array} \qquad \begin{array}{r} 17 \\ -\ \ 8 \\ \hline \mathbf{9} \end{array}$$

6.
$$\begin{array}{l} \ \ \ 11 \text{ boys} \\ +\ 16 \text{ girls} \\ \hline \ \ \ \mathbf{27\ boys\ and\ girls} \end{array}$$

7.
$$\begin{array}{r} 25 \text{ peaches} \\ +\ \ m \text{ peaches} \\ \hline 34 \text{ peaches} \end{array} \qquad \begin{array}{r} 34 \text{ peaches} \\ -\ 25 \text{ peaches} \\ \hline \mathbf{9\ peaches} \end{array}$$

8. 

9. **A. 4 × 1 = 4**

10. 3 + 3 + 3 + 3 + 3 + 3 = **6 × 3**

11.
$$\begin{array}{r} 408 \\ -\ \ \ 29 \\ \hline \mathbf{379} \end{array}$$

12.
$$\begin{array}{r} \$3.24 \\ \$1.17 \\ +\ \$7.03 \\ \hline \mathbf{\$11.44} \end{array}$$

13.
$$\begin{array}{r} 729 \\ 84 \\ +\ 654 \\ \hline \mathbf{1467} \end{array}$$

14.
$$\begin{array}{r} \$2.12 \\ -\ \$1.83 \\ \hline \mathbf{\$0.29} \end{array}$$

15.
$$\begin{array}{r} 900 \\ -\ 767 \\ \hline \mathbf{133} \end{array}$$

16. The digit showing the number of tens is in the second place from the right, so the digit is **8.**

17. Forty-seven thousand, nine hundred seventy = **47,970**

18. 3 eights = 3 × 8 = **24**

19. 2767 **<** 2776

20. The pattern is "Count up by eights":
…, 40, 48, **56**

Test 4

1. $$\begin{array}{r} 267 \text{ pages} \\ +\ 198 \text{ pages} \\ \hline \mathbf{465 \text{ pages}} \end{array}$$

2. $$\begin{array}{r} 9 \text{ players} \\ 9 \text{ players} \\ 9 \text{ players} \\ +\ 9 \text{ players} \\ \hline \mathbf{36 \text{ players}} \end{array} \qquad \begin{array}{r} 4 \\ \times\ 9 \text{ players} \\ \hline \mathbf{36 \text{ players}} \end{array}$$

3. $$\begin{array}{r} 550 \\ +\ 215 \\ \hline \mathbf{765} \end{array}$$

4. $$\begin{array}{r} 169 \text{ boys} \\ +\ \ G \text{ girls} \\ \hline 359 \text{ children} \end{array} \qquad \begin{array}{r} 359 \text{ children} \\ -\ 169 \text{ boys} \\ \hline \mathbf{190 \text{ girls}} \end{array}$$

5. 卌 ||||

6. $-4 < 4$

7. $8 + 8 + 8 + 8 \bigcirc 5 \times 8$
   $4 \times 8 \bigcirc 5 \times 8$
   $32 < 40$

8. **A.**

9. The product of any odd number and any even number is **even.**

10. $$\begin{array}{r} \$27 \\ \times\ \ \ 6 \\ \hline \mathbf{\$162} \end{array}$$

11. $$\begin{array}{r} \$4.08 \\ \times\ \ \ \ 5 \\ \hline \mathbf{\$20.40} \end{array}$$

12. $$\begin{array}{r} 4321 \\ -\ 2893 \\ \hline \mathbf{1428} \end{array}$$

13. $$\begin{array}{r} \$50.00 \\ -\ \$18.63 \\ \hline \mathbf{\$31.37} \end{array}$$

14. $2 \times 4 \times 6$
    $2 \times 24 = \mathbf{48}$

15. $3\overline{)12}$ with quotient $\mathbf{4}$

16. $12 \div 4 = \mathbf{3}$

17. $\frac{12}{6} = \mathbf{2}$

18. $m \times 10 = 90$
    $m = \mathbf{9}$

19. $$\begin{array}{r} e \\ -\ 16 \\ \hline 34 \end{array} \qquad \begin{array}{r} 34 \\ +\ 16 \\ \hline \mathbf{50} \end{array}$$

20. $$\begin{array}{r} 426 \\ +\ \ \ m \\ \hline 714 \end{array} \qquad \begin{array}{r} 714 \\ -\ 426 \\ \hline \mathbf{288} \end{array}$$

Test 5

1. The factors of 24 are the whole numbers that divide it without leaving a remainder. So **1, 2, 3, 4, 6, 8, 12,** and **24** are the factors of 24.

2. $$\begin{array}{r} N \text{ chairs in each row} \\ \times\ 5 \text{ rows} \\ \hline 25 \text{ chairs in all rows} \end{array} \qquad 5\overline{)25 \text{ chairs}} \text{ with quotient } \mathbf{5 \text{ chairs}}$$

3. One fourth means one of four equal groups.

$$\begin{array}{r} n \text{ pigs in each group} \\ \times\ 4 \text{ groups} \\ \hline 20 \text{ pigs in all groups} \end{array} \qquad 4\overline{)20 \text{ pigs}} \text{ with quotient } \mathbf{5 \text{ pigs}}$$

4. One half means one of two equal groups.

$$\begin{array}{r} n \text{ pigs in each group} \\ \times\ 2 \text{ groups} \\ \hline 20 \text{ pigs in all groups} \end{array} \qquad 2\overline{)20 \text{ pigs}} \text{ with quotient } \mathbf{10 \text{ pigs}}$$

5. $\mathbf{5 \times 6 = 30} \qquad \mathbf{30 \div 6 = 5}$
   $\mathbf{6 \times 5 = 30} \qquad \mathbf{30 \div 5 = 6}$

6. $\begin{array}{r} 270 \\ +\ 818 \\ \hline \mathbf{1088} \end{array}$

7. 3×6 ducklings $=$ **18 ducklings**

8. $3 \times (2 + 5) \bigcirc 3 + (2 \times 5)$

$3 \times 7 \bigcirc 3 + 10$

$21 > 13$

9. **D.** $\mathbf{\frac{8}{17}}$

10. $\begin{array}{r} \$20.00 \\ \$7.32 \\ +\ \$0.42 \\ \hline \mathbf{\$27.74} \end{array}$

11. $\begin{array}{r} \$7.00 \\ -\ \$3.46 \\ \hline \mathbf{\$3.54} \end{array}$

12. $15 \div (10 \div 2) = 15 \div 5 = \mathbf{3}$

13. $\begin{array}{r} \mathbf{8\ R\ 6} \\ 8\overline{)70} \\ 64 \\ \hline 6 \end{array}$

14. $\frac{42}{7} = \mathbf{6}$

15. $\begin{array}{r} \$9.04 \\ \times\quad 8 \\ \hline \mathbf{\$72.32} \end{array}$

16. $7 \times 5 \times 10$

$7 \times 50 = \mathbf{350}$

17. $\begin{array}{r} 728 \\ \times\quad 6 \\ \hline \mathbf{4368} \end{array}$

18. $\begin{array}{r} 3824 \\ 276 \\ +\ 2503 \\ \hline \mathbf{6603} \end{array}$

19. $\begin{array}{r} 1000 \\ -\quad 97 \\ \hline \mathbf{903} \end{array}$

20. **B. $\mathbf{1 + (3 + 5) = (1 + 3) + 5}$**

TEST 6

1. 1 decade $=$ 10 years
8 decades $= 8 \times 10$ years $=$ **80 years**

2. $\begin{array}{r} \$1.24 \\ \times\quad 4 \\ \hline \$4.96 \end{array}$ $\qquad \begin{array}{r} \$4.96 \\ +\ \$0.95 \\ \hline \mathbf{\$5.91} \end{array}$

3. $\begin{array}{r} 15 \\ \times\ 5 \\ \hline \mathbf{75} \end{array}$

4. $\begin{array}{r} N \text{ treats for each person} \\ \times\ 5 \text{ people} \\ \hline 20 \text{ treats} \end{array}$ $\qquad \begin{array}{r} \mathbf{4 \text{ treats}} \\ 5\overline{)20} \end{array}$

5. Since the number is even, 4 must be in the ones place. To form the greatest number possible, we place the greatest remaining digit in the hundreds place. Therefore, 7 is in the hundreds place, and 5 is in the tens place. The number is **754.**

6. The product of any two odd numbers is **odd.**

7. Student may draw a diagram (not required).

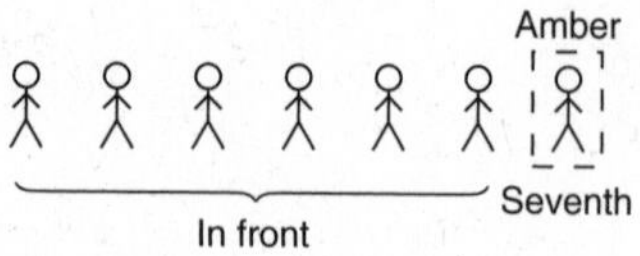

We count **6 people** in front of Amber.

8. On the scale, only every 10°F is labeled. There are 10 spaces between every 10°F, which means that every space equals 1°F. We count down from 0°F by ones to **−8°F.**

9. The factors of 27 are all the whole numbers that divide it without leaving a remainder. So **1, 3, 9,** and **27** are factors of 27.

10.
$$3738$$
$$\underline{-\ 1876}$$
$$\mathbf{1862}$$

11.
$$\$40.00$$
$$\underline{-\ \$4.87}$$
$$\mathbf{\$35.13}$$

12.
$$23$$
$$72$$
$$84$$
$$15$$
$$\underline{+\quad 2}$$
$$\mathbf{196}$$

13.
$$\overset{\mathbf{29\ R\ 5}}{8\overline{)237}}$$
$$\underline{16}$$
$$77$$
$$\underline{72}$$
$$5$$

14.
$$\overset{\mathbf{80}}{5\overline{)400}}$$
$$\underline{40}$$
$$00$$
$$\underline{\ 0}$$
$$0$$

15.
$$\overset{\mathbf{\$1.54}}{6\overline{)\$9.24}}$$
$$\underline{6}$$
$$3\,2$$
$$\underline{3\,0}$$
$$24$$
$$\underline{24}$$
$$0$$

16.
$$35$$
$$\underline{\times\quad 60}$$
$$\mathbf{2100}$$

17. $4 \times (5 + 9) = 4 \times 14 = \mathbf{56}$

18.
$$\$8.63$$
$$\underline{\times\qquad 6}$$
$$\mathbf{\$51.78}$$

19. $\dfrac{1}{2} > \dfrac{7}{16}$

20. The clock shows 28 minutes after five. Since it is afternoon, the indicated time is **5:28 p.m.**

TEST **7**

1.2. The clock shows 57 minutes after two. Since it is morning, the indicated time is **2:57 a.m.**

3.
$$1987$$
$$\underline{-\ 1966}$$
$$\mathbf{21\ years}$$

4. The factors of 36 are all the whole numbers that divide it without leaving a remainder. So **1, 2, 3, 4, 6, 9, 12, 18,** and **36** are factors of 36.

5.
$$4 \text{ quarts in each gallon}$$
$$\underline{\times\ 12 \text{ gallons}}$$
$$\mathbf{48\ quarts}$$

6. The number 34 is between 30 and 40. The number 34 is closer to **30.**

7. The number 866 is between 800 and 900. The number 866 is closer to **900.**

8. $30 - (15 - 5)\ \bigcirc\ (30 - 15) - 5$
$$30 - 10\ \bigcirc\ 15 - 5$$
$$20\ >\ 10$$

9.

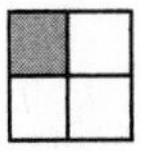

The square is divided into equal parts. One of the 4 parts is shaded. The fraction is $\frac{1}{4}$.

10.
$$\overset{\mathbf{25\%}}{4\overline{)100\%}}$$
$$\underline{8}$$
$$20$$
$$\underline{20}$$
$$0$$

11. C.

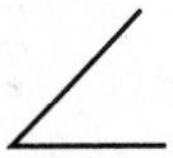

12. On the scale, only every 20 units is labeled. There are 4 spaces between every 20 units, which means that each space equals 5 units. We count up from 30 by fives to **35.**

13.
$$\begin{array}{r} \$7.26 \\ \$14.00 \\ + \;\; \$0.35 \\ \hline \mathbf{\$21.61} \end{array}$$

14.
$$\begin{array}{r} \$10.00 \\ - \;\; \$3.26 \\ \hline \mathbf{\$6.74} \end{array}$$

15. $7 \times 15 \times 20$

105×20

$$\begin{array}{r} 105 \\ \times \;\; 20 \\ \hline \mathbf{2100} \end{array}$$

16.
$$\begin{array}{r} \$3.75 \\ \times \;\; 30 \\ \hline \mathbf{\$112.50} \end{array}$$

17.
$$\begin{array}{r} \mathbf{205} \\ 7\overline{)1435} \\ 14 \\ \hline 03 \\ 00 \\ \hline 35 \\ 35 \\ \hline 0 \end{array}$$

18.
$$\begin{array}{r} \mathbf{71\ R\ 1} \\ 6\overline{)427} \\ 42 \\ \hline 07 \\ 6 \\ \hline 1 \end{array}$$

19.
$$\begin{array}{r} \mathbf{500} \\ 4\overline{)2000} \\ 20 \\ \hline 00 \\ 0 \\ \hline 00 \\ 0 \\ \hline 0 \end{array}$$

20. **B.** 

TEST 8

1. $14 + 9 =$ **23 red roses**

2. There are six spaces between 5 and 6. Each space is one sixth. The arrow points to $\mathbf{5\frac{5}{6}}$.

3. **C.** 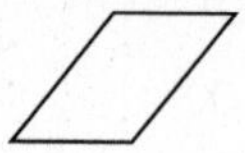

4.  (varies)

5. 

$\frac{1}{2} > \frac{1}{3}$

6. The number 329 is between 300 and 400. The number 329 is closer to **300.**

7. On this scale, only every 10°F is labeled. There are 10 spaces between every 10°F, which means that each space equals 1°F. We count up from 20°F by ones to **26°F.**

8.
$$\begin{array}{r} 238 \text{ pages} \\ - \;\; 112 \text{ pages} \\ \hline \mathbf{126 \text{ pages}} \end{array}$$

9. Seven of the ten triangles are shaded. The fraction is $\mathbf{\frac{7}{10}}$.

10.
$$\begin{array}{r} \$679.00 \\ \$0.43 \\ + \;\; \$3.80 \\ \hline \mathbf{\$683.23} \end{array}$$

11.
$$\begin{array}{r} 7732 \\ - \;\; 2857 \\ \hline \mathbf{4875} \end{array}$$

12.
$$\begin{array}{r} 249 \\ \times \;\; 400 \\ \hline \mathbf{99,600} \end{array}$$

13. $7 \times 15 \times 3$

$105 \times 3 =$ **315**

14.
$$\begin{array}{r} \mathbf{\$1.03} \\ 6\overline{)\$6.18} \\ 6 \\ \hline 0\,1 \\ 0 \\ \hline 18 \\ 18 \\ \hline 0 \end{array}$$

15. $\begin{array}{r} 23 \\ 7\overline{)161} \\ \underline{14} \\ 21 \\ \underline{21} \\ 0 \end{array}$

16. $\begin{array}{r} \mathbf{160\ R\ 1} \\ 3\overline{)481} \\ \underline{3} \\ 18 \\ \underline{18} \\ 01 \\ \underline{0} \\ 1 \end{array}$

17. $\begin{array}{r} \$100.00 \\ -\ \$27.32 \\ \hline \mathbf{\$72.68} \end{array}$

18. $200 - (20 - 2) = 200 - 18 = \mathbf{182}$

19. **B.** 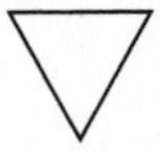 and **D.** 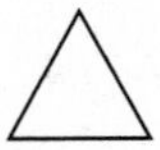

20. **May 16, 2024**

TEST 9

1. **B.** 

2. $\mathbf{1\frac{7}{8}}$ **in.**

3. $\begin{array}{r} \$15.28 \\ -\ \$11.29 \\ \hline \mathbf{\$3.99} \end{array}$

4. Forty-three can be written as $42 + 1$. Half of 42 is 21, and half of 1 is $\frac{1}{2}$, so half of 43 is $\mathbf{21\frac{1}{2}}$.

5. **D.** 

6. Two whole pentagons are shaded, and three fifths of another is shaded. The mixed number is $\mathbf{2\frac{3}{5}}$.

7. 10 millimeters = 1 centimeter
70 millimeters = **7 centimeters**

8. One quarter is twenty-five cents. Three nickels is fifteen cents. Twenty-five cents plus fifteen cents equals forty cents. Forty cents is **40%** of a dollar.

9. Student may draw a diagram (not required).

$\frac{1}{4}$ 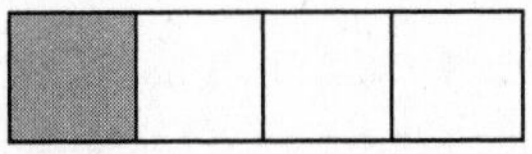

$\frac{1}{10}$ 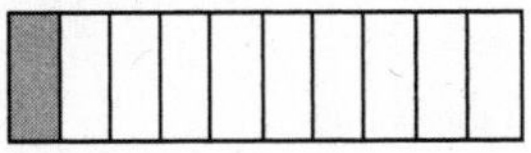

$\frac{1}{4} > \frac{1}{10}$

10. $\begin{array}{r} 3012 \\ -\ 2877 \\ \hline \mathbf{135} \end{array}$

11. $\begin{array}{r} \$8.04 \\ \times\ \ 30 \\ \hline \mathbf{\$241.20} \end{array}$

12. $7 \times 5 \times 7$
$35 \times 7 = \mathbf{245}$

13. $\begin{array}{r} \mathbf{\$12.85} \\ 4\overline{)\$51.40} \\ \underline{4} \\ 11 \\ \underline{8} \\ 3\,4 \\ \underline{3\,2} \\ 20 \\ \underline{20} \\ 0 \end{array}$

14. $\begin{array}{r} \mathbf{387} \\ 9\overline{)3483} \\ \underline{27} \\ 78 \\ \underline{72} \\ 63 \\ \underline{63} \\ 0 \end{array}$

15. $\begin{array}{r} \mathbf{7\,5\,0} \\ 7\overline{)5\,2^{3}5\,0} \end{array}$

16. $\begin{array}{r} 1\frac{1}{5} \\ +\ 4\frac{4}{5} \\ \hline 5\frac{5}{5} \end{array} = \mathbf{6}$

17. $\frac{8}{6} - \frac{3}{6} = \mathbf{\frac{5}{6}}$

18. $\frac{1}{5} - \frac{1}{5} = \mathbf{0}$

19. $40 − ($8.79 + $17 + $0.83)
= $40 − $26.62 = **$13.38**

20.
$$\begin{array}{l} \quad 30 \text{ beans in each bag} \\ \underline{\times \;\; 4 \text{ bags}\qquad\qquad\quad} \\ \textbf{120 beans} \end{array}$$

Test 10

1. Student may draw a diagram (not required).

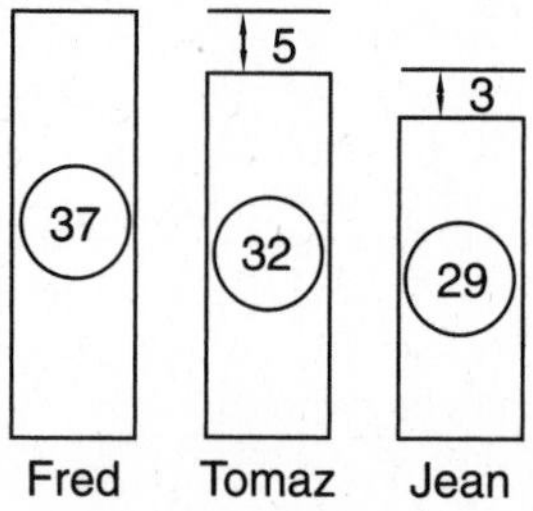

Jean is **29 years old.**

2. (8 × 1000) + (7 × 100) + (9 × 10)
= 8000 + 700 + 90 = **8790**

3. 5 feet = 5 × 12 inches = 60 inches
60 inches + 4 inches = **64 inches**

4.
$$\begin{array}{r} 14 \text{ children} \\ \underline{+\; 22 \text{ children}} \\ 36 \text{ children} \end{array} \qquad \begin{array}{l} N \text{ children in each line} \\ \underline{\times\; 2 \text{ lines}\qquad\qquad\quad} \\ \;\; 36 \text{ children} \end{array}$$

$$\begin{array}{r} \textbf{18 children} \\ 2\overline{)36} \end{array}$$

5.

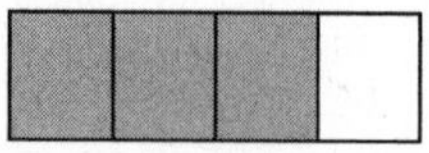

(varies)

$$\begin{array}{r} 25\% \\ 4\overline{)100\%} \\ \underline{8\;\;\;\;} \\ 20\;\; \\ \underline{20\;\;} \\ 0\;\; \end{array}$$

3 × 25% = **75%**

6. $\mathbf{1\frac{3}{4}}$ **inches**

7. 1 centimeter = 10 millimeters
7 centimeters = **70 millimeters**

8. Four hundred seventy-two thousand, three hundred thirty-eight = **472,338**

9.
$$\begin{array}{r} \$32.00 \\ \$14.32 \\ \underline{+\;\; \$6.59} \\ \mathbf{\$52.91} \end{array}$$

10.
$$\begin{array}{r} \$100.00 \\ \underline{-\;\; \$77.35} \\ \mathbf{\$22.65} \end{array}$$

11.
$$\begin{array}{r} 327 \\ \underline{\times \quad 6} \\ \mathbf{1962} \end{array}$$

12.
$$\begin{array}{r} \$9.02 \\ \underline{\times \quad\;\; 30} \\ \mathbf{\$270.60} \end{array}$$

13.
$$\begin{array}{r} 4020 \\ \underline{-\;\; 327} \\ \mathbf{3693} \end{array}$$

14.
$$\begin{array}{r} \mathbf{120\ R\ 5} \\ 6\overline{)725} \\ \underline{6\;\;\;\;} \\ 12\;\; \\ \underline{12\;\;} \\ 05 \\ \underline{0} \\ 5 \end{array}$$

15.
$$\begin{array}{r} \mathbf{428\ R\ 4} \\ 7\overline{)3000} \\ \underline{28\;\;\;\;} \\ 20\;\; \\ \underline{14\;\;} \\ 60 \\ \underline{56} \\ 4 \end{array}$$

16.
$$\begin{array}{r} \mathbf{104} \\ 4\overline{)416} \\ \underline{4\;\;\;\;} \\ 01\;\; \\ \underline{0\;\;} \\ 16 \\ \underline{16} \\ 0 \end{array}$$

17. $\frac{5}{7} - \frac{5}{7} = \mathbf{0}$

18.
$$\begin{array}{r} 4\frac{2}{5} \\ +\ 1\frac{1}{5} \\ \hline \mathbf{5\frac{3}{5}} \end{array}$$

19.
$$\begin{array}{r} 3\frac{2}{3} \\ -\ 2\frac{1}{3} \\ \hline \mathbf{1\frac{1}{3}} \end{array}$$

20. Miranda scored 9 points.
Sarita scored 4 points.
$9 - 4 = 5$
Miranda scored **5 more** points than Sarita.

Test 11

1.
$$\begin{array}{r} 3 \text{ bugs} \\ 8 \text{ bugs} \\ 7 \text{ bugs} \\ +\ 6 \text{ bugs} \\ \hline 24 \text{ bugs} \end{array} \qquad \begin{array}{l} N \text{ bugs in each jar} \\ \times\ 4 \text{ jars} \\ \hline 24 \text{ bugs} \end{array} \qquad \begin{array}{r} \mathbf{6 \text{ bugs}} \\ 4\overline{)24} \end{array}$$

2. A dozen is 12. Two dozen is 24. First divide 24 by 4 to find $\frac{1}{4}$. Then multiply by 3 to find $\frac{3}{4}$.

$$\begin{array}{r} 6 \\ 4\overline{)24} \end{array} \qquad 3 \times 6 = \mathbf{18}$$

Students may draw a diagram (not required).

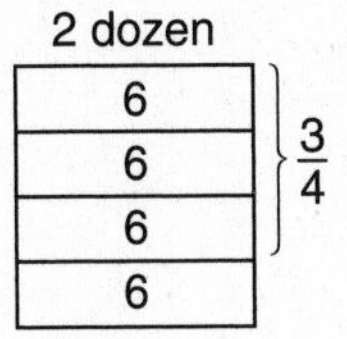

3.
$$\begin{array}{r} 1215 \\ -\ 1066 \\ \hline \mathbf{149 \text{ years}} \end{array}$$

4. The number 923 is between 920 and 930.
The number 923 is closer to **920.**

5. The diameter of a circle is always twice the radius.
$2 \times 20 \text{ mm} = 40 \text{ mm}$
$10 \text{ mm} = 1 \text{ cm}$
$40 \text{ mm} = \mathbf{4 \text{ cm}}$

6. $15 \times 1 = 15 + y$
$15 = 15 + y$
$y = \mathbf{0}$

7. Perimeter $= 8 \text{ cm} + 4 \text{ cm} + 8 \text{ cm} + 4 \text{ cm}$
$= \mathbf{24 \text{ centimeters}}$

8. **90°**

9. $(7 \times 100{,}000) + (2 \times 10{,}000) + (9 \times 100) + (1 \times 1) = \mathbf{720{,}901}$

10. **D.**

11.
$$\begin{array}{r} 2714 \\ 3217 \\ +\ \ 78 \\ \hline \mathbf{6009} \end{array}$$

12.
$$\begin{array}{r} 7623 \\ -\ 2849 \\ \hline \mathbf{4774} \end{array}$$

13.
$$\begin{array}{r} \$2.76 \\ \times\ \ 30 \\ \hline \mathbf{\$82.80} \end{array}$$

14.
$$\begin{array}{r} 53 \\ \times\ 24 \\ \hline 212 \\ 106\ \ \\ \hline \mathbf{1272} \end{array}$$

15.
$$\begin{array}{r} \mathbf{608} \\ 7\overline{)4256} \\ 42\ \ \\ \hline 05\ \\ 0\ \\ \hline 56 \\ 56 \\ \hline 0 \end{array}$$

16.
$$\begin{array}{r} \mathbf{33} \\ 20\overline{)660} \\ 60\ \\ \hline 60 \\ 60 \\ \hline 0 \end{array}$$

17. $6 + \frac{1}{4} = \mathbf{6\frac{1}{4}}$

18.
$$\begin{array}{r} 5\frac{1}{2} \\ -\ 2 \\ \hline \mathbf{3\frac{1}{2}} \end{array}$$

19. 

$$\begin{array}{r} 25\% \\ 4\overline{)100\%} \\ 8 \\ \hline 20 \\ 20 \\ \hline 0 \end{array} \qquad \begin{array}{r} 25\% \\ \times\ 1 \\ \hline \mathbf{25\%} \end{array}$$

20. First, place the commas.
32876534 → 32,876,534
32,876,534 = **thirty-two million, eight hundred seventy-six thousand, five hundred thirty-four**

Test 12

1. Three sectors have a number greater than 2: 3, 4, and 5. Each of these sectors is $\frac{1}{5}$ of the circle, so the probability that the spinner will stop on a number greater than 2 is

$$\frac{1}{5} + \frac{1}{5} + \frac{1}{5} = \mathbf{\frac{3}{5}}$$

2.
$$\begin{array}{r} \mathbf{59} \\ 6\overline{)354} \\ 30 \\ \hline 54 \\ 54 \\ \hline 0 \end{array}$$

3. Fifty-two million, six hundred eighty-seven thousand, three hundred twenty = **52,687,320**

4. **8**

5.
$$\begin{array}{r} 17 \text{ pens} \\ 7 \text{ pens} \\ 11 \text{ pens} \\ +\ 5 \text{ pens} \\ \hline 40 \text{ pens} \end{array} \qquad \begin{array}{r} N \text{ pens in box} \\ \times\ 4 \text{ boxes} \\ \hline 40 \text{ pens} \end{array} \qquad \begin{array}{r} \mathbf{10 \text{ pens}} \\ 4\overline{)40} \end{array}$$

6. Perimeter = 11 in. + 7 in. + 9 in. = **27 in.**

7. The diameter of a circle is always twice the radius.
2 × 8 cm = **16 cm**

8. $\frac{5}{5} - \frac{2}{5} = \mathbf{\frac{3}{5}}$

Student may draw a diagram (not required).

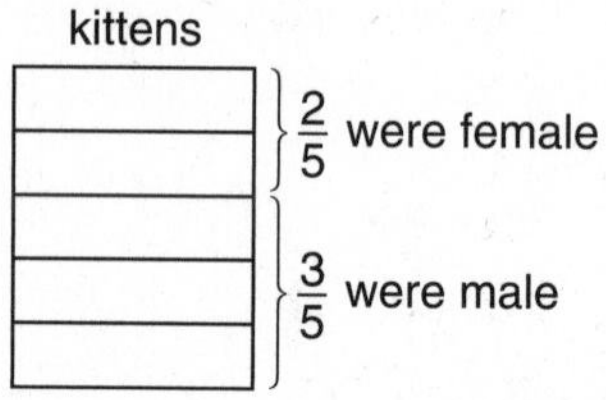

9. A 50% chance of rain would mean that it is as likely to rain as not to rain. 75% is greater than 50%, so rain is **C. likely.**

10.
$$\begin{array}{r} \mathbf{3 \text{ free throws}} \\ 5\overline{)15} \end{array}$$

11.
$$\begin{array}{r} 7832 \\ 273 \\ +\ 917 \\ \hline \mathbf{9022} \end{array}$$

12.
$$\begin{array}{r} 2420 \\ -\ 1903 \\ \hline \mathbf{517} \end{array}$$

13.
$$\begin{array}{r} 239 \\ \times\ 621 \\ \hline 239 \\ 478 \\ 1434 \\ \hline \mathbf{148{,}419} \end{array}$$

14.
$$\begin{array}{r} 805 \\ \times\ 220 \\ \hline 16100 \\ 161000 \\ \hline \mathbf{177{,}100} \end{array}$$

15.
$$\begin{array}{r} \mathbf{76\ R\ 3} \\ 10\overline{)763} \\ 70 \\ \hline 63 \\ 60 \\ \hline 3 \end{array}$$

16. $7\overline{)7023}$ = **$1003\frac{2}{7}$**

```
   1003 2/7
7)7023
  7
  00
   0
   02
    0
    23
    21
     2
```

17. $30\overline{)\$3.90}$ = **$0.13**

```
    $0.13
30)$3.90
    3 0
      90
      90
       0
```

18. $7\frac{3}{5} + 2\frac{2}{5} = 9\frac{5}{5} = \mathbf{10}$

19. $1 - \frac{1}{5} = \frac{5}{5} - \frac{1}{5} = \mathbf{\frac{4}{5}}$

20. $30 − ($17 + $9.78 + $0.23)

= $30 − $27.01

```
  $30.00
− $27.01
  $2.99
```

$2.99

Test 13

1.
```
  17 oranges        N oranges in each basket
+ 23 oranges      × 2 baskets
  40 oranges        40 oranges
```

```
  20 oranges
2)40
  4
  00
   0
   0
```

20 oranges

2. $\frac{5}{5} - \frac{1}{5} = \mathbf{\frac{4}{5}}$

Student may draw a diagram (not required).

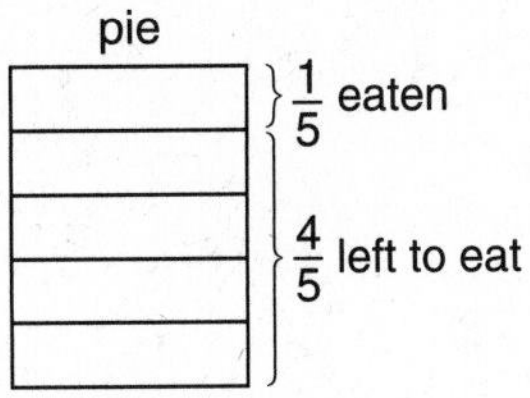

3.
```
  1956          10 years in each decade
− 1826        × N decades
  130 years     130 years
```

```
  13 decades
10)130
   10
    30
    30
     0
```

13 decades

4. The number 66 is closer to 70 than to 60.
The number 22 is closer to 20 than to 30.

```
  70
× 20
1400
```

1400

5. The number 1239 is between 1200 and 1300.
The number 1239 is closer to **1200.**

6. **A. ∠ *AMB***

7. The distance between whole numbers is divided into ten spaces, or tenths. The arrow is pointing to the mark seven spaces to the right of 7, which is the mixed number $\mathbf{7\frac{7}{10}}$.

8. Of the six equally probable outcomes of rolling a number cube, only one results in a five. Thus, the probabilitity that the cube will land with a five on top is $\mathbf{\frac{1}{6}}$.

9. (a) **5 cm**

(b) Perimeter =
5 cm + 3 cm + 5 cm + 3 cm = **16 cm**

10. $9\overline{)63}$ = **7**

11. $\frac{3}{5} + \frac{2}{5} = \frac{5}{5} = \mathbf{1}$

12. $7 - 1\frac{2}{3} = 6\frac{3}{3} - 1\frac{2}{3} = \mathbf{5\frac{1}{3}}$

13. $3 - \frac{1}{5} = 2\frac{5}{5} - \frac{1}{5} = \mathbf{2\frac{4}{5}}$

14.
$$\begin{array}{r} 3821 \\ -\ \ 930 \\ \hline \mathbf{2891} \end{array}$$

15. $20 − ($8 + $5.47) = $20 − $13.47

$$\begin{array}{r} \$20.00 \\ -\ \$13.47 \\ \hline \mathbf{\$6.53} \end{array}$$

16.
$$\begin{array}{r} 125 \\ \times\ 432 \\ \hline 250 \\ 375 \\ 500 \\ \hline \mathbf{54{,}000} \end{array}$$

17.
$$\begin{array}{r} \mathbf{704\ R\ 1} \\ 8\overline{)5633} \\ 56 \\ \hline 03 \\ 0 \\ \hline 33 \\ 32 \\ \hline 1 \end{array}$$

18.
$$\begin{array}{r} \mathbf{17} \\ 20\overline{)340} \\ 20 \\ \hline 140 \\ 140 \\ \hline 0 \end{array}$$

19. The 8 is in the first place to the right of the decimal point, the **tenths** place.

20.
$$\begin{array}{r} \mathbf{7\frac{3}{8}} \\ 8\overline{)59} \\ 56 \\ \hline 3 \end{array}$$

TEST 14

1.
$$\begin{array}{r} \$0.74 \\ \times\quad 3 \\ \hline \$2.22 \end{array} \qquad \begin{array}{r} \$5.00 \\ -\ \$2.22 \\ \hline \mathbf{\$2.78} \end{array}$$

2. Factors of 16: ①,②,④,⑧, 16
Factors of 24: ①,②, 3, ④, 6, ⑧, 12, 24
Common Factors: **1, 2, 4, 8**

3. (a) $\mathbf{\frac{27}{100}}$

(b) **0.27**

4. 76.25 = **seventy-six and twenty-five hundredths**

5. $\frac{87}{100} = \mathbf{0.87}$

6. Fourteen and twenty-three hundredths = **14.23**

7. The tenths place is the first place to the right of the decimal, so the digit is **3.**

8. **4.2 cm**

9. 0.4 > 0.04

10. The number 27 is closer to 30 than to 20.
The number 41 is closer to 40 than to 50.

$$\begin{array}{r} 30 \\ \times\ 40 \\ \hline \mathbf{1200} \end{array}$$

11.
$$\begin{array}{r} 3519 \\ 274 \\ 38 \\ 12 \\ +\quad 6 \\ \hline \mathbf{3849} \end{array}$$

12.
$$\begin{array}{r} 30{,}000 \\ -\ 21{,}325 \\ \hline \mathbf{8{,}675} \end{array}$$

13.
$$\begin{array}{r} 227 \\ \times\ 160 \\ \hline 13620 \\ 22700 \\ \hline \mathbf{36{,}320} \end{array}$$

14. $6\overline{)4248}$ = **708**

$$\begin{array}{r} 42 \\ \hline 04 \\ 0 \\ \hline 48 \\ 48 \\ \hline 0 \end{array}$$

15.
$$\begin{array}{r} 35 \\ \times\ 20 \\ \hline 700 \end{array} \qquad \begin{array}{r} 700 \\ \times\ 3 \\ \hline \mathbf{2100} \end{array}$$

16. $5\overline{)850}$ = **170**

$$\begin{array}{r} 5 \\ \hline 35 \\ 35 \\ \hline 00 \\ 0 \\ \hline 0 \end{array}$$

17. $40\overline{)\$95.60}$ = **$2.39**

$$\begin{array}{r} 80 \\ \hline 15\,6 \\ 12\,0 \\ \hline 3\,60 \\ 3\,60 \\ \hline 0 \end{array}$$

18. $7 - \left(5\frac{2}{3} - 4\right) = 7 - 1\frac{2}{3}$

$= 6\frac{3}{3} - 1\frac{2}{3} = \mathbf{5\frac{1}{3}}$

19.
$$\begin{array}{r} 1\frac{1}{5} \\ +\ 2\frac{2}{5} \\ \hline 3\frac{3}{5} \end{array} \qquad \begin{array}{r} 3\frac{3}{5} \\ +\ 3\frac{2}{5} \\ \hline 6\frac{5}{5} \end{array} = \mathbf{7}$$

20. $6\overline{)29}$ = $\mathbf{4\frac{5}{6}}$

$$\begin{array}{r} 24 \\ \hline 5 \end{array}$$

Test 15

1. One approach is to compare the fractions to $\frac{1}{2}$ and 1.

$\frac{1}{4} < \frac{1}{2} \qquad \frac{6}{6} = 1 \qquad \frac{4}{8} = \frac{1}{2} \qquad \frac{2}{3} > \frac{1}{2}$

$\mathbf{\frac{1}{4}, \frac{4}{8}, \frac{2}{3}, \frac{6}{6}}$

2. $3\overline{)8}$ = $\mathbf{2\frac{2}{3}}$

$$\begin{array}{r} 6 \\ \hline 2 \end{array}$$

3.
$$\begin{array}{rl} & 3 \text{ feet in each yard} \\ \times & 8 \text{ yards} \\ \hline & 24 \text{ feet} \end{array}$$

$$\begin{array}{rl} & 12 \text{ inches in each foot} \\ \times & 24 \text{ feet} \\ \hline & 48 \\ & 24\ \ \\ \hline & \mathbf{288 \text{ inches}} \end{array}$$

4. Thirteen and fifty-two hundredths = **13.52**

5. The hundredths place is the second place to the right of the decimal, so the digit is **9.**

6. (a) $\mathbf{\frac{79}{100}}$

(b) $\frac{79}{100}$ = **0.79**

(c) **79%**

7. The number 8734 is between 8700 and 8800. The number 8734 is closer to **8700.**

8. 8.42 **=** 8.420

9. $40\overline{)873}$ = $\mathbf{21\frac{33}{40}}$

$$\begin{array}{r} 80 \\ \hline 73 \\ 40 \\ \hline 33 \end{array}$$

10.
$$\begin{array}{r} \$9.28 \\ \$0.97 \\ +\ \$2.00 \\ \hline \mathbf{\$12.25} \end{array}$$

11. $\begin{array}{r} \$1.22 \\ -\ \$0.74 \\ \hline \mathbf{\$0.48} \end{array}$

12. $\begin{array}{r} 8.26 \\ 1.40 \\ +\ 0.72 \\ \hline \mathbf{10.38} \end{array}$

13. $\begin{array}{r} 4.892 \\ -\ 1.900 \\ \hline \mathbf{2.992} \end{array}$

14. $\begin{array}{r} \$0.46 \\ \times\ \ \ 3 \\ \hline \mathbf{\$1.38} \end{array}$

15. $8 - \left(7\frac{4}{5} - \frac{3}{5}\right) = 8 - 7\frac{1}{5}$
$= 7\frac{5}{5} - 7\frac{1}{5} = \mathbf{\frac{4}{5}}$

16. $\begin{array}{r} 2\frac{3}{5} \\ +\ 3\frac{2}{5} \\ \hline 5\frac{5}{5} \end{array} = \mathbf{6}$

17. length of $\overline{PQ}$ + length of $\overline{QR}$ = length of $\overline{PR}$

$\begin{array}{r} 50 \text{ mm} \\ +\ 20 \text{ mm} \\ \hline \mathbf{70 \text{ mm}} \end{array}$

10 mm = 1 cm
70 mm = **7 cm**

18. 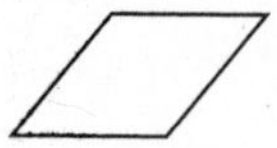 (varies)

19. Perimeter = 6 yd + 5 yd + 6 yd + 5 yd
= **22 yd**

20. Area = 6 yd × 5 yd = **30 sq. yd**

TEST 16

1. $\begin{array}{r} \$3.40 \\ \times\ \ \ 3 \\ \hline \$10.20 \end{array}$ $\begin{array}{r} \$10.20 \\ +\ \$0.85 \\ \hline \mathbf{\$11.05} \end{array}$

2. $3^4 = 3 \times 3 \times 3 \times 3$
$9 \times 9 = \mathbf{81}$

3. $\frac{9}{100} = \mathbf{0.09}$

4. 1000 meters in a kilometer
× 7.2 kilometers
7200 meters

7200 meters
× 100 centimeters in a meter
720,000 centimeters

5. 1 ton = 2000 pounds
6 tons = 6 × 2000 pounds = **12,000 pounds**

6. Both **5** and 4 are in the tenths place.

7. $\frac{1}{2} \times \frac{6}{6} \bigcirc \frac{2}{4}$
$\frac{1}{2} \mathbf{=} \frac{1}{2}$

8. $\frac{3}{5} \times \frac{2}{2} = \mathbf{\frac{6}{10}}$

9. A prime number is a whole number that has exactly two factors. The next four prime numbers are **3, 5, 7,** and **11.**

10. Each small space between each whole number equals one tenth of a centimeter. This line segment is 4 cm and 4 tenths, or **4.4 cm** long.

11. $\begin{array}{r} \$3.00 \\ \$2.87 \\ \$0.84 \\ \$16.00 \\ +\ \ \$0.06 \\ \hline \mathbf{\$22.77} \end{array}$

12. $\begin{array}{r} 21.72 \\ 2.20 \\ +\ 17.30 \\ \hline \mathbf{41.22} \end{array}$

13. $\begin{array}{r} 87.56 \\ -\ \ 8.70 \\ \hline \mathbf{78.86} \end{array}$

14.
$$\begin{array}{r} 317 \\ 30\overline{)9510} \\ \underline{90} \\ 51 \\ \underline{30} \\ 210 \\ \underline{210} \\ 0 \end{array}$$

15.
$$\begin{array}{r} \$6.39 \\ \times \quad 4 \\ \hline \mathbf{\$25.56} \end{array}$$

16.
$$\begin{array}{r} 20 \\ \times\ 40 \\ \hline 800 \end{array} \qquad \begin{array}{r} 800 \\ \times\ \ 60 \\ \hline \mathbf{48{,}000} \end{array}$$

17.
$$\begin{array}{r} \mathbf{\$14.25} \\ 4\overline{)\$57.00} \\ \underline{4} \\ 17 \\ \underline{16} \\ 1\,0 \\ \underline{8} \\ 20 \\ \underline{20} \\ 0 \end{array}$$

18. $\frac{4}{5} + \frac{2}{5} = \frac{6}{5} = \mathbf{1\frac{1}{5}}$

19. $\frac{2}{3} \times \frac{1}{3} = \mathbf{\frac{2}{9}}$

20.
$$\begin{array}{r} \text{length of } \overline{EF} \\ +\ \text{length of } \overline{FG} \\ \hline \text{length of } \overline{EG} \end{array} \qquad \begin{array}{r} 34 \text{ mm} \\ -\ N \text{ mm} \\ \hline 95 \text{ mm} \end{array} \qquad \begin{array}{r} 95 \text{ mm} \\ -\ 34 \text{ mm} \\ \hline \mathbf{61 \text{ mm}} \end{array}$$

TEST 17

1. **B. 3**

2. An obtuse angle is larger than a right angle, so **∠*ADC*** (or **∠*CDA***) is obtuse.

3. Perpendicular lines are always at a right angle to each other. $\overline{CB}$ (or $\overline{BC}$) is perpendicular to $\overline{CD}$.

4. **B. cone**

5. (a)
$$\begin{array}{r} N \text{ in each group} \\ \times\ 3 \text{ groups} \\ \hline 24 \end{array} \qquad \begin{array}{r} \mathbf{8} \\ 3\overline{)24} \\ \underline{24} \\ 0 \end{array}$$

(b) $\frac{1}{3}$ of 24 = 8

$\frac{2}{3}$ of 24 = 2 × 8 = **16**

6.
$$\begin{array}{r} N \text{ houses in each group} \\ \times\ 6 \text{ groups} \\ \hline 18 \text{ houses} \end{array}$$

$$\begin{array}{r} \mathbf{3 \text{ houses were brick}} \\ 6\overline{)18} \\ \underline{18} \\ 0 \end{array}$$

7. (a) Factors of 16: ①,②,④,⑧, 16

Factors of 24: ①,②, 3, ④, 6, ⑧, 12, 24

Greatest Common Factor: **8**

(b) $\frac{16 \div 8}{24 \div 8} = \mathbf{\frac{2}{3}}$

8.
$$\begin{array}{r} 4 \text{ quarts in a gallon} \\ \times\ 3 \text{ gallons} \\ \hline \mathbf{12 \text{ quarts}} \end{array}$$

9. (a) $\frac{4 \div 4}{16 \div 4} = \mathbf{\frac{1}{4}}$

(b) $\frac{14 \div 7}{21 \div 7} = \mathbf{\frac{2}{3}}$

(c) $\frac{3 \div 3}{6 \div 3} = \mathbf{\frac{1}{2}}$

10.
$$\begin{array}{r} 37.850 \\ 6.629 \\ +\ 263.400 \\ \hline \mathbf{307.879} \end{array}$$

11.
$$\begin{array}{r} 29.684 \\ -\ 17.790 \\ \hline \mathbf{11.894} \end{array}$$

12. $6^3 = 6 \times 6 \times 6$

$36 \times 6 = \mathbf{216}$

13. $4\overline{)\$35.00}$ = **\$8.75**

$$\begin{array}{r} 32 \\ \hline 3\,0 \\ 2\,8 \\ \hline 20 \\ 20 \\ \hline 0 \end{array}$$

14. $$\begin{array}{r} 7010 \\ -\ 3976 \\ \hline \mathbf{3034} \end{array}$$

15. $$\begin{array}{r} 450 \\ \times\ 803 \\ \hline 1350 \\ 360000 \\ \hline \mathbf{361{,}350} \end{array}$$

16. $60\overline{)5220}$ = **87**

$$\begin{array}{r} 480 \\ \hline 420 \\ 420 \\ \hline 0 \end{array}$$

17. $$\begin{array}{r} 5\frac{4}{5} \\ +\ 2\frac{4}{5} \\ \hline 7\frac{8}{5} \end{array} = 7 + \frac{8}{5} = 7 + 1\frac{3}{5} = \mathbf{8\frac{3}{5}}$$

18. $5 - \left(\frac{1}{4} + 2\right) = 5 - 2\frac{1}{4}$

$= 4\frac{4}{4} - 2\frac{1}{4} = \mathbf{2\frac{3}{4}}$

19. $\frac{2}{5} \times \frac{4}{4} \bigcirc \frac{2}{5} \times \frac{2}{2}$

$\frac{2}{5} \; \mathbf{=} \; \frac{2}{5}$

20. The sides of a square are all the same length.
Area = 2 cm × 2 cm = **4 sq. cm**

TEST 18

1. Jermael should start for the theater 25 minutes before 7:45 p.m., which is **7:20 p.m.**

2. $5\overline{)12}$ = $\mathbf{2\frac{2}{5}}$

$$\begin{array}{r} 10 \\ \hline 2 \end{array}$$

3. $\frac{12 \div 3}{15 \div 3} = \mathbf{\frac{4}{5}}$

4. There are 2 pints in 1 quart, 2 cups in 1 pint, and 8 ounces in 1 cup, so there are
2 × 2 × 8 ounces = **32 ounces** in one quart.

5. Parallel lines are always the same distance apart, $\overline{CB}$ (or $\overline{BC}$) is parallel to $\overline{AD}$.

6. An acute angle is smaller than a right angle, so ∠***DCB*** (or ∠***BCD***) is an acute angle.

7. **B. cylinder**

8. $\frac{4}{5}$ of $30 = \frac{4}{5} \times \frac{30}{1} = \frac{120}{5}$ $\quad 5\overline{)120}$ = **24**

$$\begin{array}{r} 10 \\ \hline 20 \\ 20 \\ \hline 0 \end{array}$$

9. Arranging the scores in numerical order gives us:
80, 80, 80, 90, 95, 95, 100
The middle number, or median, of the seven test scores is **90.**

10. $\frac{3}{4} \times \frac{3}{3} = \mathbf{\frac{9}{12}}$

11. 0.625 = **six hundred twenty-five thousandths**

12. $$\begin{array}{r} 84{,}016 \\ -\ 8{,}127 \\ \hline \mathbf{75{,}889} \end{array}$$

13. $$\begin{array}{r} 907 \\ \times\ 120 \\ \hline 18140 \\ 90700 \\ \hline \mathbf{108{,}840} \end{array}$$

14. $\begin{array}{r} \$3.81 \\ 4\overline{)\$15.24} \\ \underline{12} \\ 3\,2 \\ \underline{3\,2} \\ 04 \\ \underline{4} \\ 0 \end{array}$

15. $\frac{2}{5} \times \frac{1}{3} = \mathbf{\frac{2}{15}}$

16. The square root of 49 is **7,** because $7 \times 7 = 49$.

17. $\frac{3}{5} \div \frac{3}{5} = \mathbf{1}$

18. $\frac{7}{8} - \frac{3}{8} = \frac{4 \div 4}{8 \div 4} = \mathbf{\frac{1}{2}}$

19. $2\frac{1}{6} + 3\frac{1}{6} = 5\frac{2}{6}$

$\frac{2 \div 2}{6 \div 2} = \frac{1}{3}$

$5\frac{2}{6} = \mathbf{5\frac{1}{3}}$

20. The sides of a square are all the same length.
Perimeter = 20 mm + 20 mm + 20 mm + 20 mm = **80 mm**

Test 19

1. $\mathbf{\frac{4}{5}}$

2. $\frac{1}{3} \times \frac{5}{5} = \mathbf{\frac{5}{15}}$

$\frac{1}{5} \times \frac{3}{3} = \mathbf{\frac{3}{15}}$

$\frac{5}{15} + \frac{3}{15} = \mathbf{\frac{8}{15}}$

3. 2 dozen cookies = 24 cookies

$\begin{array}{l} N \text{ cookies in each row} \\ \underline{\times \ \ 6 \text{ rows}} \\ 24 \text{ cookies} \end{array}$

4 cookies in each row

$6\overline{)24}$

4. (a) $\frac{1}{5}$ of 70 $= \frac{1}{5} \times \frac{70}{1} = \frac{70}{5} = \mathbf{14}$

(b) $\frac{3}{5}$ of 70 $= 3 \times 14 = \mathbf{42}$

5. (a) $\frac{5 \div 5}{10 \div 5} = \mathbf{\frac{1}{2}}$

(b) $\frac{5}{10} = \mathbf{0.5}$

(c) $\frac{5}{10}$ of 100% = **50%**

6. One approach is to compare each number to the first number listed, three tenths.

$0.3 = \frac{3}{10}$ $\quad \frac{2}{3} > \frac{3}{10}$ $\quad 0 < \frac{3}{10}$

0, 0.3, $\mathbf{\frac{2}{3}}$

7. (a) **1.7 centimeters**

(b) **17 millimeters**

8. $\begin{array}{r} 1\frac{10}{14} \\ 14\overline{)24} \\ \underline{14} \\ 10 \end{array}$ $\quad \frac{10 \div 2}{14 \div 2} = \frac{5}{7}$

$1\frac{10}{14} = \mathbf{1\frac{5}{7}}$

9. $\frac{2}{3} = \frac{x}{12}$

$\frac{2}{3} \times \frac{4}{4} = \frac{8}{12}$

$x = \mathbf{8}$

10. $\begin{array}{r} 27.20 \\ 9.43 \\ \underline{+\ 17.50} \\ \mathbf{54.13} \end{array}$

11. $\begin{array}{r} 5201 \\ \underline{-\ 3839} \\ \mathbf{1362} \end{array}$

12. $\begin{array}{r} \$6.25 \\ \underline{\times \quad 24} \\ 2500 \\ \underline{1250} \\ \mathbf{\$150.00} \end{array}$

13.
$$\begin{array}{r} \mathbf{\$2.20} \\ 10\overline{)\$22.00} \\ \underline{20} \\ 2\,0 \\ \underline{2\,0} \\ 00 \\ \underline{0} \\ 0 \end{array}$$

14. The square root of 36 is **6,** because $6 \times 6 = 36$.

15.
$$\begin{array}{r} \mathbf{52} \\ 16\overline{)832} \\ \underline{80} \\ 32 \\ \underline{32} \\ 0 \end{array}$$

16.
$$\begin{array}{r} 7\frac{7}{8} \\ -\ 2\frac{5}{8} \\ \hline 5\frac{2}{8} \end{array} = \mathbf{5\frac{1}{4}}$$

17.
$$\begin{array}{r} 2\frac{4}{5} \\ +\ 3\frac{4}{5} \\ \hline 5\frac{8}{5} \end{array} = 5 + 1\frac{3}{5} = \mathbf{6\frac{3}{5}}$$

18. $\frac{3}{4} \times \frac{5}{1} = \frac{15}{4}$

$$\begin{array}{r} \mathbf{3\frac{3}{4}} \\ 4\overline{)15} \\ \underline{12} \\ 3 \end{array}$$

19.
$$\begin{array}{r} \$7.83 \\ \$0.59 \\ +\ \$14.00 \\ \hline \mathbf{\$22.42} \end{array}$$

20.
$$\begin{array}{r} \$0.35 \\ \times\ \ \ 14 \\ \hline 140 \\ 35 \\ \hline \mathbf{\$4.90} \end{array}$$

TEST 20

1. $24 \div 3 = 8$

$12 \div 3 = 4$

$3 \div 3 = 1$

D. It divides by 3.

2. $9 \div 3 = \mathbf{3}$

3. $\frac{\text{girls}}{\text{boys}} = \frac{12}{15} = \mathbf{\frac{4}{5}}$

4. 1 centimeter = 10 millimeters

12 centimeters = 12×10 millimeters

= **120 millimeters**

5.
$$\begin{array}{r} 3.4 \text{ cm} \\ 1.4 \text{ cm} \\ +\ 1.4 \text{ cm} \\ \hline \mathbf{6.2\ cm} \end{array}$$

6. (a) 16 ounces = 1 pint **1 ounce =** $\mathbf{\frac{1}{16}}$ **pint**

(b) 2 pints = 1 quart **1 pint =** $\mathbf{\frac{1}{2}}$ **quart**

(c) $\frac{1}{16} \times \frac{1}{2} = \frac{1}{32}$ **1 ounce =** $\mathbf{\frac{1}{32}}$ **quart**

7.
$$\begin{array}{r} 300 \\ \times\ \ 700 \\ \hline \mathbf{210{,}000} \end{array}$$

8. 00021.090 = **21.09**

9. $\mathbf{\frac{4}{9}}$

10. $\frac{12 \div 3}{21 \div 3} = \frac{4}{7}$

$8\frac{12}{21} = \mathbf{8\frac{4}{7}}$

11.
$$\begin{array}{r} 8.50 \\ 0.31 \\ +\ 2.00 \\ \hline \mathbf{10.81} \end{array}$$

12. $2^6 = 2 \times 2 \times 2 \times 2 \times 2 \times 2$

$4 \times 4 \times 4$

$16 \times 4 = \mathbf{64}$

13. $\begin{array}{r} 7.982 \\ -\ 0.350 \\ \hline \mathbf{7.632} \end{array}$

14. $\begin{array}{r} \$0.97 \\ \times\ \ \ 10 \\ \hline \mathbf{\$9.70} \end{array}$

15. The square root of 64 is **8,** because $8 \times 8 = 64$.

16. $34\overline{)700}$ = **20 R 20**

$\begin{array}{r} 68 \\ \hline 20 \\ 0 \\ \hline 20 \end{array}$

17. $9 - \left(3\frac{2}{5} + 1\frac{1}{5}\right) = 9 - 4\frac{3}{5}$

$= 8\frac{5}{5} - 4\frac{3}{5} = \mathbf{4\frac{2}{5}}$

18. $\frac{3}{5} \div \frac{2}{3} = \frac{3}{5} \times \frac{3}{2} = \mathbf{\frac{9}{10}}$

19. $\frac{5}{6} + \frac{5}{6} = \frac{10}{6} = 1\frac{4}{6} = \mathbf{1\frac{2}{3}}$

20. $\frac{2}{3} \times \frac{2}{1} = \frac{4}{3} = \mathbf{1\frac{1}{3}}$

TEST 21

1. The value \$12.64 is between \$12 and \$13. Since \$0.64 is greater than \$0.50, \$12.64 is closer to **\$13.**

2. (a) The number 18.3 is between 18 and 19. Since 0.3 is less than 0.5, 18.3 is closer to **18.**

(b) The number $7\frac{2}{3}$ is between 7 and 8. Since $\frac{2}{3}$ is greater than $\frac{1}{2}$, $7\frac{2}{3}$ is closer to **8.**

3. $0 < 0.2 < 0.8$

0, 0.2, 0.8

4. $3\overline{)24}$ = 8 $\quad 2 \times 8$ books = **16 books**

Student may draw a diagram (not required).

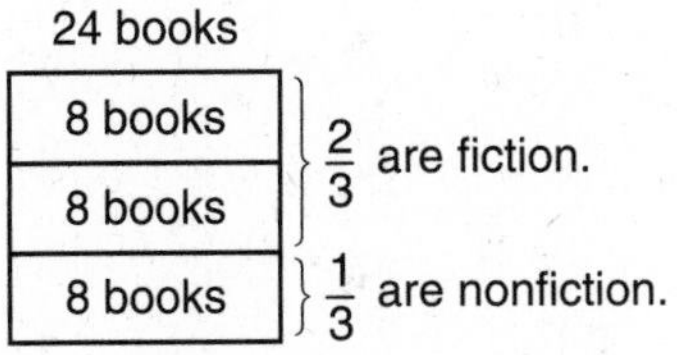

5. $2.0 + 3.8 + CD = 7.9$

$5.8 + CD = 7.9$

$\begin{array}{r} 7.9\text{ cm} \\ -\ 5.8\text{ cm} \\ \hline \mathbf{2.1\text{ cm}} \end{array}$

6. A basketball is shaped like a **sphere.** 

7. From the origin go to the left 3 and up 2 to reach **point *D*.**

8. $\begin{array}{r} 9.75 \\ 2.30 \\ +\ 4.00 \\ \hline \mathbf{16.05} \end{array}$

9. $\begin{array}{r} 5.17 \\ -\ 0.60 \\ \hline \mathbf{4.57} \end{array}$

10. $\begin{array}{r} 5.0 \\ -\ 3.4 \\ \hline \mathbf{1.6} \end{array}$

11. $\begin{array}{r} 144 \\ \times\ 24 \\ \hline 576 \\ 288\ \\ \hline \mathbf{3456} \end{array}$

12. $5 - \left(2\frac{7}{8} - \frac{5}{8}\right) = 5 - 2\frac{2}{8}$

$= 4\frac{8}{8} - 2\frac{2}{8} = 2\frac{6}{8} = \mathbf{2\frac{3}{4}}$

13. $2\frac{6}{8} + 1\frac{6}{8} = 3\frac{12}{8} = 4\frac{4}{8} = \mathbf{4\frac{1}{2}}$

14. $7\overline{)7014}$ = **1002**

$$\begin{array}{r} 7 \\ \hline 00 \\ 0 \\ \hline 01 \\ 0 \\ \hline 14 \\ 14 \\ \hline 0 \end{array}$$

15. $30\overline{)426}$ = **14 R 6**

$$\begin{array}{r} 30 \\ \hline 126 \\ 120 \\ \hline 6 \end{array}$$

16. $\frac{1}{4} \div \frac{1}{3} = \frac{1}{4} \times \frac{3}{1} = \mathbf{\frac{3}{4}}$

17. $\frac{3}{4} \times \frac{4}{1} = \frac{12}{4} = \mathbf{3}$

18. Volume = 4 in. × 2 in. × 2 in. = **16 cu. in.**

19. 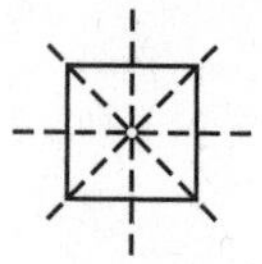

20. $\frac{1}{2} \times \frac{5}{5} = \frac{5}{10}$

$\frac{9}{10} - \frac{5}{10} = \frac{4}{10} = \mathbf{\frac{2}{5}}$

TEST 22

1. The thousandths place is the third place to the right of the decimal, so the digit is **8.**

2. 1 yard = 3 feet
25 yards = 25 × 3 feet = **75 feet**

3. **A.** 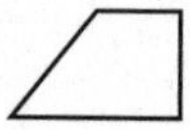

4. $\frac{\text{girls}}{\text{boys}} = \frac{8}{16} = \mathbf{\frac{1}{2}}$

5. $10\% = \frac{10}{100}$

$\frac{10 \div 10}{100 \div 10} = \mathbf{\frac{1}{10}}$

6. $\frac{16}{50} \times \frac{2}{2} = \frac{32}{100} = \mathbf{32\%}$

7. (a) The number $7\frac{4}{9}$ is between 7 and 8. Since $\frac{4}{9}$ is less than $\frac{1}{2}$, $7\frac{4}{9}$ is closer to **7.**

(b) The number 32.701 is between 32 and 33. Since 0.701 is greater than 0.5, 32.701 is closer to **33.**

8. 0.036 **<** 0.36

9. Perimeter =
13 in. + 13 in. + 13 in. + 13 in. = **52 in.**

10. Area = 13 in. × 13 in. = **169 sq. in.**

11. $$\begin{array}{r} 14.000 \\ 6.820 \\ 0.927 \\ +\ \ 3.000 \\ \hline \mathbf{24.747} \end{array}$$

12. $$\begin{array}{r} 4.6 \\ -\ 1.0 \\ \hline \mathbf{3.6} \end{array}$$

13. $$\begin{array}{r} 7.00 \\ -\ 2.91 \\ \hline \mathbf{4.09} \end{array}$$

14. $$\begin{array}{r} 2.4 \\ \times\ 0.3 \\ \hline \mathbf{0.72} \end{array}$$

15. $$\begin{array}{r} 0.16 \\ \times\ \ 0.6 \\ \hline \mathbf{0.096} \end{array}$$

16. $2\frac{2}{5} + \left(7 - 3\frac{2}{5}\right)$

$7 - 3\frac{2}{5} = 6\frac{5}{5} - 3\frac{2}{5} = 3\frac{3}{5}$

$2\frac{2}{5} + 3\frac{3}{5} = 5\frac{5}{5} = \mathbf{6}$

17. $\frac{4}{5} \times \left(5 \times \frac{1}{4}\right)$

$\frac{5}{1} \times \frac{1}{4} = \frac{5}{4}$

$\frac{4}{5} \times \frac{5}{4} = \frac{20}{20} = \mathbf{1}$

18. $2\frac{2}{3} + 2\frac{2}{3} = 4\frac{4}{3} = \mathbf{5\frac{1}{3}}$

19. $\frac{9}{10} \div \frac{3}{1} = \frac{9}{10} \times \frac{1}{3} = \frac{9}{30} = \mathbf{\frac{3}{10}}$

20. Volume =
8 in. × 3 in. × 10 in. = **240 cu. in.**

Test 23

1. The number 6.38 is closer to 6 than 7.
The number 5.83 is closer to 6 than 5.
The number 7.66 is closer to 8 than 7.
$6 + 6 + 8 = \mathbf{20}$

2. **C. 40%**

3. On the scale, only every 10°F is labeled. There are 10 spaces between every 10°F, which means that every space equals 1°F. We count down from 0°F by ones to **–6°F.**

4. Go to the left 1 and down 2 to reach **point *E*.**

5. *G* is 2 to the left and up 1 at **(–2, 1).**

6. $13 + 10 + 9 + 5 + 8 = 45$
$45 \div 5 = \mathbf{9}$

7. $\frac{\text{3 odd numbers}}{\text{6 numbers total}} = \frac{3}{6} = \mathbf{\frac{1}{2}}$

8. (a) Perimeter = 2 cm + 3 cm + 3 cm + 3 cm + 5 cm + 6 cm = **22 cm**

(b)

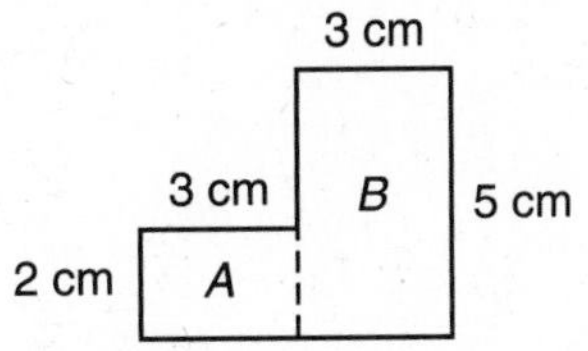

	Area *A*	= 2 cm × 3 cm	=	6 sq. cm
+	Area *B*	= 3 cm × 5 cm	=	15 sq. cm
	Combined area		=	**21 sq. cm**

9.

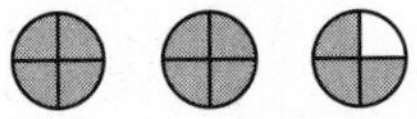

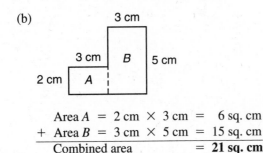

$\frac{4}{4} + \frac{4}{4} + \frac{3}{4} = \mathbf{\frac{11}{4}}$

10. Multiples of 8:
8, 16, 24, 32, (40), 48, 56, 64, 72, (80), ...
Multiples of 10:
10, 20, 30, (40), 50, 60, 70, (80), ...
The least common multiple of 8 and 10 is **40.**

11.
$$\begin{array}{r} 5.820 \\ 17.000 \\ 0.186 \\ +\ 13.000 \\ \hline \mathbf{36.006} \end{array}$$

12.
$$\begin{array}{r} 8.69 \\ -\ 4.00 \\ \hline 4.69 \end{array} \qquad \begin{array}{r} 10.00 \\ -\ 4.69 \\ \hline \mathbf{5.31} \end{array}$$

13.
$$\begin{array}{r} 4.7 \\ \times\ 12 \\ \hline 94 \\ 47\ \ \\ \hline \mathbf{56.4} \end{array}$$

14.
$$\begin{array}{r} \mathbf{29\ R\ 12} \\ 18\overline{)534} \\ \underline{36}\ \ \\ 174 \\ \underline{162} \\ 12 \end{array}$$

15. To multiply by 100, shift the decimal two places to the right.

$0.5432 \times 100 = 54.32$

The product is **54.32.**

16.
$$\begin{array}{r} 0.01 \\ \times\ \ 0.2 \\ \hline \mathbf{0.002} \end{array}$$

17. $2\frac{7}{8} + 2\frac{7}{8} = 4\frac{14}{8} = 5\frac{6}{8} = \mathbf{5\frac{3}{4}}$

18.
$$\begin{array}{r} 5\frac{5}{8} \\ -\ 1\frac{3}{8} \\ \hline 4\frac{2}{8} \end{array} = \mathbf{4\frac{1}{4}}$$

19. $\frac{3}{4} \times \frac{2}{3} = \frac{6}{12} = \mathbf{\frac{1}{2}}$

20. $\frac{5}{6} \div \frac{2}{3} = \frac{5}{6} \times \frac{3}{2} = \frac{15}{12} = 1\frac{3}{12} = \mathbf{1\frac{1}{4}}$